METHODS IN MOLECULAR BIOLOGY

For further volumes:
http://www.springer.com/series/7651

For over 35 years, biological scientists have come to rely on the research protocols and methodologies in the critically acclaimed *Methods in Molecular Biology series*. The series was the first to introduce the step-by-step protocols approach that has become the standard in all biomedical protocol publishing. Each protocol is provided in readily-reproducible step-by-step fashion, opening with an introductory overview, a list of the materials and reagents needed to complete the experiment, and followed by a detailed procedure that is supported with a helpful notes section offering tips and tricks of the trade as well as troubleshooting advice. These hallmark features were introduced by series editor Dr. John Walker and constitute the key ingredient in each and every volume of the *Methods in Molecular Biology* series. Tested and trusted, comprehensive and reliable, all protocols from the series are indexed in Pub Med.

Vaccinia Virus

Methods and Protocols

Edited by

Jason Mercer

MRC-Laboratory for Molecular Cell Biology, University College London, London, UK

Humana Press

Editor
Jason Mercer
MRC-Laboratory for Molecular Cell Biology
University College London
London, UK

ISSN 1064-3745 ISSN 1940-6029 (electronic)
Methods in Molecular Biology
ISBN 978-1-4939-9592-9 ISBN 978-1-4939-9593-6 (eBook)
https://doi.org/10.1007/978-1-4939-9593-6

Cover illustration: Pictured is an electron micrograph of a cell infected with vaccinia virus. The red objects are called viral crescents, which grow to form the violet circles called immature virions, when the viral genome is encapsidated into these they become known as immature virions with nucleoid (the genome is the very dense structure within), the green objects are fully assembled infectious vaccinia viruses, which are called mature virions. Image credit: Ian J. White; MRC-LMCB UCL

This Humana imprint is published by the registered company Springer Science+Business Media, LLC part of Springer Nature.
The registered company address is: 233 Spring Street, New York, NY 10013, U.S.A.

Foreword

Poxvirology has a long and storied history. Part of that history includes a significant list of "firsts." The first vaccine was the smallpox vaccine. Smallpox was the first and only human disease that has been eradicated from afflicting humankind. A new first has been added to this illustrious list. A new poxvirus therapeutic has been FDA approved. It is the first drug to gain approval through the FDA's "Animal Rule" because efficacy studies in humans were not ethical or feasible. The drug that was known as ST-246 is now named tecovirimat (commercial name TPOXX), the first drug that specifically treats smallpox.

The scientific understanding and advances in poxvirology are based on the work of generations of scientists. Since the publication of the last vaccinia virus methods book in this series, we have lost a number of contributors to the field of poxvirology. We note the passage of Donald (D.A.) Henderson (1928–2016), R. Mark Buller (1949–2017), Enzo Paoletti (1943–2018), and Keith R. Dumbell (1922–2018).

This scientific understanding of vaccinia virus and poxviruses continues to grow. This new volume of Methods in Molecular Biology provides a resource for scientists to bring new methods and procedures into their lab to make exciting discoveries that will continue to deepen our understanding of this fascinating virus family.

Stuart N. Isaacs
Perelman School of Medicine
at the University of Pennsylvania
Philadelphia, PA, USA

Preface

Vaccinia Virus: Methods and Protocols provides a resource of practical information for the laboratory that can be applied to study vaccinia and other poxviruses. This methods edition looks to emphasize long-standing field standards and complement the previous two editions (Volumes 269 and 890 in the Methods in Molecular Biology series). This includes a range of chapters focused on emerging new technologies applied in the field of poxvirology. The methods and protocols have been designed with the bench scientist in mind, being presented in a fashion that makes them useful for both starting and veteran poxvirus researchers.

London, UK *Jason Mercer*

Acknowledgments

I would like to acknowledge the poxvirus community for all the great meetings and discussions over the years (science and non-science). I want to specifically thank Paula Traktman, who convinced me that poxviruses are amazing, and the fab five, Rich Condit, Nissan Moussatche, Grant Mcfadden, Ed Niles, and Stewart Shuman, who dared to let a first-year graduate student compete in the famed Poxvirus Open (PVO), if only to FROOB. I want to thank all of the authors for their outstanding contributions to this book and all the members of my lab for remembering to keep poxvirology fun.

Contents

Contributors

DAVID ALBRECHT • *MRC Laboratory for Molecular Cell Biology, University College London, London, UK*

JEFFREY L. AMERICO • *Genetic Engineering Section, Laboratory of Viral Diseases (LVD), NIAID, NIH, Bethesda, MD, USA*

SHUAI CAO • *Division of Biology, Kansas State University, Manhattan, KS, USA*

CARMELA DI GIOIA • *Barts Cancer Institute, Queen Mary University, London, UK*

MARIANO ESTEBAN • *Department of Molecular and Cellular Biology, Centro Nacional de Biotecnología, CSIC, Madrid, Spain*

DAVID EVANS • *Department of Medical Microbiology and Immunology and Li Ka Shing Institute of Virology, 6020 Katz Group Centre, University of Alberta, Edmonton, AB, Canada*

JAMES GIBBS • *Cell Biology Section, LVD, NIAID, NIH, Bethesda, MD, USA*

CARMEN ELENA GÓMEZ • *Department of Molecular and Cellular Biology, Centro Nacional de Biotecnología, CSIC, Madrid, Spain*

ROBERT GRAY • *MRC Laboratory for Molecular Cell Biology, University College London, London, UK*

DOUGLAS W. GROSENBACH • *SIGA Technologies, Inc., Corvallis, OR, USA*

HEATHER D. HICKMAN • *Viral Immunity and Pathogenesis Unit, Laboratory of Clinical Immunology and Microbiology (LCIM), National Institute of Allergy and Infectious Diseases (NIAID), National Institutes of Health (NIH), Bethesda, MD, USA*

DENNIS E. HRUBY • *SIGA Technologies, Inc., Corvallis, OR, USA*

MOONA HUTTUNEN • *MRC Laboratory for Molecular Cell Biology, University College London, London, UK*

NOUHOU IBRAHIM • *Department of Biochemistry and Molecular Biology, Medical University of South Carolina, Charleston, SC, USA; Department of Microbiology and Immunology, Medical University of South Carolina, Charleston, SC, USA*

STUART N. ISAACS • *Division of Infectious Diseases, Perelman School of Medicine at the University of Pennsylvania and the Corporal Michael J. Crescenz VA Medical Center, Philadelphia, PA, USA*

OLENA KAMENYEVA • *Biological Imaging Section, Research Technology Branch, NIAID, NIH, Bethesda, MD, USA*

QUINTEN KIESER • *Department of Medical Microbiology and Immunology and Li Ka Shing Institute of Virology, 6020 Katz Group Centre, University of Alberta, Edmonton, AB, Canada*

SAMUEL KILCHER • *Institute of Food, Nutrition, and Health, ETH Zurich, Zurich, Switzerland*

JAMES LIN • *Department of Medical Microbiology and Immunology and Li Ka Shing Institute of Virology, 6020 Katz Group Centre, University of Alberta, Edmonton, AB, Canada*

YONGQUAN LIN • *Division of Biology, Kansas State University, Manhattan, KS, USA; Key Laboratory of Molecular Microbiology and Technology, Ministry of Education, College of Life Sciences, Nankai University, Tianjin, China*
CAROLINE MARTIN • *MRC Laboratory for Molecular Cell Biology, University College London, London, UK*
N. BISHARA MARZOOK • *The University of Sydney, School of Life and Environmental Sciences, Sydney, NSW, Australia*
JASON MERCER • *MRC-Laboratory for Molecular Cell Biology, University College London, London, UK*
HARRIET MOK • *MRC-Laboratory for Molecular Cell Biology, University College London, London, UK*
TIMOTHY P. NEWSOME • *The University of Sydney, School of Life and Environmental Sciences, Sydney, NSW, Australia*
RYAN NOYCE • *Department of Medical Microbiology and Immunology and Li Ka Shing Institute of Virology, 6020 Katz Group Centre, University of Alberta, Edmonton, AB, Canada*
ANNABEL T. OLSON • *Nebraska Center for Virology, University of Nebraska, Lincoln, NE, USA; School of Biological Sciences, University of Nebraska, Lincoln, NE, USA*
PATRICK PASZKOWSKI • *Department of Medical Microbiology and Immunology and Li Ka Shing Institute of Virology, 6020 Katz Group Centre, University of Alberta, Edmonton, AB, Canada*
BEATRIZ PERDIGUERO • *Department of Molecular and Cellular Biology, Centro Nacional de Biotecnología, CSIC, Madrid, Spain*
WENTAO QIAO • *Key Laboratory of Molecular Microbiology and Technology, Ministry of Education, College of Life Sciences, Nankai University, Tianjin, China*
GLENNYS V. REYNOSO • *Viral Immunity and Pathogenesis Unit, Laboratory of Clinical Immunology and Microbiology (LCIM), National Institute of Allergy and Infectious Diseases (NIAID), National Institutes of Health (NIH), Bethesda, MD, USA*
AMBER B. RICO • *School of Veterinary Medicine and Biomedical Sciences, University of Nebraska, Lincoln, NE, USA; Nebraska Center for Virology, University of Nebraska, Lincoln, NE, USA*
RACHEL L. ROPER • *Department of Microbiology and Immunology, Brody School of Medicine, East Carolina University, Greenville, NC, USA*
JOHN P. SHANNON • *Viral Immunity and Pathogenesis Unit, Laboratory of Clinical Immunology and Microbiology (LCIM), National Institute of Allergy and Infectious Diseases (NIAID), National Institutes of Health (NIH), Bethesda, MD, USA*
PAULA TRAKTMAN • *Department of Biochemistry and Molecular Biology, Medical University of South Carolina, Charleston, SC, USA; Department of Microbiology and Immunology, Medical University of South Carolina, Charleston, SC, USA*
SHIN-LIN TU • *Department of Biochemistry and Microbiology, University of Victoria, Victoria, BC, Canada*
CHRIS UPTON • *Department of Biochemistry and Microbiology, University of Victoria, Victoria, BC, Canada*
YAOHE WANG • *Barts Cancer Institute, Queen Mary University, London, UK*

MATTHEW S. WIEBE • *School of Veterinary Medicine and Biomedical Sciences, University of Nebraska, Lincoln, NE, USA; Nebraska Center for Virology, University of Nebraska, Lincoln, NE, USA*
ARTUR YAKIMOVICH • *MRC-Laboratory for Molecular Cell Biology, University College London, London, UK*
ZHILONG YANG • *Division of Biology, Kansas State University, Manhattan, KS, USA*
MING YUAN • *Barts Cancer Institute, Queen Mary University, London, UK*

Chapter 1

Working Safely with Vaccinia Virus: Laboratory Technique and Review of Published Cases of Accidental Laboratory Infections with Poxviruses

Stuart N. Isaacs

Abstract

Vaccinia virus, the prototype Orthopoxvirus, is widely used in the laboratory as a model system to study various aspects of viral biology and virus-host interactions, as a protein expression system, as a vaccine vector, and as an oncolytic agent. The ubiquitous use of vaccinia viruses in laboratories around the world raises certain safety concerns because the virus can be a pathogen in individuals with immunological and dermatological abnormalities, and on occasion can cause serious problems in normal hosts. This chapter reviews standard operating procedures when working with vaccinia virus and reviews published cases of accidental laboratory infections with poxviruses.

Key words Vaccinia virus, Biosafety Level 2, Class II Biological Safety Cabinet, Personal protective equipment, Smallpox vaccine, Complications from vaccination, Laboratory accidents

1 Introduction

Poxviruses are large DNA viruses with genomes of nearly 200 kilobases. Their unique site of DNA replication and transcription [1], the fascinating immune evasion strategies employed by the virus [2, 3], and the relative ease of generating recombinant viruses that express foreign proteins in eukaryotic cells [4, 5] have made poxviruses an exciting system to study and a common laboratory tool. Variola virus, the causative agent of smallpox, is the most famous member of the poxvirus family. It was eradicated as a human disease by the late 1970s and now work with that virus is confined to only two World Health Organization-sanctioned sites under Biosafety Level 4 conditions. It is interesting to note in the context of this chapter that the last human case of smallpox was due to a

The views expressed in this chapter are solely those of the author and do not necessarily reflect the position or policy of the Department of Veterans Affairs or the University of Pennsylvania.

Jason Mercer (ed.), *Vaccinia Virus: Methods and Protocols*, Methods in Molecular Biology, vol. 2023,
https://doi.org/10.1007/978-1-4939-9593-6_1,

laboratory mishap that led to the tragic death of a researcher who was working in an adjacent area [6]. While variola virus was an important human pathogen, vaccinia virus (VACV) is more widely studied and has become the prototype member of the Orthopoxvirus genus. VACV was used as the vaccine to confer immunity to variola virus and helped in the eradication of smallpox. In the United States, routine vaccination with the smallpox vaccine ended in the early 1970s. Since then, the Advisory Committee on Immunization Practices (ACIP) and the CDC have recommended that people working with poxviruses continue to get vaccinated [7–12]. This recommendation for those working with VACV is based mainly on the potential problems that an unintentional infection due to a laboratory accident may cause. Rationale for this recommendation is furthered by the understanding that the strains of VACV used in the laboratory setting (e.g., WR; *see* **Note 1**) are more virulent than the vaccine strain. Also, lab workers frequently handle virus at much higher titers than the dose given in the vaccine (*see* **Note 2**). There have been reports of laboratory accidents involving VACV (discussed later in this chapter), but a much greater number of such incidents likely go unreported. The total number of people working with VACV and the frequency with which they work with the virus is also unknown. Thus, for laboratory workers both the full extent of the problem and potential benefit from the vaccine are not known. The ACIP/CDC recommendations were recently updated [12]. New to the recommendations is the use of the "Grading of Recommendations Assessment, Development and Evaluation (GRADE)" methodology. While the inclusion of this methodology gives a more scientific feel to the recommendations, it is surprising that this method gave the same level of evidence to support smallpox vaccination to prevent accidental infection by vaccinia virus as it does for the more virulent viruses, variola and monkeypox. This chapter discusses laboratory procedures, personal safety equipment, and published laboratory accidents, all of which will serve as aides in preventing accidental laboratory infections and highlight the need to work safely with the virus.

2 Materials and Equipment

1. Class II Biological Safety Cabinet (BSC).
2. Personal protective equipment.
3. Autoclave.
4. Disinfectants: 1% Sodium hypochlorite, 2% glutaraldehyde, formaldehyde, 10% bleach, Spor-klenz, Expor, 70% alcohol.
5. Sharps container disposal unit.
6. Centrifuge bucket safety caps.
7. Occupational medicine access to the smallpox vaccine (*see* **Note 3**).

3 Methods

3.1 Laboratory and Personal Protective Equipment

The following section describes safety practices when working with fully replication-competent live VACVs that produce infectious virus in humans. Table 1 summarizes some published cases of laboratory accidents and will be used to highlight various aspects of working safely with the virus. In addition to fully replication-competent VACVs, there are highly attenuated strains of VACV (e.g., MVA and NYVAC) that are unable to form infectious progeny virus in mammalian cells ("non-replicating virus," *see* **Note 4**). These highly attenuated, non-replicating VACVs are considered Biosafety Level (BSL)-1 agents [30]. With that said, labs that work with both replication-competent and non-replicating viruses should be wary of potential contamination of stocks of avirulent virus with replication-competent poxviruses. This could result in an accidental laboratory infection. Since unintentional VACV infections most commonly occur through direct contact with the skin or eyes, the most important aspect of working safely with VACV is to use proper laboratory and personal protective equipment to help prevent accidental exposure to the virus. One of the first lines of defense against an accidental exposure is to always work with infectious virus in a biological safety cabinet (BSC). A BSC is a requirement when working with VACV. Not only the cabinet confines the virus to a work area that is easily defined and cleaned, but also the glass shield on the front of the BSC serves as an excellent barrier against splashes into the face. A BSC draws room air through the front grille, circulates HEPA-filtered air within the cabinet area, and also HEPA filters the air that is exhausted. Thus, working in a BSC protects the worker and the room where VACV is being handled from the unlikely event of aerosolization of the virus (*see* **Note 5**).

An equally important line of defense against accidental exposure to the virus is wearing proper personal protective equipment. This includes gloves, lab coat, and eye protection. VACV does not enter intact skin but can gain access through breaks in the skin. Thus, gloves are critical (*see* **Note 6**). Accidental infections due to breaks in skin are highlighted by Cases 1, 5–7, 18, and 19 in Table 1 and Fig. 1a–c. Some of these accidents could have been prevented by use of personal protective equipment. While the front shield of the BSC serves as a first line of protection against splashes into the eye, it is also recommended that safety glasses with solid side shields be worn when working with VACV. Depending upon the work being done (e.g., handling high-titer purified stocks of VACV), one should consider additional eye and/or protection like goggles or a full-face shield. This is important to consider because, as an immunologically privileged site [31], the eye can be susceptible to a serious infection even in those previously vaccinated [32]. Finally, a lab coat or some other type of outer protective garment

Table 1
Published cases of accidental laboratory infections with Orthopoxviruses

Case no.	Journal, year (reference)	Age (years) and/or state (year) and underlying medical conditions	Exposure activity	Virus	Site and cause of infection	Prior vaccination status	Illness	Antibiotics/surgery/antivirals	Resolution and follow-up
1	Nature (1986) [13]	>31	Injecting mice	TK-minus WR strain (2×10^6 pfu/50 μL)	Cut on right ring finger	Vaccinated 30 years prior to exposure	4 Days after exposure finger was red and swollen and it progressed from base of finger nail to first joint; day 8 right axillary LN became swollen; no fever or malaise		10 Days; worker developed antibodies to the recombinant VACV-expressed protein

2	Lancet (1991) [14]	London (1990)	Injecting mice	TK-minus WR strain	Needle sticks into the left thumb and left forefinger	Vaccinated 1 year prior to exposure	3 Days after the needle stick. Regions became itchy and by day 4 were red and papular. Days 5–6 the lesions were discharging serous fluid and reached a max diameter of 1 cm; kept in occlusive dressing and healed spontaneously		No antibody response to protein expressed by recombinant VACV, but potential evidence of T-cell response
3	NEJM (2001) [15]	28 (15 weeks pregnant w/h/o epidermolytic hyperkeratosis)	Dog bite	Copenhagen strain-based rabies vaccine	Technically not a lab accident, but unintentional exposure to a recombinant virus via a dog bite	Reportedly no prior smallpox vaccination (born 1971)	3 Days after exposure developed blisters on her forearm; 8 days after bite, hospitalized for progressive pain, erythema, and swelling of left forearm; 10 days after bite, swelling and erythema worsened, left axillary LN	Antibiotics and went to OR for incision and drainage of the forearm	30 Days; developed antibodies to the recombinant VACV expressed protein; no pregnancy complications and delivered a healthy baby

(continued)

Table 1
(continued)

Case no.	Journal, year (reference)	Age (years) and/or state (year) and underlying medical conditions	Exposure activity	Virus	Site and cause of infection	Prior vaccination status	Illness	Antibiotics/ surgery/ antivirals	Resolution and follow-up
4	EID (2003) [16]	26	Needle stick during virus purification step	WR (~10^8 pfu)	Needle stick into the left thumb	Previously vaccinated in childhood (>20 years earlier)	Developed erythema and pain 3 days after inoculation; additional pustules on fourth and fifth fingers developed days 5 and 6; day 6 axillary LN; day 8 necrotic areas around lesion and a large erythematous lesion on left forearm	Day 9 began on antibiotics because of concern of bacterial superinfection; went to OR for surgical excision of necrotic tissue	Improved and lesions healed over ~3 weeks; evidence of increased anti-VACV antibodies

5	J Invest Dermatol (2003) [17]	40	Contact exposure through broken skin	TK-minus WR strain (10^9 pfu/mL)	Working with high titer in tissue culture with evidence of small erosions on both hands (from working in cold temperatures)	Prior vaccinations 28 and 39 years prior to exposure	Middle inner side of right second finger; second lesion developed (large nodule with central necrosis) on third finger of left hand 2 days later; no LN	Unsuccessful surgical incision followed by topical disinfectants (e.g., polyvidone iodine)	Two weeks and then healed; evidence of increased anti-VACV antibodies
6	Can Commun Dis Rep (2003) [18]	48 (history of eczema)	Contact exposure through broken skin	TK-minus virus	Chronic eczema on both hands and a cut on her finger, usually did NOT wear gloves when working with the virus	Vaccinated as a child	First developed pain and redness over dorsal aspect of her index finger; 5 days later admitted with blistering lesions on right index finger; also noted to have swollen axillary LN	Did not respond to antibiotics; treated with occlusive dressing	Spontaneously resolved

(continued)

Table 1
(continued)

Case no.	Journal, year (reference)	Age (years) and/or state (year) and underlying medical conditions	Exposure activity	Virus	Site and cause of infection	Prior vaccination status	Illness	Antibiotics/surgery/antivirals	Resolution and follow-up
7	J Clin Virol (2004) [19]	25	Contact exposure through broken skin		Cut on finger with secondary spread by contact to another site	Never vaccinated	Developed a pustule at the site of a cut on the finger. Squeezed pus that squirted on to her face. 2 days later a lesion formed on her chin; axillary and submental LN, malaise, fever; on day 20, four other lesions were noted on her palms, back of the knee, and upper back and felt to be generalized vaccinia	Did not respond to antibiotics	By day 28 lesions were fading, but continued to have fatigability; by day 36 just a scab on her finger but felt back to full strength; developed anti-VACV antibodies ~1 month after presentation

8	EID (2006) [20]	Graduate student PA (2004)	Unknown mechanism of infection	Recombinant WR strain	Unknown, but question of hand to eye or microscope eyepiece to eye or aerosol exposure	Never vaccinated	Painful eye infection (no keratitis or orbital cellulitis) requiring hospitalization	Antibiotics and then antiviral eye drops; VIG	Improvement 24 h after starting VIG; no sequelae, but recovery took a few weeks. Developed anti-VACV antibodies ~2 month after presentation; no secondary VACV infections of contacts were identified
9	Military Medicine (2007) [21]	28	Splash into eye		Sprayed ~1 mL of fluid containing virus in eye; washed eye for 2 min	Never vaccinated	Developed eye burning several hours after exposure		No infection occurred
10	J Viral Hepatitis (2007) [22]	30	Needlestick injury	Recombinant non-TK-minus WR strain (10^8 pfu/mL)	Needlestick into the left thumb	Never vaccinated	8 Days after needlestick developed pain and erythema of the thumb and axillary LN; painful swelling of thumb worsened	15 Days after injury necrosis at the injection site was surgically removed	Developed antibodies and T-cell responses to the recombinant VACV-expressed protein

(continued)

Table 1
(continued)

Case no.	Journal, year (reference)	Age (years) and/or state (year) and underlying medical conditions	Exposure activity	Virus	Site and cause of infection	Prior vaccination status	Illness	Antibiotics/surgery/antivirals	Resolution and follow-up
11	MMWR (2008) [23]	CT (2005)	Injecting mice	TK-minus WR strain	Needlestick in finger	Vaccinated as a child and ~10 years before the accident	3 Days after accident developed fever, LN, and bulla at the inoculation site	Hospitalized for 1 day	Symptoms improved rapidly
12	MMWR (2008) [23]	PA (2006)	Injecting mice	TK-minus WR strain	Needlestick in thumb	Never vaccinated	6 Days after accident sought medical attention for a lesion at the site of inoculation and a secondary lesion near the nail; 9 days after accident had malaise, fever, and LN	Finger surgically debrided 14 days after the accident	Began feeling better 13 days after the accident

13	MMWR (2008) [23]	IA (2007)	Needlestick	TK-minus WR strain (3×10^6 pfu)	Needlestick to finger while unsheathing a sterile needle	Never vaccinated	11 Days after injury developed fever, chills, and lesions with swelling at the inoculation site		Recovered fully
14	MMWR (2008) [23]	MD (2007)	Injecting animals	TK-minus WR strain (10^4 pfu in 5 μL)	Needlestick to finger	Unsuccessful immunization ~6 years prior to the accident	No infection	After accident put finger into disinfectant containing hypochlorite and then was vaccinated on the day of accident	No infection
15	MMWR (2008) [23]	NH (2007)	Needle scratch while working with mice	WR strain (5×10^4 pfu/mL)	Needle scratch to finger	Never vaccinated	7 Days after the accident developed a pustule; afebrile	Hospitalized with streaking up arm	Recovered

(continued)

Table 1
(continued)

Case no.	Journal, year (reference)	Age (years) and/or state (year) and underlying medical conditions	Exposure activity	Virus	Site and cause of infection	Prior vaccination status	Illness	Antibiotics/surgery/antivirals	Resolution and follow-up
16	MMWR (2009) [24]	20s	Unknown mechanism of infection	WR strain (a contaminating virus in a stock of recombinant virus the lab usually works with)	Ear and eye, with additional lesions on chest, shoulder, arm, and leg	Never vaccinated	Pain and swelling of right earlobe and cervical LN and fevers developed 4–6 days after working with vaccinia virus; 4 days after onset of symptoms, pustular lesions were on right ear, left eye, chest, shoulder, left arm, and right leg	Symptoms worsened on antibiotics and steroids; hospitalized; acyclovir given	Full recovery and returned to work ~1 month after infection; no secondary VACV infections of contacts were identified

17	MJA (2009) [25]	26	Injecting mice	WR strain	Needlestick into the left second finger	Vaccinated within 5 years of accident	2 Days after injury developed a cloudy vesicle; 5 days after injury finger became inflamed with streaking up the arm and axillary LN		All symptoms resolved after 10 days
18	MMWR (2009) [26]	35 (taking immuno suppressive medication for inflammatory bowel disease)	Skin abrasion	Copenhagen strain-based rabies vaccine	Technically not a lab accident, but exposed to recombinant VACV while handling raccoon rabies vaccine bait	Never vaccinated	4 Days after exposure developed some red papules that then increased in number; day 9 had 26 lesions on her arm with edema; afebrile	Hospitalized and treated with VIG on day 6 after exposure and then a repeat dose on day 12. Started on investigational antiviral agent, ST-246 × 14 days	Discharged on day 19 after exposure. By day 28, all lesion scabs had separated
19	JID (2012) [27][a]	Researcher, IL (2010)	Possible small cut on finger	Recombinant cowpox virus contaminating a stock of another virus used in the lab	Finger. Working with a material in lab and mice that turned out to be contaminated with a recombinant cowpox virus	Never vaccinated (lab not actively working with poxviruses for 5 years)	Painful, ulcerated lesion on a finger that lasted 3 months; 5 days of axillary LN; fever; body aches; malaise; headache	Antibiotics and then debridement of necrotic tissue	Lesion resolved with scarring, residual pain, and loss of range of movement at the site of the lesion

(continued)

Table 1
(continued)

Case no.	Journal, year (reference)	Age (years) and/ or state (year) and underlying medical conditions	Exposure activity	Virus	Site and cause of infection	Prior vaccination status	Illness	Antibiotics/ surgery/ antivirals	Resolution and follow-up
20	EID (2014) [28]	28, Researcher, India, (2014)	Cut caused by a broken glass ampule during freeze-drying	Buffalopox	Palm. Cut caused by a broken glass ampule during freeze-drying	Never vaccinated	Lesion appeared on day 3 post-injury; pain, high fever, malaise on day 9 post-injury; went to surgery on day 11 post-injury	Site immediately cleaned with 70% ethanol and treated with antibiotic ointment. Surgical excision of necrotic tissue, followed by antibiotics	Healing by post-injury day 85, with scarring
21	MMWR (2015) [29]	27, MA (2013)	Needlestick while recapping the needle	VACV	Needlestick resulting in a necrotic thumb lesion	Recently vaccinated (10 months prior to injury)	Initial red streaks up arm and small fluid collection in thumb; thumb lesion becomes necrotic	Antibiotics then surgical debridement 23 days post-injury	Skin lesion resolved ~6 weeks post-injury

Abbreviations: *LN* lymph node, *OR* operating room, *pfu* plaque-forming units, *TK* thymidine kinase, *VACV* vaccinia virus, *VIG* vaccinia immune globulin, *WR* western reserve

[a]First reported in Medscape News, February 8, 2011 (http://www.medscape.com/viewarticle/737030)

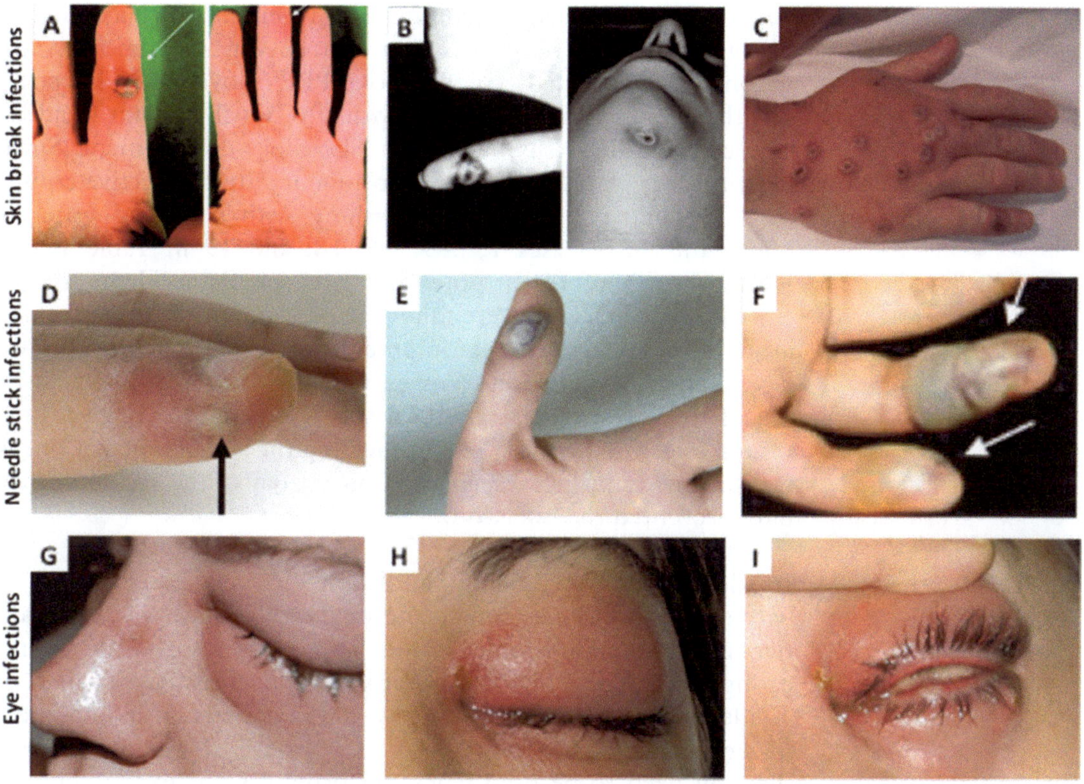

Fig. 1 (**a**) Photograph of non-needlestick infection of fingers ~5–7 days after the onset of symptoms. Reprinted by permission from Macmillan Publishers Ltd: The Journal of Investigative Dermatology, reference [17], copyright © 2003. (**b**) Photographs of primary and secondary lesions 18 days after the onset of symptoms. Reprinted from the Journal of Clinical Virology, reference [19], copyright © 2004, with permission from Elsevier. (**c**) Photograph of the right hand of a woman 11 days after contact with a raccoon rabies vaccine bait. Figure and legend reproduced from [26]. These published materials are in the public domain. (**d**) Photograph of a finger 2 days after a needlestick injury. Arrow points to cloudy vesicle forming at the site of inoculation. Figure reproduced with permission from reference [25]: Senanayake SN. Needlestick injury with smallpox vaccine. Med J Aust 2009; 191 (11): 657. © Copyright 2009 *The Medical Journal of Australia*. (**e**) Lesion on left thumb 9 days postinoculation. Figure and legend reproduced from [29]. These published materials are in the public domain. (**f**) Local reaction on the left hand after accidental needlestick inoculation with VACV 11 days postinoculation. Arrows indicate the lesion areas. Figure and legend reproduced from [16]. These published materials are in the public domain. (**g**) Left eye of a man with laboratory-acquired VACV infection ~4 days after the onset of symptoms. Figures and legend reproduced from [24]. These published materials are in the public domain. (**h**) VACV eye infection 5 days after the onset of symptoms. The primary pox lesion is located at the inner canthus. (**i**) A satellite lesion on lower conjunctiva developing 7 days after the onset of symptoms. (**h** and **i**) Photographs by E. Claire Newbern. Figures and legend reproduced from [20]. These published materials are in the public domain. For color versions of images, see electronic version of this chapter or web links. (**c**) Link: https://www.cdc.gov/mmwr/preview/mmwrhtml/mm5843a2.htm. (**e**) Link: https://www.cdc.gov/mmwr/preview/mmwrhtml/mm6416a2.htm. (**f**) Link: http://www.cdc.gov/ncidod/EID/vol9no6/images/02-0732_1b.jpg. (**g**) Link: http://www.cdc.gov/mmwr/preview/mmwrhtml/figures/m829a1f.gif. (**h**) Link: https://wwwnc.cdc.gov/eid/article/12/1/05-1126-f1. (**i**) Link: https://wwwnc.cdc.gov/eid/article/12/1/05-1126-f2

decreases the chance of contaminating clothing. If such a contamination occurs, an outer garment can be quickly removed and decontaminated. Furthermore, use of PPE (gloves and a lab coat) and their removal after working with VACV will prevent accidentally carrying the virus out of the laboratory environment. Since the virus can be stable in the environment, after protective equipment is removed, good hand-washing with soap and water is important [33]. Cases 1, 5–9, 16, 18, and 19 in Table 1 and Fig. 1a–c, g, h represent potentially preventable accidents if general biosafety practices were followed. Also note that in some published accidents, prompt interventions may have prevented development of infections. For example, in Case 9, prompt flushing of the eye with water after a splash exposure may have prevented infection. In Case 14, disinfecting the site of inoculation, as well as active smallpox vaccination on the day of the accident, may have prevented the infection.

3.2 Laboratory Safety

In addition to VACV being handled at Biosafety Level-2 [34], as with all biohazardous agents, routine good laboratory safety practices need to be fully implemented. This includes such things as no eating or drinking in the laboratory. To decrease the chance of accidental infections the use of any sharps or glass should be minimized while working with VACV (*see* **Note 7**). Syringes and needles still need to be used when performing some experiments in animals, and thus personnel who need to be working with needles to inject animals with the virus would be considered to be performing a higher risk procedure (*see* **Note 8** and Subheading 3.3). If sharps or disposable glassware are necessary, the proper leak-proof, puncture-resistant sharps disposal container needs to be conveniently located close to the work area to prevent accidents during disposal of needles and glass (*see* **Note 9**). Multiple cases of laboratory accidents due to needlesticks have been reported. Cases 2, 4, 10–15, 17, and 21 are examples of needlestick accidents while working with VACV (Table 1 and Fig. 1d–f). Case 21 also serves as a reminder that needles should never be recapped.

As discussed earlier, a BSC is a requirement when working with VACV. As a virus that has an outer membrane envelope, VACV is susceptible to inactivation by a variety of detergents and disinfectants [35]. Thus, after working with the virus in the BSC, surfaces should be wiped down with freshly prepared 1% bleach. Since experiments with VACV in tissue culture frequently involve aspirating and discarding virus-contaminated growth medium, one must properly inactivate the virus in the media prior to disposal. For aspirating off media from infected cells using the lab vacuum system, one must include a trap to collect the aspirated media and a vacu-guard to prevent contamination of the house vacuum system with the virus (*see* **Note 10**). The virus contained in the liquid must be inactivated by the addition of disinfectant (*see* **Note 11**).

The virus is also susceptible to heat, and thus autoclaving contaminated instruments, dryware, animal cages, and bedding exposed to the virus is also required. Properly packaged disposable plasticware used for culturing the virus should be sterilized in a humidified autoclave. While small samples containing virus have been shown to be inactivated in autoclaves in as little as 15 min [36], infectious waste should be autoclaved at 121 °C for at least 60 min at 15 pounds per square inch and then disposed of according to the institutional guidelines (*see* **Note 11**). When centrifuging large volumes of media containing virus, it is best to use centrifuge buckets that have safety lids to contain a spill and contain potential aerosolization of the virus if a tube should leak during centrifugation. Sonication of infected cells, which is frequently done in poxvirus protocols as a means of releasing virus from cells and breaking up virus clumps, can also cause aerosolization of the virus (*see* **Note 12**). Therefore, sonication should be performed in a cup sonicator with the virus or virus-infected cells remaining in a closed tube. Larger virus preparations may need to be sonicated with a probe sonicator, which should only be performed if the sonicator device is contained in a proper BSC that will properly filter the air and remove any potential aerosolized virus (*see* **Note 5**). As highlighted in Case 19, labs should also create systems in the lab to segregate the use and storage of pathogenic and nonpathogenic viruses to prevent the accidental use of the wrong stock of virus.

3.3 Occupational Smallpox Vaccination

In the USA, the currently licensed smallpox vaccine (ACAM 2000 [37]) is recommended for people who work with poxviruses [12] (*see* **Note 13**). While there is no debate that this policy should be followed for researchers who work with such poxviruses as variola virus, camelpoxvirus, and monkeypox virus (*see* **Note 14**), there is considerable debate whether smallpox vaccination should be carried out in all people who have contact with VACV [38–43]. The Advisory Committee on Immunization Practices (ACIP) and the CDC continue to recommend vaccination every 10 years of all people who have contact with VACV [12]. As mentioned previously, this recommendation does not apply to those working with highly attenuated strains of VACV (i.e., MVA, NYVAC, ALVAC, and TROVAC). These viruses are considered to be extremely safe because they do not generate progeny virus in mammalian cells and are avirulent in normal and immunosuppressed animal models (*see* **Note 15**). However, ACIP recommendations about smallpox vaccination of laboratory workers handling replication-competent strains of VACV remain unchanged from earlier decades [7, 8]. The issue is that these older recommendations were at a time when essentially all adults working with VACV had previously been vaccinated at least once in childhood. Since routine vaccination in the USA ended in the early 1970s, there is now a growing population of workers who have never been vaccinated and thus the

Table 2
Incidence of reported VACV vaccine complications in adults (number of cases/million vaccinations)

Vaccination group	Accidental transfer	Generalized vaccinia	Eczema vaccinatum	Progressive vaccinia	Post-vaccinial encephalitis
Primary[a]	606	212	30	–[b]	–[b]
Secondary[a]	25	9	4	7	4
US Military (2005)[c]	178	59[d]	0	0	1[e]
Civilian (2004)[f]	569	74	0	0	25

[a]Data from [45] for individuals ≥20 years of age. Receiving the vaccine for the first time (primary); receiving the vaccine in those who had previously been vaccinated (secondary)
[b]No report of this complication was included in the 1968 10-state survey [45]
[c]Data adapted from [46] reporting on 730,580 vaccinations as of January 4, 2005 (71% were primary vaccinations)
[d]All were in primary vaccinees
[e]One case in a primary vaccinee and another case in a secondary vaccinee
[f]Data adapted from [46] reporting on 40,422 civilian healthcare and public health workers vaccinated between January 24, 2003, and January 31, 2004 (36% were primary vaccinations)

recommendation to vaccinate such workers would represent primary vaccinations. The rate of complications from primary vaccination is 10 to 20 times greater than the rate of complications in those who had previously been vaccinated [44, 45] (*see* Table 2 and **Note 16**).

Thus, the risk-benefit ratio for routine vaccination of all lab workers who handle VACV in the USA has significantly changed from the 1980s and 1990s. Also, a recent survey of lab workers revealed that previously unvaccinated people experience more post-vaccination symptoms than previously vaccinated people [47]. In addition, since vaccinated individuals can accidentally transmit the virus to close contacts, the potential infection of unvaccinated contacts is more problematic now than in previous times when most of the population had been vaccinated (*see* **Note 17**). Recent cases of accidental transmission of the vaccine have been documented [48–53] and at least one of these accidental transmissions resulted in significant morbidity [49]. In contrast to the ACIP recommendation to vaccinate lab workers, advisory committees in other countries have reached different conclusions than the ACIP and do not recommend routine vaccination, but a vaccination based on risk assessment [54]. These committees concluded that the risk of vaccination of all workers (i.e., knowingly infecting an individual with VACV) outweighs the potential benefit of protecting them from an accidental exposure that results in an infection.

Some feel strongly that vaccination should be mandatory [55, 56]. Another approach that has been implemented at some institutions in the USA is to offer mandatory counseling by occupational medicine regarding vaccination, but to then allow each individual to make an informed personal decision whether or not to be vaccinated [38–43]. The types of procedures being done with VACV by the worker should enter into this decision-making process (*see* **Notes 8** and **18**). Since accidental infections have occurred in workers with distant past vaccination (Cases 1, 4–6; Table 1, Fig. 1a, f) or recent vaccination (Cases 2, 11, 17, 21; Table 1, Fig. 1d, e), the role prior vaccination may have played is unclear. With that said, the majority of published cases are in workers who have never been vaccinated (Cases 3, 7–10, 12–16, 18, and 19–20; Table 1, Fig. 1a–c, f, g). Some would argue that prior vaccination prevents accidental infections from becoming more serious. Others would argue that it is unclear if the cases of accidental infection in the setting of no prior vaccination were any different than those who had prior vaccination. However, it has been pointed out that more workers with accidental infections who had never been vaccinated were hospitalized (Cases 3, 8, 15, 16, 18; Table 1) than those who had prior vaccination (Case 11; Table 1 and reference [56]). But the problem with such data is that not all lab accidents are reported. Furthermore, it is more likely that serious infections resulting in hospitalization would come to the attention of the CDC and thus further skew the data.

Another important issue related to the counseling session is that the worker should have a complete medical assessment to determine whether the smallpox vaccine can safely be given. Thus, workers may be identified who have medical conditions that would preclude them from vaccination. Such conditions are listed in Table 3 (*see* **Note 16**). Importantly, some feel that the presence of

Table 3
Contraindications for nonemergency use of ACAM2000[a]

History or presence of atopic dermatitis (*see* **Note 19**)
Other active exfoliative skin conditions (e.g., eczema, burns, impetigo, varicella zoster virus infection, herpes simplex virus infection, severe acne)
Immunosuppression
Persons aged <1 year
Women who are pregnant or breastfeeding
Serious allergy to any component of ACAM2000
Underlying heart disease with or without symptoms (e.g., coronary artery disease or cardiomyopathy)
Primary vaccinees with three or more known major cardiac risk factors (i.e., hypertension, diabetes, hypercholesterolemia, heart disease at age 50 years in a first-degree relative, and smoking)

[a]*See* [12] for details

these conditions should preclude someone from working with VACV. Others think that for a properly trained researcher, working with VACV while pregnant or with mild eczema can be allowed, as long as the individual is not performing high-risk procedures with the virus (*see* **Note 8**). Laboratory directors and workers need to be made aware of the risks and preventative measures available in order to make educated decisions. Each institution should develop a policy that addresses these concerns.

3.4 Future Considerations

Over the past decade, the concerns about smallpox bioterrorism led to the development of a number of new products including a newly licensed vaccine grown in tissue culture [57], next-generation smallpox vaccines with better safety profiles [58], and a number of promising new anti-poxvirus therapeutics like tecovirimat (ST-246) [59, 60] and brincidofovir (CMX001) [61]. As these new therapeutics and safer next-generation vaccines move forward they will certainly remedy some safety concerns of working with the non-attenuated laboratory strains of VACV. For example, the future availability of a smallpox vaccine that is safer than what is currently FDA approved in the USA will likely end the controversy of whether or not laboratory workers handling VACV should be vaccinated. A vaccine with few to no side effects (e.g., MVA) would clearly shift the risk-benefit ratio in favor of vaccination to prevent infections resulting from a laboratory accident. In addition, the future availability of anti-poxvirus therapeutics will be beneficial for treatment of serious infections after a lab accident.

4 Notes

1. VACV strain WR (Western Reserve) is the most commonly used laboratory strain of VACV and was selected for its neurovirulence in mice.
2. The smallpox vaccine is administered by scarification with a bifurcated needle. When the needle is dipped into the vaccine stock (~10^8 pfu/mL) approximately 2.5 μL of solution is taken up and thus approximately 2.5×10^5 pfu is delivered.
3. Institutions can obtain the smallpox vaccine from the CDC, Drug Services, National Center for Infectious Diseases (404-639-3670; drugservice@cdc.gov).
4. One should be aware of the nuance in use of the terms "replicating" and "non-replicating" virus since this may be confused with viral DNA replication. Many of the highly attenuated VACVs are able to replicate the viral DNA, but do not assemble to form infectious progeny virus, and are thus referred to as non-replicating virus.

5. If aerosol-generating procedures are to be performed outside of a BSC, they should be done under some type of containment, for example, using tubes with caps during sonication or centrifugation. One should also remember that after centrifugation, tubes should only be opened within a BSC. If such procedures cannot be contained, respiratory protection is required. Individuals requiring respiratory protection should be enrolled in the institution's respiratory protection program.
6. Nitrile or powder-free latex gloves should be used when working with virus.
7. Disposable plasticware that is now commonly used has significantly decreased the need for glassware while working with the virus. For example, instead of using sterilized glass Pasteur pipets to aspirate medium while culturing the virus in cell culture plates, one can use plastic pipet tips attached to vacuum tubing. A simple system used by many labs is to have a P-1000 pipet tip attached to the vacuum tubing. One can then use P-200 pipet tips over the P-1000 tip for direct contact with various culture wells. This allows easy changing of the tip to prevent any cross contamination of specimens. But some procedures still require glass that can result in accidents. Case 20 highlights the need to carefully inspect glass for cracks and try to minimize conditions that could cause glass to shatter.
8. Workers performing higher risk procedures like vaccinating animals, direct handling of infected animals (*see* **Note 18**), or preparing and purifying high-titer virus are individuals who might benefit the most from prophylactic smallpox vaccination. Also included in the group for whom vaccination should be strongly considered are those generating or working with recombinant VACVs that express potentially toxic proteins, proteins that might enhance virulence of the recombinant VACV, or proteins to which seroconversion may be problematic (e.g., HIV, dengue). However, to successfully vaccinate such individuals, workers need additional education about the potential benefits of vaccination to overcome the fear of adverse effects of smallpox vaccination [62].
9. Sharps containers (when 2/3 full) should be autoclaved prior to disposal in accordance with the institution's infectious waste policy.
10. Labs frequently use a large-volume Erlenmeyer flask placed underneath the BSL-2 cabinet. This flask can be filled with detergent or bleach to inactivate the virus. Bleach should be at a final concentration of ~10%. Decontaminated liquid waste should be disposed in accordance with the institution's regulatory requirements. The level of fluid in the waste collection

flask should be monitored to prevent overfilling. As a further safety measure to prevent contamination of the vacuum system if the collection trap accidentally becomes overfilled, it is recommended to have two forms of line protection placed before the vacuum outlet: an additional trap as well as a vent/gas filter with a hydrophobic membrane.

11. Special consideration must be made when working with virus and radioactivity. The waste stream for all radioactivity must be accounted for and disposed of following institutional guidelines. Thus, virus-contaminated radioactive waste must first be inactivated. Radioactive liquid waste should be treated with disinfectant and then disposed of with the normal radioactive liquid waste. Since one should *not* autoclave radioactive waste, radioactive contaminated disposable plasticware should be washed down with disinfectant solution to inactivate virus prior to disposing it as radioactive solid waste.

12. During sonication, use of proper ear protection is recommended.

13. Since the publication of the first and second editions of vaccinia virus methods [63, 64], relatively large numbers of people in the US military have continued to be vaccinated with VACV. With careful screening of potential vaccine recipients, the experience with US military found that the rate of the most serious adverse events was less than previously reported (*see* Table 2). However, despite careful screening measures there have still been serious adverse events. There have been reports of progressive vaccinia in a military recruit [65] as well as a life-threatening case of eczema vaccinatum in an infant that was exposed to his father who was vaccinated prior to deployment [49, 66]. Data for the vaccination of US civilians revealed that the rate of the most serious adverse events was similar to that previously reported (*see* Table 2).

14. For those working with these highly virulent poxviruses (e.g., variola virus, monkeypox virus), vaccination every 3 years is recommended [9].

15. In fact, these highly attenuated strains of VACV can now be handled in the laboratory at Biosafety Level-1 [30].

16. The potential serious complications from smallpox vaccine include eczema vaccinatum, post-vaccinial encephalitis, and progressive vaccinia in immunocompromised hosts who were inadvertently vaccinated or were accidentally infected by exposure to a vaccinated individual (*see* Table 2). Photographs of these types of complications are available [9, 67]. In the course of the pre-event smallpox vaccination programs during the first decade of the twenty-first century, myocarditis/pericarditis has

also been reidentified as a potential complication from the vaccine [11, 68, 69].

17. Because of potential spread of the virus from the vaccination site, close contact with individuals who have the conditions listed in Table 3 is considered to be a relative contraindication to elective vaccination. However, a worker who is not vaccinated because of potential contact with a person with a contraindication should still be able to work with VACV following guidelines of personal protection equipment and good handwashing. Alternatively, an individual who still wants to be vaccinated can choose to isolate themselves from contact with such at-risk people until the vaccination site scab has fallen off (typically anywhere from 2 to 4 weeks). These issues affect the risk-benefit ratio decision.
18. The recent ACIP recommendations [12] have adjusted the recommendation for animal care personnel. Routine vaccination is not recommended for animal care personnel (or healthcare personnel) whose contact with VACV is limited to contaminated materials (e.g., cages or dressings). While VACV can spread from infected mice to naïve mice within a cage [70–72], animal care workers usually have no direct contact with animals (e.g., use of tongs to transfer animals from cage to cage). This new recommendation is supported by studies that showed no transmission of virus between subcutaneously vaccinated and unvaccinated guinea pigs [73], and a study [72] that looked for VACV transmission to sentinel mice by the soiled bedding from cages that housed mice infected by the subcutaneous or intrarectal route with VACV. None of the mice exposed to the soiled bedding seroconverted.
19. There is increased likelihood of serious complications in those with skin conditions like atopic dermatitis or eczema whether the skin disease is active or non-active at the time of vaccination. While it is not known why such individuals have increased complications, it is theorized that there is some type of cutaneous immune abnormality [74].

Acknowledgment

The author is supported by the Philadelphia Veterans Affairs Medical Center (recently renamed, Corporal Michael J. Crescenz VA Medical Center) and by NIH grants R21 AI117100, R44 AI115759, and R44 AI125005.

References

1. Moss B (2013) Poxviridae: the viruses and their replication. In: Knipe DM, Howley PM, Cohen JI et al (eds) Fields virology, 6th edn. Wolters Kluwer/Lippincott Williams & Wilkins Health, Philadelphia, PA, pp 2129–2159
2. Seet BT, Johnston JB, Brunetti CR, Barrett JW, Everett H, Cameron C, Sypula J, Nazarian SH, Lucas A, McFadden G (2003) Poxviruses and immune evasion. Annu Rev Immunol 21:377–423
3. Haga IR, Bowie AG (2005) Evasion of innate immunity by vaccinia virus. Parasitology 130(Suppl):S11–S25
4. Moss B (1996) Genetically engineered poxviruses for recombinant gene expression, vaccination, and safety. Proc Natl Acad Sci U S A 93(21):11341–11348
5. Carroll MW, Moss B (1997) Poxviruses as expression vectors. Curr Opin Biotechnol 8(5):573–577
6. Hawkes N (1979) Smallpox death in Britain challenges presumption of laboratory safety. Science 203(4383):855–856
7. CDC (1985) Recommendations of the immunization practices advisory committee smallpox vaccine. MMWR Morb Mortal Wkly Rep 34:341–342
8. CDC (1991) Vaccinia (smallpox) vaccine. Recommendations of the Immunization Practices Advisory Committee (ACIP). MMWR Morb Mortal Wkly Rep 40(RR-14):1–10
9. CDC (2001) Vaccinia (smallpox) vaccine Recommendations of the Advisory Committee on Immunization Practices (ACIP), 2001. MMWR Morb Mortal Wkly Rep 50(RR10):1–25
10. CDC (2003) Recommendations for using smallpox vaccine in a pre-event vaccination program. Supplemental recommendations of the Advisory Committee on Immunization Practices (ACIP) and the Healthcare Infection Control Practices Advisory Committee (HICPAC). MMWR Recomm Rep 52(RR-7):1–16
11. CDC (2003) Supplemental recommendations on adverse events following smallpox vaccine in the pre-event vaccination program: recommendations of the Advisory Committee on Immunization Practices. MMWR Morb Mortal Wkly Rep 52(13):282–284
12. Petersen BW, Harms TJ, Reynolds MG, Harrison LH (2016) Use of vaccinia virus smallpox vaccine in laboratory and health care personnel at risk for occupational exposure to orthopoxviruses—recommendations of the Advisory Committee on Immunization Practices (ACIP), 2015. MMWR Morb Mortal Wkly Rep 65(10):257–262. https://doi.org/10.15585/mmwr.mm6510a2
13. Jones L, Ristow S, Yilma T, Moss B (1986) Accidental human vaccination with vaccinia virus expressing nucleoprotein gene. Nature 319:543
14. Openshaw PJ, Alwan WH, Cherrie AH, Record FM (1991) Accidental infection of laboratory worker with recombinant vaccinia virus. Lancet 338:459
15. Rupprecht CE, Blass L, Smith K, Orciari LA, Niezgoda M, Whitfield SG, Gibbons RV, Guerra M, Hanlon CA (2001) Human infection due to recombinant vaccinia-rabies glycoprotein virus. N Engl J Med 345(8):582–586
16. Moussatche N, Tuyama M, Kato SE, Castro AP, Njaine B, Peralta RH, Peralta JM, Damaso CR, Barroso PF (2003) Accidental infection of laboratory worker with vaccinia virus. Emerg Infect Dis 9(6):724–726
17. Mempel M, Isa G, Klugbauer N, Meyer H, Wildi G, Ring J, Hofmann F, Hofmann H (2003) Laboratory acquired infection with recombinant vaccinia virus containing an immunomodulating construct. J Invest Dermatol 120(3):356–358
18. Loeb M, Zando I, Orvidas MC, Bialachowski A, Groves D, Mahoney J (2003) Laboratory-acquired vaccinia infection. Can Commun Dis Rep 29(15):134–136
19. Wlodaver CG, Palumbo GJ, Waner JL (2004) Laboratory-acquired vaccinia infection. J Clin Virol 29(3):167–170
20. Lewis FM, Chernak E, Goldman E, Li Y, Karem K, Damon IK, Henkel R, Newbern EC, Ross P, Johnson CC (2006) Ocular vaccinia infection in laboratory worker, Philadelphia, 2004. Emerg Infect Dis 12(1):134–137
21. Peate WF (2007) Prevention of vaccinia infection in a laboratory worker. Mil Med 172(10):1117–1118
22. Eisenbach C, Neumann-Haefelin C, Freyse A, Korsukewitz T, Hoyler B, Stremmel W, Thimme R, Encke J (2007) Immune responses against HCV-NS3 after accidental infection with HCV-NS3 recombinant vaccinia virus. J Viral Hepat 14(11):817–819
23. CDC (2008) Laboratory-acquired vaccinia exposures and infections—United States, 2005–2007. MMWR Morb Mortal Wkly Rep 57(15):401–404

24. CDC (2009) Laboratory-acquired vaccinia virus infection—Virginia, 2008. MMWR Morb Mortal Wkly Rep 58(29):797–800
25. Senanayake SN (2009) Needlestick injury with smallpox vaccine. Med J Aust 191(11-12):657
26. CDC (2009) Human vaccinia infection after contact with a raccoon rabies vaccine bait—Pennsylvania, 2009. MMWR Morb Mortal Wkly Rep 58(43):1204–1207
27. McCollum AM, Austin C, Nawrocki J, Howland J, Pryde J, Vaid A, Holmes D, Weil MR, Li Y, Wilkins K, Zhao H, Smith SK, Karem K, Reynolds MG, Damon IK (2012) Investigation of the first laboratory-acquired human cowpox virus infection in the United States. J Infect Dis 206(1):63–68. https://doi.org/10.1093/infdis/jis302
28. Riyesh T, Karuppusamy S, Bera BC, Barua S, Virmani N, Yadav S, Vaid RK, Anand T, Bansal M, Malik P, Pahuja I, Singh RK (2014) Laboratory-acquired buffalopox virus infection, India. Emerg Infect Dis 20(2):324–326. https://doi.org/10.3201/eid2002.130358
29. Hsu CH, Farland J, Winters T, Gunn J, Caron D, Evans J, Osadebe L, Bethune L, McCollum AM, Patel N, Wilkins K, Davidson W, Petersen B, Barry MA, Centers for Disease Control and Prevention (2015) Laboratory-acquired vaccinia virus infection in a recently immunized person—Massachusetts, 2013. MMWR Morb Mortal Wkly Rep 64(16):435–438
30. Office of Recombinant DNA Activities; recombinant DNA research: action under the guidelines. National Institutes of Health (NIH), PHS, DHHS (2016) Notice of action under the NIH Guidelines for Research Involving Recombinant DNA Molecules (NIH Guidelines). Fed Regist 81 (https://osp.od.nih.gov/wp-content/uploads/NIH_Guidelines.html)
31. Niederkorn JY (2002) Immune privilege in the anterior chamber of the eye. Crit Rev Immunol 22(1):13–46
32. Ruben FL, Lane JM (1970) Ocular vaccinia. An epidemiologic analysis of 348 cases. Arch Ophthalmol 84:45–48
33. Jonczy EA, Daly J, Kotwal GJ (2000) A novel approach using an attenuated recombinant vaccinia virus to test the antipoxviral effects of hand soaps. Antiviral Res 45(2):149–153
34. Richmond JY, RW MK (eds) (1999) Biosafety in microbiological and biomedical laboratories. HHS publication; no. (CDC) 93-8395, 4th edn. U.S. Department of Health and Human Services, PHS, CDC, NIH, Washington DC
35. Block SS (ed) (2001) Disinfection, sterilization, and preservation, 5th edn. Lippincott Williams & Wilkins, Philadelphia
36. Espy MJ, Uhl JR, Sloan LM, Rosenblatt JE, Cockerill FR 3rd, Smith TF (2002) Detection of vaccinia virus, herpes simplex virus, varicella-zoster virus, and Bacillus anthracis DNA by LightCycler polymerase chain reaction after autoclaving: implications for biosafety of bioterrorism agents. Mayo Clin Proc 77(7):624–628
37. CDC (2008) Notice to readers: newly licensed smallpox vaccine to replace old smallpox vaccine. MMWR Morb Mortal Wkly Rep 57(8):207–208
38. Baxby D (1989) Smallpox vaccination for investigators. Lancet 2:919
39. Wenzel RP, Nettleman MD (1989) Smallpox vaccination for investigators using vaccinia recombinants. Lancet 2(8663):630–631
40. Perry GF (1992) Occupational medicine forum. J Occup Med 34:757
41. Baxby D (1993) Indications for smallpox vaccination: policies still differ. Vaccine 11(4):395–396
42. Williams NR, Cooper BM (1993) Counselling of workers handling vaccinia virus. Occup Med (Oxf) 43:125–127
43. Isaacs SN (2002) Critical evaluation of smallpox vaccination for laboratory workers. Occup Environ Med 59(9):573–574
44. Lane JM, Ruben FL, Neff JM, Millar JD (1969) Complications of smallpox vaccination, 1968. National surveillance in the United States. New Engl J Med 281:1201–1208
45. Lane JM, Ruben FL, Neff JM, Millar JD (1970) Complications of smallpox vaccination, 1968: results of ten statewide surveys. J Infect Dis 122:303–309
46. Poland GA, Grabenstein JD, Neff JM (2005) The US smallpox vaccination program: a review of a large modern era smallpox vaccination implementation program. Vaccine 23(17–18):2078–2081
47. Baggs J, Chen RT, Damon IK, Rotz L, Allen C, Fullerton KE, Casey C, Nordenberg D, Mootrey G (2005) Safety profile of smallpox vaccine: insights from the laboratory worker smallpox vaccination program. Clin Infect Dis 40(8):1133–1140
48. CDC (2004) Secondary and tertiary transfer of vaccinia virus among U.S. military personnel—United States and worldwide, 2002–2004. MMWR Morb Mortal Wkly Rep 53(5):103–105
49. CDC (2007) Household transmission of vaccinia virus from contact with a military smallpox vaccinee–Illinois and Indiana, 2007. MMWR Morb Mortal Wkly Rep 56(19):478–481

50. CDC (2007) Vulvar vaccinia infection after sexual contact with a military smallpox vaccinee–Alaska, 2006. MMWR Morb Mortal Wkly Rep 56(17):417–419
51. CDC (2010) Vaccinia virus infection after sexual contact with a military smallpox vaccinee–Washington, 2010. MMWR Morb Mortal Wkly Rep 59(25):773–775
52. Young GE, Hidalgo CM, Sullivan-Frohm A, Schult C, Davis S, Kelly-Cirino C, Egan C, Wilkins K, Emerson GL, Noyes K, Blog D (2011) Secondary and tertiary transmission of vaccinia virus from US military service member. Emerg Infect Dis 17(4):718–721
53. Hughes CM, Blythe D, Li Y, Reddy R, Jordan C, Edwards C, Adams C, Conners H, Rasa C, Wilby S, Russell J, Russo KS, Somsel P, Wiedbrauk DL, Dougherty C, Allen C, Frace M, Emerson G, Olson VA, Smith SK, Braden Z, Abel J, Davidson W, Reynolds M, Damon IK (2011) Vaccinia virus infections in martial arts gym, Maryland, USA, 2008. Emerg Infect Dis 17(4):730–733
54. Advisory Committee on Dangerous Pathogens and Advisory Committee on Genetic Modifications (1990; re-issued 2011) Vaccination of laboratory workers handling vaccinia and related poxviruses infectious for humans. HMSO Publications Center, London (http://www.hse.gov.uk/pUbns/priced/acdp-acgm-vaccine.pdf)
55. Fulginiti VA (2003) The risks of vaccinia in laboratory workers. J Invest Dermatol 120(3):viii
56. MacNeil A, Reynolds MG, Damon IK (2009) Risks associated with vaccinia virus in the laboratory. Virology 385(1):1–4
57. Greenberg RN, Kennedy JS (2008) ACAM2000: a newly licensed cell culture-based live vaccinia smallpox vaccine. Expert Opin Investig Drugs 17(4):555–564
58. Kennedy JS, Greenberg RN (2009) IMVAMUNE: modified vaccinia Ankara strain as an attenuated smallpox vaccine. Expert Rev Vaccines 8(1):13–24
59. Jordan R, Goff A, Frimm A, Corrado ML, Hensley LE, Byrd CM, Mucker E, Shamblin J, Bolken TC, Wlazlowski C, Johnson W, Chapman J, Twenhafel N, Tyavanagimatt S, Amantana A, Chinsangaram J, Hruby DE, Huggins J (2009) ST-246 antiviral efficacy in a nonhuman primate monkeypox model: determination of the minimal effective dose and human dose justification. Antimicrob Agents Chemother 53(5):1817–1822
60. Chinsangaram J, Honeychurch KM, Tyavanagimatt SR, Leeds JM, Bolken TC, Jones KF, Jordan R, Marbury T, Ruckle J, Mee-Lee D, Ross E, Lichtenstein I, Pickens M, Corrado M, Clarke JM, Frimm AM, Hruby DE (2012) Safety and pharmacokinetics of the anti-orthopoxvirus compound ST-246 following a single daily oral dose for 14 days in human volunteers. Antimicrob Agents Chemother 56(9):4900–4905. https://doi.org/10.1128/AAC.00904-12
61. Lanier R, Trost L, Tippin T, Lampert B, Robertson A, Foster S, Rose M, Painter W, O'Mahony R, Almond M, Painter G (2010) Development of CMX001 for the treatment of poxvirus infections. Viruses 1(12):2740–2762
62. Benzekri N, Goldman E, Lewis F, Johnson CC, Reynolds SM, Reynolds MG, Damon IK (2010) Laboratory worker knowledge, attitudes and practices towards smallpox vaccine. Occup Med (Lond) 60(1):75–77
63. Isaacs SN (2004) Vaccinia virus and poxvirology: Methods and protocols, Methods in Molecular Biology. Humana Press, c2004, Totowa, NJ
64. Isaacs SN (2012) Working safely with vaccinia virus: laboratory technique and review of published cases of accidental laboratory infections. Methods Mol Biol 890:1–22. https://doi.org/10.1007/978-1-61779-876-4_1
65. CDC (2009) Progressive vaccinia in a military smallpox vaccinee—United States, 2009. MMWR Morb Mortal Wkly Rep 58(19):532–536
66. Vora S, Damon I, Fulginiti V, Weber SG, Kahana M, Stein SL, Gerber SI, Garcia-Houchins S, Lederman E, Hruby D, Collins L, Scott D, Thompson K, Barson JV, Regnery R, Hughes C, Daum RS, Li Y, Zhao H, Smith S, Braden Z, Karem K, Olson V, Davidson W, Trindade G, Bolken T, Jordan R, Tien D, Marcinak J (2008) Severe eczema vaccinatum in a household contact of a smallpox vaccinee. Clin Infect Dis 46(10):1555–1561. https://doi.org/10.1086/587668
67. Cono J, Casey CG, Bell DM (2003) Smallpox vaccination and adverse reactions. Guidance for clinicians. MMWR Recomm Rep 52(RR-4):1–28
68. CDC (2003) Update: cardiac-related events during the civilian smallpox vaccination program—United States, 2003. MMWR Morb Mortal Wkly Rep 52(21):492–496
69. Halsell JS, Riddle JR, Atwood JE, Gardner P, Shope R, Poland GA, Gray GC, Ostroff S, Eckart RE, Hospenthal DR, Gibson RL, Grabenstein JD, Arness MK, Tornberg DN (2003) Myopericarditis following smallpox

vaccination among vaccinia-naive US military personnel. JAMA 289(24):3283–3289
70. Briody BA (1959) Response of mice to ectromelia and vaccinia viruses. Bacteriol Rev 23:61–95
71. Lee SL, Roos JM, McGuigan LC, Smith KA, Cormier N, Cohen LK, Roberts BE, Payne LG (1992) Molecular attenuation of vaccinia virus: mutant generation and animal characterization. J Virol 66:2617–2630
72. Gaertner DJ, Batchelder M, Herbst LH, Kaufman HL (2003) Administration of vaccinia virus to mice may cause contact or bedding sentinel mice to test positive for orthopoxvirus antibodies: case report and follow-up investigation. Comp Med 53(1):85–88
73. Holt RK, Walker BK, Ruff AJ (2002) Horizontal transmission of recombinant vaccinia virus in strain 13 guinea pigs. Contemp Top Lab Anim Sci 41(2):57–60
74. Engler RJ, Kenner J, Leung DY (2002) Smallpox vaccination: risk considerations for patients with atopic dermatitis. J Allergy Clin Immunol 110(3):357–365

Chapter 2

Bioinformatics for Analysis of Poxvirus Genomes

Shin-Lin Tu and Chris Upton

Abstract

In recent years, there have been numerous technological advances in the field of molecular biology; these include next- and third-generation sequencing of DNA genomes and mRNA transcripts and mass spectrometry of proteins. Perhaps, however, it is genome sequencing that impacts a virologist the most. In 2017, more than 480 complete genome sequences of poxviruses have been generated, and are constantly used in many different ways by almost all molecular virologists. Matching this growth in data acquisition is an explosion of the relatively new field of bioinformatics, providing databases to store and organize this valuable/expensive data and algorithms to analyze it. For the bench virologist, access to intuitive, easy-to-use, software is often critical for performing bioinformatics-based experiments. Three common hurdles for the researcher are (1) selection, retrieval, and reformatting genomics data from large databases; (2) use of tools to compare/analyze the genomics data; and (3) display and interpretation of complex sets of results. This chapter is directed at the *bench virologist* and describes the software that helps overcome these obstacles, with a focus on the comparison and analysis of poxvirus genomes. Although poxvirus genomes are stored in public databases such as GenBank, this resource can be cumbersome and tedious to use if large amounts of data must to be collected. Therefore, we also highlight our Viral Orthologous Clusters database system and integrated tools that we developed specifically for the management and analysis of complete viral genomes.

Key words Poxvirus, Vaccinia virus, Smallpox, Bioinformatics, Genomics, Dotplot, Multiple sequence alignment, MSA, VOCs, VGO, BBB, BLAST, JDotter

1 Introduction

It is almost 30 years since the first complete poxvirus genome, from a vaccinia virus (VACV), was reported [1]. In 2017, more than 480 complete poxvirus genomes are available from public databases (*see* **Note 1**). Although for a few species such as variola virus (VARV), VACV, monkeypox virus (MPXV), and myxoma virus (MYXV) many isolates have been sequenced, there are a surprising number of viruses with no or very few close relatives. In these examples, one or a small number of viruses comprise a genus, frequently with a very long branch to the backbone of the phylogenetic tree. An interesting consequence of this is when these new viruses are examined, they are often found to encode a relatively

Jason Mercer (ed.), *Vaccinia Virus: Methods and Protocols*, Methods in Molecular Biology, vol. 2023,
https://doi.org/10.1007/978-1-4939-9593-6_2,

large proportion of novel genes. For example, we were recently involved in the annotation of two closely related poxviruses isolated from kangaroos [2] and of the 162 and 165 genes contained by these viruses, 81 genes were common to all chordopoxviruses that infect vertebrates, but about 40 genes were unique to these viruses. Furthermore, almost all of these "novel" genes had no counterpart in any public database. Even the prototypic poxvirus, VACV, contains genes for which no function has yet been ascribed. Thus, the annotation of poxvirus genomes is likely to remain an ongoing process for many years to come. Such annotation is not merely a "stamp-collecting" endeavor; without an understanding of all the processes encoded or controlled by such complex viruses as the poxviruses, it is not possible to get close to a full understanding of their biology and pathology.

Running parallel to the rapid advances in DNA sequencing technologies has been an equally explosive growth in bioinformatics as evidenced by the number of journals dedicated to the science and others more loosely affiliated with the topic (https://en.wikipedia.org/wiki/List_of_bioinformatics_journals). There are a great many subtopics within bioinformatics; some of those that are most likely to interest virologists are genome assembly, similarity searches, genome annotation, multiple sequence alignment (MSA), gene/protein analysis, and visualization. This chapter, which touches on all of these areas, also highlights the importance of tools for data management because as the volume of -omic data grows it has become increasingly important to have special tools to organize, display, and provide access to that data. For many of the examples presented here, we have used tools developed in our own lab as these were created specifically to help with the annotation of newly sequenced genomes from large DNA viruses. These tools, which are collectively known as the Viral Bioinformatics Resource Center (VBRC; 4virology.net), were also developed with the virologist in mind, providing graphical user interfaces (GUIs) to allow users to easily access and analyze data and then display results in easily interpreted presentations.

For this chapter, most of the processes and analyses discussed fall into the "comparative genomics" category and our primary goal is to provide readers with easy-to-follow demonstrations of data retrieval and management as well as several common bioinformatics methods for analyzing poxvirus genomes. The latter include (1) comparison of entire poxvirus genomes; (2) comparison of poxvirus genes and proteins within orthologous families; (3) similarity searching of poxvirus proteins against databases for prediction of function; and (4) searching of poxvirus proteins for motifs and functional domains. Although it is not possible to go into great detail for all the software mentioned, help files and manuals are available and detailed instructions will be provided on how to access the software that is used through the VBRC web site (*see* **Note 2**).

Finally, although we endeavor to make the database of poxvirus genomes as accurate as possible, we view one of the greatest values of this Bioinformatics Resource as providing the ability to quickly and easily investigate the data behind particular results. For example, a genome GenBank file might list a gene as being 20% shorter than most of its orthologs: to understand the biology that might follow from this, it is necessary to determine whether a mutation has truncated the ORF at the 3′ end (perhaps little effect on the gene product) or 5′ end (perhaps inactivating the gene product) or whether the apparent mutation could be a sequencing error (no change to gene product). Hopefully, the following examples will show readers the importance of reviewing the raw data, and that with the right tools it doesn't have to take all day.

2 Materials

1. To use the VBRC tools, researchers need a reasonably up-to-date desktop computer and operating system that can support the installation of Java Runtime Environment 8 (*see* **Note 3**).
2. In developing our own tools, we have preferred the Java client-server format (*see* **Note 3**). This design was chosen because of the greater flexibility and functionality it allows in the client software and the use of Java also overcomes the problem of supporting a large variety of WWW browsers and different operating systems. The download is easy and quick, and each time a user opens a Java program, Java Web Start automatically checks the version of the client on the user's machine and downloads an updated version if required. However, other systems are available: Virus Pathogen Resource (VIPR; https://www.viprbrc.org/) uses a web interface that supports a tool workbench and improvements to JavaScript libraries have led to a variety of new Web-based tools that use GenomeD3Plot to produce rich graphical outputs [3].
3. Key software:
   (a) *Virus Orthologous Clusters* (VOCs): A MySQL database (*see* **Note 4**) of information derived from all complete annotated poxvirus genomes. A powerful, but easy-to-use, interface has been built to query the database. VOCs [4] can be accessed via the VBRC web site at https://4virology.net/virology-ca-tools/vocs/.
   (b) *Viral Genome Organizer* (VGO): Software for visualizing and comparing the organization of genes within complete poxvirus genomes. The program allows custom retrieval of

protein/DNA sequence regions from a genome and can display preprocessed result files from a variety of analyses alongside the genomic data. A variety of searches can be performed on genome-sized DNA sequences. Although the genomes are not aligned in this tool, because VGO "talks" to the VOCs database, it understands and can highlight orthologs to display the similarity between genomes. VGO [5] can be accessed via the VBRC web site at https://4virology.net/virology-ca-tools/vgo/.

(c) *JDotter*: We developed a Java client-server version of DOTTER [6] called JDotter that can use DNA sequences in FASTA files or import genomes/sequences from the VOCs database. JDotter can also produce dotplots for protein sequences, and multiple DNA or protein sequences. JDotter [7] can be accessed via the VBRC web site at https://4virology.net/virology-ca-tools/jdotter/.

(d) *Base-By-Base* (BBB): Software for generating, visualizing, and editing MSAs. The program can use DNA and protein sequences in FASTA files or import genomes/sequences from the VOCs database. BBB uses standard alignment algorithms, such as ClustalO [8], MAFFT [9], and MUSCLE [10]. It has the ability to display annotations from GenBank files and allows users to add their own comments, including primer-binding sites, to the sequences. BBB [11] can be accessed via the VBRC web site at https://4virology.net/virology-ca-tools/base-by-base/.

(e) *Genome Annotation Transfer Utility* (GATU): Software for annotating genomes using the information from a reference genome, and writing GenBank files. This tool provides for interactive annotation; it automatically annotates genes that are very similar between a reference virus and leaves others for a human decision. GATU [12] can be accessed via the VBRC web site at https://4virology.net/virology-ca-tools/gatu/.

(f) *HHpred*: A Web-based interface to a search tool that uses Hidden Markov Model (HMM) profile matching to detect distantly related proteins to the query protein. It is available at https://toolkit.tuebingen.mpg.de/#/tools/hhpred [13].

(g) *Command Line Tools*: MIRA [14], SPAdes [15] (assembly), Burrows-Wheeler Aligner (BWA) [16], Tanoti (http://www.bioinformatics.cvr.ac.uk/tanoti.php) (sequence mapping), SAMtools [17], NGSUtils [18] (raw sequence data manipulation), RAxML [19] (phylogeny). These tools are invoked from the command line window (the Terminal in macOS, or the command line prompt in Windows).

3 Methods

3.1 Connecting to the VBRC Web Site

The VBRC web site is available at https://4virology.net/. It is well worth spending a little time browsing the *Help* documents on this site, but most of the software discussed here can be found as follows:

1. Select tool from the *VBRC Tools* menu.
2. Click on the *Click to Launch* button.
3. Some programs require Java Web Start to be installed on the local computer. This is built into the macOS operating system; on other operating systems, you will be alerted if you need to install Java Web Start (*see* **Note 5**).

3.2 Organizing Poxvirus Genome Data

Although a 127–350 kb genome is small by current sequencing capacities, detailed investigation of the information contained within a single poxvirus genome, never mind 350+ such genomes, is only realistic with specialized bioinformatics software. All of the various programs that we have developed over the last 20 years for characterization of the poxvirus genomes have the same goals (1) to provide a dedicated SQL (structured query language) database to organize and store poxvirus genomic sequence data together with various types of sequence annotation, and (2) to give virologists intuitive common interfaces to this database and the other analysis software, thereby reducing the need to connect to and learn multiple web sites or programs. VOCs and VGO are described first because several other sections of this review refer to analyses that they perform.

3.3 Viral Orthologous Clusters

VOCs is a Java client-server application that accesses a large, up-to-date, MySQL database containing all complete and fully annotated poxvirus genomes. The database is administered by the corresponding author's laboratory and is provided free to all researchers via the Internet at the VBRC web site (https://4virology.net/). The VOCs database stores complete genome sequences, which are parsed from GenBank files, along with the DNA and protein sequences, promoter region sequences, annotations, molecular weights, predicted isoelectric points, calculated nucleotide (nt) and amino acid (aa) frequencies, and calculated codon usages of the individual genes. Thus, the researcher can interact with the data at multiple levels, for example: (1) complete genomic DNAs: genomes can be searched for subsequences; (2) gene and protein sequences: these can be retrieved from the database and then aligned or searched; and (3) comparatively: determine which genes are common to all poxviruses in the database. Unfortunately, some of the poxvirus GenBank sequence files contain annotation and DNA sequence errors. When we become aware of these, we update our database; similarly, we annotate genomes in the database with new information found in the literature. Therefore, VOCs is the most current source of poxvirus genomic information.

Besides providing access to all of the poxvirus sequences, probably the most important and useful feature of the VOCs database is that each poxvirus gene/protein is assigned to an orthologous gene family based on BLASTP (protein query search against a protein database) similarity scores. Currently, the database contains 350+ complete poxvirus genomes with ~70,000 predicted genes or gene fragments. These predicted genes have been grouped into ~1700 ortholog families that have been given names based on function (if known). Many of the ortholog families contain genes from only one or a few viruses; these represent genes that are unique to a virus species or genus. This feature allows the user to collect all examples of a poxvirus ortholog (gene or protein sequence) in a matter of seconds and also determine which viruses do not possess a particular gene.

Among the applications integrated into VOCs are position-specific iterative BLAST (PSI-BLAST) [20], BLASTP, BLASTX (DNA query translated and searched against a protein database), TBLASTN (protein query search against a DNA database translated to six reading frames on the fly) [21], VGO [5], BBB [11], and JDotter [7]. Thus, the user is able to select the sequences they want and then send them directly to the application for analysis. If external tools are needed, then VOCs can supply the selected data as a text file.

When a user opens the VOCs program, Java Web Start automatically checks the version of the client on the user's machine and downloads an updated version if required. VOCs was specifically designed to be used by virologists and queries are GUI menu driven. Three windows are used to navigate through the VOCs program: the Sequence Filter, Ortholog Group Filter, and Genome Filter. The Sequence Filter window allows users to select genomes and to search the database for genes based on such parameters as gene name, size, pI, NCBI protein ID, as well as nucleotide and amino acid sequences and constraints. The Ortholog Group Filter window allows users to search the database for gene families based on family name, ID, numbers of genes per family, and specific annotations. Although the Virus Selector section of the VOCs interface windows allows the user to select or exclude particular genomes for subsequent queries, the Genome Filter window allows users to restrict the set of genomes that will be used in the search processes and then actually updates the Virus Selector section of the windows; for example, the Genome Filter window can be used to limit the view to the 16 Capripoxvirus genomes instead of showing all 350+ genomes. Table 1 shows some example queries that can easily be made within VOCs.

Table 1
Some of the functions that can be performed with VOCs

Window	Query
SF	Find genes with names that match *A10*
SF	Find genes with DNA sequence containing a given sequence of nucleotides (ACGATCGATT)
SF	Find proteins with $4.5 < pI < 6.5$
SF	Find proteins with $25{,}000 < MW < 30{,}000$ kDa
SF	Find proteins with *serine* content greater than *13%*
SF	Find proteins with *leu + ile + val + ala gt; 40%*
SF	Draw a map of genes in ECTV genome
OGF	Find the gene family containing ectromelia virus gene 108
OGF	Find gene families containing a single (unique) gene
OGF	Find gene families that contain a gene from all orthopoxviruses
OGF	Find gene families present in myxoma virus but absent from SFV

SF sequence filter window, *OGF* ortholog group filter window

Once a database query has been selected, the results of the query can be seen as a Gene Count or Ortholog Group Count in which case only the number of hits, of that particular type, in the database is displayed. The list of genes and families can be viewed in a new window by clicking on the appropriate *View* button. The *count* function is used to check the query and prevent the accidental selection of the wrong gene set or very large numbers of genes. For an additional safeguard, VOCs asks users to confirm that they want to display more than 1000 items. The Gene and Ortholog Results Tables can be sorted by clicking on a column header; for example, *sort by A+T%* instantly reveals that there is considerable variation among the genes. Within these tables, the genes and Ortholog Groups are also available for display or further analysis using the aforementioned tools. It is very simple to select a gene, click on *Family* (bottom panel of the results panel), and then view all genes within that Ortholog Group. The predicted protein sequences, DNA, or promoter region sequences can be selected for input into a tool to generate a multiple alignment, or particular sequences can be used to search the VOCs database using any of the BLAST programs [21]. Similarly, genes can be viewed in individual virus genomes instead of being categorized by the ortholog groups. Rather than manually sifting through a GenBank file, the user can simply select a virus in VOCs, and click on *GeneView* to see the Gene Results Table; every gene number, function, ORF positions, ORF strand, A+T%, pI, and amino acid length are available to the user in one glance. The positioning of the columns can be changed by dragging. Columns denoting individual properties can be hidden or displayed via top panel Column gt; Selection. Likewise, multiple genome statistics and properties can be viewed, sorted, and compared using the *GenomeView* function.

An often-used feature provided in VOCs is the ability to create a gene content (ortholog) comparison table for multiple viruses compared to a reference genome (Table 2). Here, multiple poxvirus genomes are compared to a reference genome to highlight the presence/absence of genes among multiple viruses. To compare orthologs to a reference, select the subject virus in Virus Selector, click on GeneView on the bottom right panel, click *Select a genome to compare*, choose the GenBank name of the comparison genome, and confirm by clicking "select." The information from your comparison genome is now filled out in the last column of the table. The table of results is also useful for determining the names of the orthologs in different viruses.

3.3.1 VOCs Use Example #1: Alignment of Poxvirus Uracil-DNA Glycosylase (UNG) Proteins

Within VOCs, there is often several ways to achieve the same result when trying to select a particular set of genes. Here, we will just give the most intuitive and simplest method.

1. Open VOCs as described in Subheading 3.1.
2. Select your favorite poxvirus in Virus Selector panel (most virologists will have a particular virus for which they are familiar with gene names/numbers).
3. Click on *Gene View* button. This lists all the genes in the genome.
4. Scroll through the gene list to find the gene of interest. Since we happen to know that in VACV-Copenhagen UNG is D4R and in approximately the middle of the genome, this is quicker than setting up a more detailed database search. Select the UNG gene.
5. Click the family button to show all UNG genes. For the table that is produced, all columns can be sorted, for example, virus name, A+T%, and protein length.
6. Select genes of interest.
7. From *Alignment* menu, choose *Protein Sequence Alignment* and the type of alignment software (use MUSCLE or ClustalO).
8. The alignment is returned in a BBB MSA editor window.

3.3.2 VOCs Use Example #2: Display Those Ortholog Groups that Have a Protein Member Present in Myxoma Virus but Not in Rabbit Fibroma Virus

1. Open VOCs as described in Subheading 3.1; if the tool is already open, go to the *select* menu and *clear all* to remove any previous search parameters. Ensure that all elements in the VOCs windows can be seen.
2. From the Virus Selector, unclick *Enable Selection*.
3. Click on *Ortholog GroupFilter* tab at the top of the window.

Table 2
Part of the table generated from VOCs gene results table comparing genomes

CPXV-BR GB gene #	Ortholog group name	CMLV-CMS GB gene #	ECTV-Mos GB gene #
116	CPV-B-116	No gene found	No gene found
117	Crescent membrane and immature virion formation	102R	89
118	mRNA capping enzyme large subunit	103R	90
119	Virion core	104L	91
119A	VV_Cop-D ORF B	104.5R	No Gene Found
120	Virion core	105R	92
121	Uracil-DNA glycosylase, DNA polymerase processivity factor	106R	93
122	NTPase, DNA primase	107R	94
123	Morphogenesis, VETF-s (early transcription factor small)	108R	95
124	RNA polymerase (RPO18)	109R	96
125	Carbonic anhydrase, GAG-binding IMV membrane protein	110L	97
126	mRNA-decapping enzyme	111R	98
127	mRNA-decapping enzyme	112R	99
128	ATPase, NPH1	113L	100
129	mRNA-capping enzyme small subunit	114L	101

CPXV-Brighton Red was chosen as the reference genome for comparison to camelpox CMLV-CMS and ectromelia ECTV-Moscow; orthologous genes are displayed; users should note that VOCs gene numbers *may* be different to GenBank gene numbers (GB gene #) in order to standardize the convention used across the VOCs database; however, both numbering schemes are presented in the VOCs window for reference

4. Scroll down to the *Ortholog Group Query* section, and select checkbox for *Select Ortholog Groups that ....*
5. Select *Contain* radio button; select *MYXV-Lau from* menu, and click *Add Criteria* button below this.
6. Select *AND* operator radio button at the right side, and click *Add Operator* button.
7. Select *Do NOT Contain* radio button; select *RFV-Kas* from menu; and click *Add Criteria* button. Your query will be shown in the window.
8. Click on *OrtGrpCnt* or *OrtGrpView* button. The search requires only a few seconds.
9. The result reveals that myxoma virus has only four genes that are absent from rabbit fibroma virus.

3.3.3 VOCs Use Example #3: Find Ortholog Groups that Have Gene Present in All Poxvirus Genomes

1. Open VOCs as described in Subheading 3.1; if the tool is already open, go to the *select* menu and *clear all* to remove any previous search parameters.
2. Click on *Ortholog Group Filter* tab at the top of the window.
3. Scroll down to the *Ortholog Group Size* section, and click on the checkbox "#viruses is between" blank and a max number of viruses in the database. Fill in the blank with max number of viruses. The result will show ortholog families in which all viruses are represented.
4. Click on *OrtGrpCnt* or *OrtGrpView* button. The search requires only a few seconds.

This type of query can also be used to find out which viruses do not have a particular gene. To find any missing viruses in a selected ortholog group, simply go to the results table generated from *OrtGrpView*, click on the ortholog group (row), and then go to the top panel to select "View > View Missing Viruses." Furthermore, the tools within VOCs can then be used to determine why a gene appears to be missing from a virus, e.g., missed annotation, DNA deleted, ORF fragmented, and possible sequencing error. The ability to make these determinations is important because whether a gene is actually absent from a virus can have significant ramifications for other research projects.

3.3.4 VOCs Use Example #4: Draw Genome Map

1. Open VOCs as described in Subheading 3.1; if the tool is already open, go to the *select* menu and *clear all* to remove any previous search parameters.
2. Select a virus of interest.
3. From the *Draw* menu, select *Genome Map*.
4. Modify *ORF Labels* and *ORF Shape* as needed.
5. Modify *Colors* for ORFs to reflect coding direction, gene type, or ortholog conservation on the genera level, or import personal setting.

This tool is not intended for the preparation of publication quality images. Rather, it was added to VOCs to allow rapid visualization and editing/coloring of genome maps, with a quality suitable for slide presentations.

3.3.5 VOCs Use Example #5: Create Sets of Concatenated Genes from Multiple Genomes

1. Open VOCs as described in Subheading 3.1; if the tool is already open, go to the *select* menu and *clear all* to remove any previous search parameters.
2. Select viruses of interest, and click *OrtGrpView*.
3. Multi-select the orthologs of your choice.

4. From the *Analysis* menu, select whether to concatenate the DNA or amino acid sequences. This feature concatenates all the orthologs from one virus into a single sequence with a series of Xs to separate each gene.
5. Results open in Base-By-Base genome editor tool. Align concatenated sequences by clicking *Edit* > *Select Whole Sequences,* and then *Tools* > *Align Selection.* During the process, the Xs align together and force the neighboring ortholog groups to be aligned separately. Without the Xs, sequences might align across the concatenated junctions between orthologs, and create errors for any extended or truncated genes.
6. Export the alignment under *File* > *Save As* and select FASTA format.
7. Open the FASTA file in a text editor, remove "uppercase X" via simple search, and replace.

This feature was added to VOCs to facilitate the creation of MSAs for building phylogenetic trees with concatenated gene sets (*see* Subheading 3.11), where the use of single genes may not provide a sufficient signal but the use of complete genomes would introduce errors into alignments because of the indels at the termini. It overcomes the very tedious process of collecting multiple genes from multiple viruses into concatenated sets.

3.4 The Viral Genome Organizer

VGO [5] provides a simple-to-use, but powerful, graphical interface to interact with complete poxvirus genomic sequences. Because VGO also understands and accesses the VOCs database, it manages large amounts of information including the complete genome sequence, all gene and protein sequences, and can display (1) genes designated in the genome GenBank file; (2) computer-predicted open reading frames (ORFs) of any user-selected size; (3) all start and stop codons; (4) search results for restriction sites or any other subsequence defined in a regular expression (e.g., TTTTTNT); (5) graphs of nucleotide composition; and (6) results from a user-defined input file (*see* **Note 6**). VGO also permits the user to quickly display gene and protein sequences by clicking on their graphical representation in its genome map; these sequences are then available in windows that allow the user to copy/paste sequences into other analysis tools. Particularly useful features of VGO are that it (1) allows a user to select and retrieve any region (DNA sequence) of a genome by simple interactions with the graphical genome map; (2) highlights orthologs in multiple genomes, thereby providing a visual way to view synteny between genomes; and (3) displays all start and stop codons in all six reading frames at once, allowing the user to scan for potential frameshift mutations.

VGO was designed with *comparative genomics* in mind, and can display multiple genomes in a single window (limited only by screen "real estate"). As a *stand-alone* tool, VGO can read GenBank files; however, additional powerful features are available if genome files are read from the VOCs database. For example, the *Auto-highlight Related Genes* feature is then able to highlight orthologs of any selected gene in the other viral genomes being displayed. When comparing multiple genomes, it is convenient to label the genes either by their gene numbers (for easy reference with the VOCs database) or with VOCs *Family Number* (Ortholog Group). The latter allows orthologous genes in the multiple viruses to appear with the same number. For example, this solves the problem that the GenBank names for DNA ligase are A50R, J4R, K4R, 168R, 148, 171, and 188 in a series of Orthopoxviruses, whereas they are all in the same VOCs Ortholog Group, #292. VGO is also capable of generating graphs for nucleotide composition across genome(s).

Since VGO performs comparative analysis and similarity searches with ease, the next exercise describes a protocol that can determine homology across genomic regions. We use this to validate a gene (<65 codons) in myxoma virus that has no annotation in other viruses (Fig. 1).

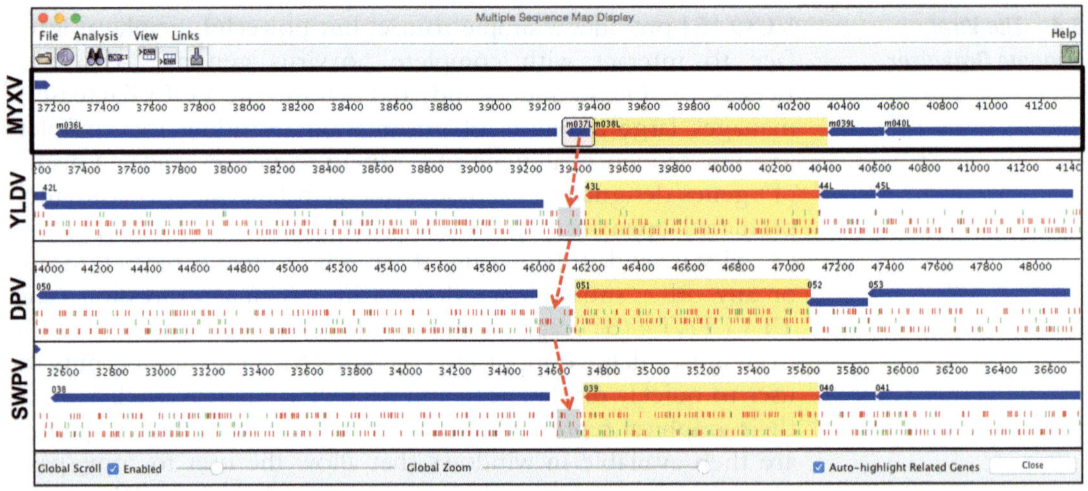

Fig. 1 Viral Genome Organizer (VGO), a depiction of the analysis from Subheading 3.4.1. Rectangles are genes; red blocks are auto-highlighted orthologs; red dashes represent small ORFs that could be potential orthologs of myxoma-m037L gene. On the computer screen (and the electronic version of this chapter), start/stop codons are green and red, respectively; however, we are evaluating color changes to accommodate users with color-blindness

3.4.1 VGO Use Example #1: Validating a Small ORF as a Real Gene (See **Note 7***)*

1. Open VGO as described in Subheading 3.1.
2. From *File* > *Open* > *Open from VOCs DB*, select myxoma (MYXV-Lau), and the other clade II poxvirus representatives like yaba-like disease virus (YLDV-Davis), deerpox (DPV-83), and swinepox (SWPV-Nebraska) (*see* **Note 8**); click OK.
3. Select all four viruses (*see* **Note 8**) from the *Working List* (pink bars in VGO window), and from the *View* menu choose *Sequence Map*.
4. From the *Viewing Options* button for one virus, check boxes for *Show Start/Stop Codons* and *Show Gene Labels: GenBank name* and then click *Apply to All* (*see* **Note 9**).
5. Stretch the windows vertically to see all frames for the viruses. Use the *Global Zoom scroll* bar to zoom until genes and gene names are legible (*see* **Notes 9** and **10**). Select the *Auto-highlight Related Genes* box at the bottom of the window.
6. In the *MYXV* window, scroll to find and select gene m038L (color changes to red).
7. Scroll along all the other genomes until the orthologous highlighted gene (red) appears.
8. Manually align the m038L orthologs in the windows using scroll bars.
9. Click on MYXV-m036L. Thus, flanking genes are syntenic for these viruses. However, notice that in between the genes, MYXV has a very small gene (m037L; annotated with no known function) not found in the others. However, review of the *Start/Stop* codons indicates that the ORF is conserved. So, the question is this: *Is this small gene (32aa) a wrong annotation in MYXV or did it get missed in all the other viruses?*
10. Select *Genome Subsequence* from the VGO *View* menu. A new window appears.
11. Use your cursor to drag a box over the region of the DPV genome between genes 050 and 051 on the bottom strand (this fills in the correct coordinates into the *Genome Subsequence* window).
12. Click the *Display button* in the *Subsequence Grabber* window; on the top panel, click *Analysis* and run BLASTx against the VOCs database (NCBI database also available).
13. BLASTx shows a small ORF in this DPV region that is similar to VACV-Cop-O3L orthologs as well as m037L orthologs from the MYXV species.

14. Repeat **steps 10–13** with the other viruses to find unannotated conserved sequences similar to MYXV-m037L. Additionally, BLASTx results suggest that both the missed ORFs and the m037L orthologs are orthologs/relatives of the O3L gene. The small O3L ORFs were characterized in 2009 [22], and chordopoxvirus orthologs were confirmed to participate in entry-fusion complex in 2012 [23].

3.4.2 VGO Use Example #2: Search ECTV-Moscow Genome for All TTTTTNT Sequences (See **Note 11***)*

1. Open VGO and the ECTV-Moscow genome as described in Subheading 3.4.1.
2. From *Analysis* menu, select *Search Selected Sequence/Reg. Expression search* (nucleotide pattern is represented as a regular expression; https://en.wikipedia.org/wiki/Regular_expression).
3. Type *TTTTTNT or TTTTT.T* (signals for termination of early transcription) into the box, and click OK.
4. Results are displayed above (forward strand) and below (reverse strand) the gene representations.

3.5 Comparison of Poxvirus Genomic Sequences in JDotter

One of the indispensable tools for the comparison of large DNA virus sequences in pairwise fashion is the dotplot. Each nucleotide, or small window of nucleotides, of one sequence is compared to every nucleotide of another and the results are visually displayed in an easy-to-understand plot. The two sequences form the horizontal and vertical axes of a matrix. Each cell in the matrix records whether the residues from the horizontal and vertical axial positions are identical/similar. For analysis of poxvirus genomes, the software must be able to handle DNA sequences in excess of 300 kb. We have found that DOTTER [6] is an extremely effective tool because *after* the plot is calculated, the user can change scoring parameters in real time as they are viewing the plot. It is also possible to zoom into particular regions of the dotplot by selecting an area with the cursor. This results in recalculation of the plot in that smaller region. We have created a Java interface for DOTTER (JDotter) [7] that permits it to be used as a graphical display in the VOCs interface with complete genomes and gene or protein sequences. The regions of high similarity, as well as gaps in the alignment, are immediately obvious on a dotplot. The *GreyMap Tool* (Fig. 2; inset) is used to rapidly change scoring parameters without the need for recalculation of the complete dotplot. An alignment tool is also available (not shown) which displays a continuously scrollable window showing the alignment of the two DNA sequences at any point in the plot chosen by the user. Dotplots are particularly useful for detecting direct and inverted repeats; the poxviral inverted terminal repeats (ITRs) are visible in the top right and bottom left of the dotplot in Fig. 2. Users of dotplots should be aware that the resolution of the plot is low (related to the number of nucleotides/screen pixel) when large sequences are compared; therefore, to get a good sense of fine sequence similarity from a dotplot, it is necessary to "*zoom-in.*"

3.5.1 Visualizing Potential Horizontal Gene Transfer (HGT)

1. Load the GenBank file of *Molluscum contagiosum* virus strain subtype1 (MOCV-st1) as both the horizontal and vertical sequence input.
2. Click *Run Dot Plot* and use default settings.
3. Forward and reverse genes are displayed as red and blue rectangles, respectively, on the top and left edges of the Dot Plot Results.
4. Notice the "*stripes*" in the background (regions with less dots; Fig. 2).
5. For better contrast, try adjusting the scoring parameters using the GreyMap Tool (Fig. 2).
6. Reviewing the sequence in VOCs and VGO indicates that the sequences in the "*stripes*" have distinct nucleotide composition

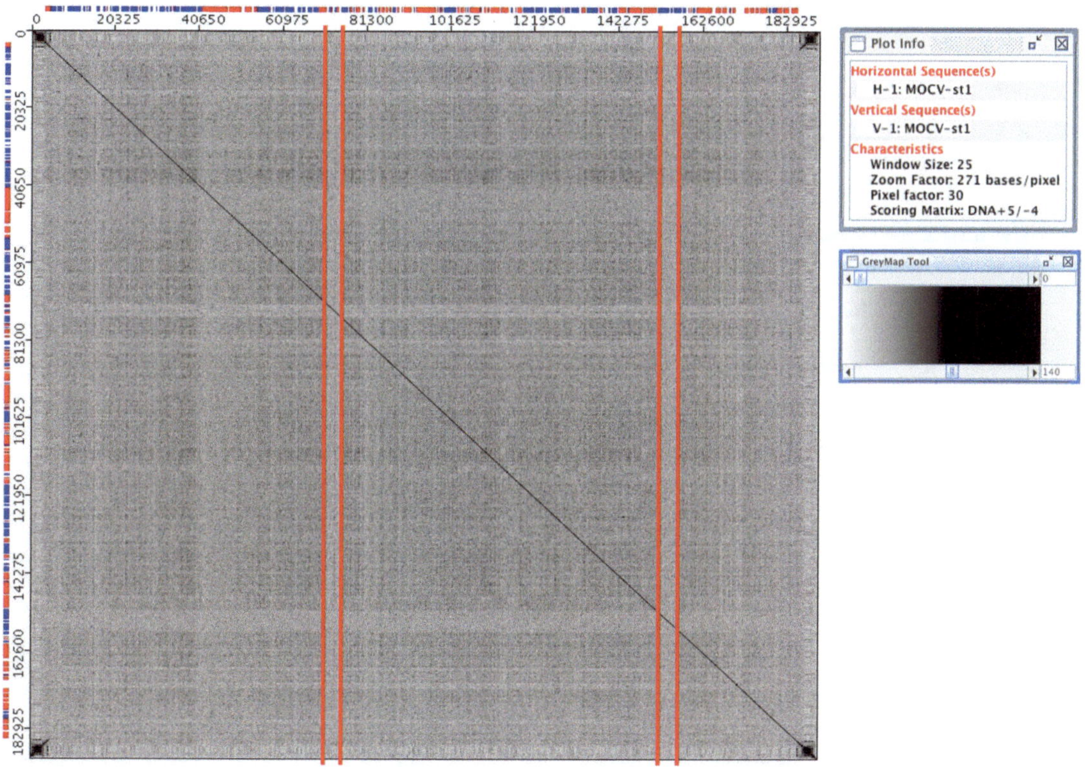

Fig. 2 A dotplot of complete DNA genome of molluscum contagiosum virus (subtype 1 strain) compared against itself (a selfplot). *Insets* show (1) *GreyMap Tool* for changing scoring parameters and (2) plot information. *Small blocks* along axes are usually colored to represent genes with transcription orientation. Red lines highlight genomic regions of distinct nucleotide composition (suggested as potential HGT regions)

compared to the MOCV-stl genome average. This suggests that they may be pathogenicity islands in MOCV with potential HGT origins [24].

3.6 Annotation of Poxvirus Genomes in GATU

Once a poxvirus genome has been sequenced there remains the, sometimes challenging but always time-consuming, problem of annotating the genome. The level of difficulty is inversely proportional to the similarity of the genome to other annotated genomes that can be used as a reference. Although poxvirus genes do not contain introns, which complicate the prediction of many eukaryotic genes, deciding which ORFs should be designated as a (potential) gene and annotated in the GenBank file is still difficult; the problem is exacerbated in GC-rich genomes because of the low frequency of stop codons in noncoding sequence.

Many ORFs are simple to annotate because they have been shown to be expressed or are conserved in a number of diverse poxviruses. To take advantage of the growing resource of annotated poxvirus genomes, we developed GATU [12] (Fig. 3), a tool that transfers annotations from a reference genome to the new target genome. Depending on the similarity between the reference and target genomes, 70–100% of the genes may be annotated by

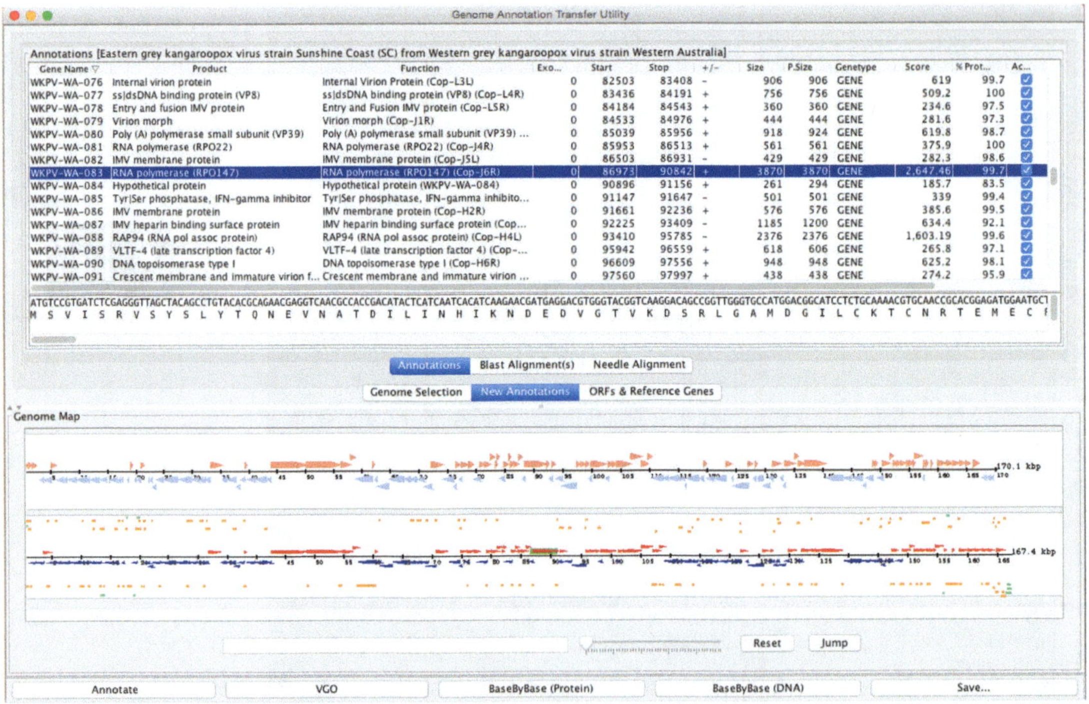

Fig. 3 Gene Annotation Transfer Utility (GATU). *Top panel* shows list of genes predicted to be located in the genome requiring annotation. *Bottom panel* shows a genome map of the reference genome (*top line*) and predicted genes of the genome to be annotated (*bottom line*). On the computer screen, gene symbols are colored to provide clarity

GATU with essentially no effort by the user. An important feature of GATU is that the tool leaves final control of the annotation process in the hands of the user (Fig. 3). Although obvious annotations suggested by GATU are preselected in checkboxes for acceptance, the user can reject these if required. Similarly, novel ORFs or those that significantly vary from orthologs in the reference genome are provided to the user as *suggested ORFs*. When the user is happy with the final annotations, these can be written out either as a GenBank file or as a Sequin table file for submission to GenBank.

However, small ORFs that have no obvious ortholog within the reference, or other poxvirus genomes, are especially problematic for annotators. The VGO genome display tool, which can be used to examine colinearity between the target and reference genomes, may help in these situations; it also provides A+T% plots for searching for potential AT-rich promoter regions and a convenient display of start/stop codons in the six possible coding frames to allow the annotator to look for potential sequencing errors that may have broken otherwise complete orthologous genes. A number of criteria have been applied to poxvirus gene prediction, the simplest being ORF size; others include potential overlap with other genes, presence of promoter-like elements, isoelectric point, amino acid composition [25], and codon usage. For vaccinia virus, and other AT-rich poxviruses, the coding strand of genes is positively correlated with purine content [26].

For genome annotation, our philosophy is that "less is more" because it is easier to add annotations than take them away once a genome has been accepted by GenBank. For example, vaccinia virus strain Copenhagen [1] was initially annotated with major ORFs (genes) and 65 minor ORFs; most of the latter substantially overlap with larger genes, usually on the opposite DNA strand. These extra ORFs were named "X-ORF-Y," where X represents the *HindIII* genome fragment and Y is a letter representing the rank of the ORF from left to right. It is unlikely that any of these minor ORFs are functional genes, but to the inexperienced eye it appears that this virus has a large series of unique genes. It would, therefore, be useful if annotation systems could include the option for *probably nonfunctional ORF* descriptions since relatively minor mutations can easily destroy gene function without changing the ORF very much.

3.7 Searching for Distantly Related Proteins

A common question in bioinformatics is as follows: "What is my protein (or DNA) sequence similar to?" Similarity searches in which a protein or DNA sequence is searched against a database of all known sequences are most often performed with one of the BLAST programs [21]. There are various search algorithm strategies, and an important design factor is how they balance search sensitivity and search speed. In this regard, it is important to note that default BLAST search parameters are not set to their most sensitive; for example, *WORD SIZE*, the match length that triggers an extended alignment of sequence regions, should be adjusted to the lowest possible for the most sensitive searches. Sometimes, to increase speed it is useful to limit the size of the database. For example, if one is only interested in searching for poxvirus orthologs, a BLAST search can be restricted to Poxviridae (taxid:10240) and unclassified Poxviridae (taxid:40069) in the browser interface (Choose Search Set > Organism). Similarly, this parameter can be set to limit searches to specific poxvirus species, genera, or other organisms (for example a suspected host). For "routine" protein database searches, BLASTP is sufficient to find database matches at >30% identity. However, a more sensitive database search program is PSI-BLAST [20] which automatically constructs a new position-specific scoring matrix (PSSM), using a multiple alignment of the highest scoring matches to be used in the next round of an iterative series of BLAST searches. Thus, on each round of searching, the program uses a modified scoring matrix that reflects the most conserved residues in sequences that have already been identified as similar to the query. Users can also perform their searches using the Domain Enhanced Lookup Time Accelerated BLAST algorithm (DELTA-BLAST) [27]; this creates a PSSM with the annotated domains in Conserved Domain Database (discussed in Subheading 3.8) and searches it against the protein nr sequence database to enhance similarity detection. Another addition to the NCBI PSI-BLAST [20] utility is the constraint-based multiple alignment tool which is enhanced for creating multiple alignments of distantly related proteins (COBALT) [28]. Another variation is to use a profile-based database search created from a set of related sequences rather than a single sequence. These can be more sensitive than sequence-based approaches such as PSI-BLAST [20]. One example of a profile-based search is HHpred (https://toolkit.tuebingen.mpg.de/#/tools/hhpred), which uses a profile based on a hidden Markov model (HMM) to perform the search [13]. HHpred initially runs several iterations of PSI-BLAST to create a multiple alignment of the query plus related protein sequences. This alignment is then turned into an HMM that is used to search HMMs previously created from the protein structure databases (PDB, SCOP, and CATH). The HMMs of both the query and the proteins in each of the databases are created based on secondary structure either from a prediction of secondary structure using PSI-PRED or from known 3D structures. Results (hits) are displayed with a probability that the query sequence is a true match to the hit sequence.

Often, poxvirus proteins are given predicted functions based on low similarity scores to presumed homologs; if the proteins can be shown to be structurally related, this type of prediction is greatly strengthened. However, since determining a protein's structure is difficult and time consuming, a bioinformatics solution is often used, that is, to determine if it is possible to fold the unknown protein sequence such that it creates a structure similar to that of the known structure. A number of tools are available to model tertiary structure: (1) Protein Homology/analogY Recognition Engine (PHYRE) [29] produces results relatively quickly (within an hour), which models only the subsection of the protein that aligns with a hit in the database; (2) Robetta [30], which will try to model difficult regions using ab initio approaches, has the disadvantage that it can take weeks to produce a result; and (3) I-TASSER (Iterative Threading ASSEmbly Refinement) [31], which is newer high-performance 3D protein modeler. Recently, we used I-TASSER to show that a novel protein from the bat-derived Pteropox virus (PTPV-Aus-040), which has similarity to a TNF-like ligand domain, could be modeled onto a TRAIL protein [32], thereby providing support for the predicted function. The UCSF Chimera [33] is an excellent downstream interactive visualization tool for 3D structures; the *Match->Align* function (under *Tools > Sequence*) generates structure-based sequence alignment.

3.8 Motif Searches

A protein motif, often described by a regular expression, can be defined as a short series of amino acids that is characteristic of a functional/structural unit (domain) in a series of proteins. One of the most commonly used motif databases is PROSITE (http://prosite.expasy.org/) [34]. In Sept 2017, PROSITE contained more than 1700 different entries. Motif searches are frequently useful in identifying domains within a protein when no other large areas of similarity exist with other proteins; that is, it acts as a "*fingerprint.*" For example, [KR]-[LIVA]-[LIVC]-[LIVM]-x-G-[QI]-D-P-Y is the PROSITE (PS00130) motif for uracil-DNA glycosylases (UNG). It is important that PROSITE motifs are continually updated, as new members of a protein family are recognized. For example, since the first edition of this book, the number of acceptable amino acids at the second position of this motif was increased to four with the inclusion of alanine. Similarly, one needs to appreciate that although this motif detects over 400 UNG proteins, there are also 6 UNGs that are not found by searching with this particular motif; these are designated *false negatives* in the PROSITE documentation. These *false negatives* are a consequence of maintaining this strict motif that minimizes false-positive hits. This particular PROSITE motif is written as a regular expression (a *pattern* in PROSITE), a format that allows mismatches and variability in spacing between residues in the motif. Software tools within VOCs allow users to search sequences with any regular expressions; at https://prosite.expasy.org/, software can be used to search (1) a protein sequence against all motifs or (2) a PROSITE or user-created motif against all protein sequences. Advantages of performing the search through VOCs are the speed and the fact that only poxvirus sequences are searched. It should be noted, however, that the PROSITE (PS00130) motif (omit the dashes) only matches 334 of the 353 UNGs present in VOCs, whereas a modified motif [KLNR]-[LIV]-[LIVC]-[LIVM]-x-G-[QIY]-[D]-[SP]-[YF] (omit the dashes) will find 350 of 353 UNGs in VOCs but matches several *false-negative* proteins in SWISS-PROT. Some PROSITE motifs (*profiles* in PROSITE) are scored using an amino acid match scoring matrix but are included in the ScanProsite [35] search at https://prosite.expasy.org/scanprosite/.

In addition to PROSITE, there are several systems that integrate multiple motif/profile searching tools. The Conserved Domain Database (CDD) Search [36] (https://www.ncbi.nlm.nih.gov/Structure/cdd/wrpsb.cgi) utilizes the RPS-BLAST program, a variant of PSI-BLAST, which quickly scans the query sequence against the position-specific scoring matrices (PSSMs) of annotated conserved protein domains. These PSSMs have adjusted scores for conserved residues specific to each protein profiles (from MSA), and provide a useful supplement to hone in and confirm BLASTP alignments with low % identity scores. The MotifScan Server (https://myhits.isb-sib.ch/cgi-bin/motif_scan) searches user-supplied protein sequences for Pfam (*P*rotein *Fam*ilies database of alignments and HMMs) and PROSITE motifs. The Pfam Web site (http://pfam.xfam.org/) lists families for many of the poxvirus proteins that are common to all poxviruses and is an excellent source of information, although the output can be rather overwhelming and difficult to interpret. InterPro (https://www.ebi.ac.uk/interpro/) is another comprehensive assembly of motif and sequence databases that is in turn connected to a variety of other databases; it is especially useful for combined searches.

A further application for motif searches is for the researcher to develop their own regular expressions from a MSA of the most varied of a set of orthologous proteins. This pattern, which is first designed to be specific to the poxvirus orthologs, is gradually made less specific by adding more matching options to the regular expression and the results of protein database searches are monitored for matches that may be significant to poxvirus biology. This process requires that the investigator understands both protein biochemistry and poxvirus biology as it does not give a black and white answer.

3.9 Multiple Sequence Alignments in Base-By-Base

The generation of MSAs is a "bread and butter" technique in comparative molecular biology. Determining which amino acids are conserved at particular locations in a group of proteins may help predict which residues are likely to have important roles in the structure or biochemistry of the proteins. A variety of computer programs are available to align DNA and protein sequences, attempting to maximize a score from matching amino acids or nucleotides and also minimize penalty scores due to insertions and deletions (indels). One of the best known alignment tools is ClustalO [8], but other algorithms include T-COFFEE [37] and MUSCLE [10]. DNA alignments of short- and medium-length sequences, such as promoters and genes, respectively, can be achieved with the same software, but sequences the length of poxvirus genomes benefit from specialized tools such as MAFFT [9], DIALIGN [38], and MAUVE [39].

However, whatever software is used, MSAs should be carefully reviewed as they often need final adjustments to be made *by-hand*, especially around gaps, using a sequence alignment editor to produce final accurate alignments. A final hurdle is the production of a publication quality figure, if required. We developed the Base-By-Base (BBB) tool [11] for creating, editing, and viewing not only MSAs of proteins and short DNA sequences, but also complete poxvirus genomes. BBB was developed by our group as a Java MSA editor to interface with VOCs, but it also functions as a stand-alone tool and can save files to be used at later times, in the same way that word processors do. Over the years, BBB has been enhanced with several unique features, for example (1) easy display of differences between adjacent sequences, or between the top sequence in an alignment and the others; (2) display of three-frame translation of DNA sequences; (3) display of *top* or *bottom* strand for a DNA sequence; (4) ability to read in GenBank files including annotation on DNA sequences; (5) ability to read sequences from VOCs database; (6) ability to search for internal sequences using regular expressions or fuzzy (allowing mismatches) motifs; (7) ability to add *user-comments* to sequence regions; (8) ability to realign internal regions of an MSA with several algorithms and import results to the existing alignment; (9) generate % identity table for sequences in an alignment; (10) map primer sequences to a DNA sequence; and (11) save alignment pictures for publication. Finally, one of the most powerful features of BBB is its ability to summarize differences between genomes in an MSA. When provided with two or more aligned and annotated genomes from closely related viruses (e.g., isolates of variola virus), the "*View CDS Statistics*" and the "*View Multi Genome Comparisone Statistics*" features (under the *Reports* menu bar) in BBB can detect all nucleotide differences and present this information with an analysis of the consequences of the differences (truncation of ORF, silent mutation, coding change, in predicted promoter region).

The detection of SNPs across poxvirus genomes allows virologists to visualize patterns of SNPs associated with particular sets of genomes. For example, it is easy to show which SNPs are uniquely associated with the clade that contains smallpox [40, 41]. The "*Find Differences*" feature allows the user to select one or many viruses to be the reference or target sets. The query is "*find SNPs present in virus group A, but absent from virus group B.*" The searches can be set to "*tolerate*" errors in the form of false-negative/positive matches, which can occur via additional recent spontaneous mutations. The results are viewed in tables and charts that help highlight patterns between the groups of SNPs that are associated with different evolutionary events. The following exercise describes the identification of SNPs in the O1L gene that may be associated with VARV host range [41].

3.9.1 Identification of SNP Patterns in VARV O1L Gene

1. Open VOCs as described in Subheading 3.1; if the tool is already open, go to the *select* menu and *clear all* to remove any previous search parameters.
2. Multi-select a representative set of orthopoxviruses; ensure that TATV-DAH68 (NC_008291), CMLV-CMS (AY009089), and VARV-UK1946-Harvey (DQ441444) are included. Display these sequences by clicking *Base-By-Base* under the *Tools* menu.
3. In BBB, click *Edit > Mark All Sequences* to highlight all sequences. Then click on *Tools > Align selection with MAFFT*. Manually edit the alignment in BBB as needed.
4. Once sequence alignment is done (relatively slow), select *Advanced > See Advanced/Experimental Tools > Find Differences*. Multi-select TATV-CMLV-VARV as a group under the list that looks for "all the same" SNPs in the chosen group; multi-select everything else under the other list to look for "all different" SNPs in the remainder viruses. Click "Ok."
5. A log for the positions of all the desired SNPs is generated. Add these positions as comments to highlight SNPs in the BBB window.
6. In the BBB window, under *Reports > Visual Summary*, zoom and scan to find a cluster of SNPs at ~21,000 bp in the alignment. Scroll to this region in the main BBB window. Toggle on the "show/hide translation" (left-side bar) to find that these SNPs correspond the left transcribing O1L genes in the TATV-CMLV-VARV.
7. Repeat another round of *Find Differences* from **step 4**; this time set a tolerance parameter of 2. Mark the SNPs in another color. The number of SNPs in the O1L sequences should increase by >20 SNPs.
8. Repeat another round of *Find Differences* from **step 4**; this time exclusively compare TATV, CMLV, and VARV against each other to elucidate event post-speciation of these viruses. Mark the SNPs in another color.

3.10 Poxvirus Whole-Genome Assembly

Genome sequencing and assembly serves as the basis for all comparative genomics work. With the dramatic growth of next-generation sequencing (NGS) (low costs, ease of sample preparation, and increased access to sequencers) in recent years, researchers can sequence complete poxvirus genomes with relative ease. Similarly, the sequencing reads from poxvirus genomes can now be processed and assembled on an average desktop computer (e.g., the work described here was performed with 3.2 GHz Intel core i5; 16 GB memory). Below we describe methods to assemble poxvirus genomes using a selection of free software (a combination of command line (CL) and GUI programs).

A three-part method for the assembly of poxvirus genomes is described. First, we describe a pre-assembly quality control (QC) protocol to identify and remove contaminants. Second, we provide two sample scripts for assembly with two different assemblers: SPAdes (St. Petersburg genome Assembler) [15] and MIRA (Mimicking Intelligent Read Assembly) [14]; the two assemblers can be used exclusive to each other, or in parallel to compare the contigs that are generated. Third, we provide suggestions to troubleshoot failed assemblies and check the validity of the assembled genome. The key steps are listed in Subheading 3.10.1 and the associated command-line commands are referenced in Table 3.

3.10.1 Pre-assembly Quality Control of Raw Sequence Reads

1. Identify any major source(s) of DNA contamination by submitting the raw read FASTQ sequence file(s) to Taxonomer (http://taxonomer.iobio.io/) [42]. This metagenomic Web tool traces raw reads back to their source organisms/viruses, and illustrates the distribution in a pie chart.
2. For the major contaminant source identified in **step 1**, download the corresponding reference genome FASTA file from the NCBI (*see* **Note 12**). The goal is to remove these contaminating reads before attempting assembly.
   - The human reference genome can be downloaded here: https://www.ncbi.nlm.nih.gov/projects/genome/guide/human/index.shtml.
   - Other organisms can be searched under the NCBI database for "Genome" here: https://www.ncbi.nlm.nih.gov/genome. Depending on the organism, reference genomes could exist as a complete genome, or a scaffold of the genome, or there could be no available genome at all. In the latter case, find the genome for the closest relative.
3. Create a database from the contaminant genome using the Burrows-Wheeler Aligner program. Corresponding CL commands are listed in Table 3, and should be cross-referenced with the alphabetical substeps below.
   A. Index the contaminant FASTA sequence into a database (a quick process). A series of database files will be created.
   B. Map the raw reads to the database (a slow process) to generate a SAM file (*see* **Notes 13** and **14**).

Table 3
List of CL commands and tools used, cross-referenced with the steps listed under Subheading 3.10.1

A	`bwa index RefGenome.fasta`
B	`bwa mem RefGenome.fasta R1.fastq R2.fastq > ref_aln-pe.sam`
C	`head ref_aln-pe.sam`
D	`samtools view -bS ref_aln-pe.sam > ref_aln-pe.bam`
E	`samtools view -bT RefGenome.fasta ref-pe.sam > ref_aln-pe.bam`
F	`samtools flagstat ref_aln-pe.bam`
G	`samtools sort ref_aln-pe.bam ref_sorted`
H	`samtools view -S -f0x4 ref_sorted.bam | wc -l`
I	`samtools view -b -f 4 ref_sorted.bam > ref_unmapped.bam`
J	`bamutils tofastq -unmapped -read1 ref_unmapped.bam >& ref_unmapped_R1.fastq` `bamutils tofastq -unmapped -read2 ref_unmapped.bam >& ref_unmapped_R2.fastq`
K	`expr $(cat ref_unmapped_R1.fastq | wc -l) / 4` `expr $(cat ref_unmapped_R2.fastq | wc -l) / 4`
L	`fastqutils properpairs ref_unmapped_R1.fastq ref_unmapped_R2.fastq processed_R1.fastq processed_R2.fastq`
M	`expr $(cat processed_R1.fastq | wc -l) / 4` `expr $(cat processed_R2.fastq | wc -l) / 4`
N	`[(From step N: R1 # + R2 #) x average read length) / (expected virus length)]`

Note that the provided scripts use arbitrary file names; modify your files accordingly (*see* **Notes 15** and **16**)

4. From the SAM file, isolate poxviral reads by extracting *only* reads that are *unmapped* to the contaminant source.
   C. Check if a header exists for the input SAM file. A header exists if the sampled read has @ at the start of its sequence name.
   D. If a header exists, then convert to a BAM file using command D in Table 3.
   E. If a header does not exist, then convert to a BAM file using command E.
   F. Retrieve basic statistics for the mapped BAM file (*see* **Note 14**). A focus is put on the number of reads to crosscheck pre-QC file with the processed file in **substep K** below (fast).
   G. Sort the BAM file. This is a technical step needed for downstream processing (relatively slow).

H. From the sorted BAM file, count how many reads are unmapped to the reference genome.
I. Extract only the *unmapped* reads.
J. Convert the file from BAM to FASTQ.
K. Count the number of extracted reads in the processed file.
L. If working with paired-end data, pair the extracted reads and remove singletons using the fastqutils feature from NGSUtils tool.
M. Count the number of reads in the final post-QC file to get an idea of the extent of the decontamination process.
N. Calculate the average read coverage to get an estimate of the extent of the QC process.

3.10.2 De Novo Sequence Assembly

In this section, we describe two free CL genome assemblers, and provide sample protocols for running the most basic *de novo* genome assemblies from a set of arbitrary paired-end Illumina reads. SPAdes is based on the De-Bruijn Graph (DBG) algorithm, whereas MIRA uses the Overlap Layout Consensus (OLC) algorithm. The assemblers can use reads sequenced by Illumina, IonTorrent, and PacBio. MIRA additionally runs Roche (454) sequencing reads, and performs its own quality control on base-call qualities, as well as trimming of sequence tags. The authors have found that SPAdes generally runs slightly faster than MIRA and uses less memory. Although benchmarking has not shown one assembler to be better than the other [43], DBG-based SPAdes might work slightly better with Illumina data, whereas OLC-based MIRA works better with IonTorrent data. Unfortunately, for now both SPAdes and MIRA run only on macOS computers. Official assembler manuals outline more thorough descriptions of the various parameters, options, and customizations.

Sample commands for running poxvirus genome assembly on different assemblers:

1. Install the latest versions of SPAdes (3.13) from http://cab.spbu.ru/software/spades/, and MIRA (4.0.2) from https://sourceforge.net/projects/mira-assembler/.
2. Optional: Put the paired-end read files into the same folder as your assembler tool (required, if you do not add the tool to the user's system path) (*see* **Note 17**).
3. In the CL window, use "cd" to change directory from the root to the folder where the read files are located.

On SPAdes (St. Petersburg genome Assembler)

(a) In the CL window, run SPAdes assembly to completion using the command:

```
spades.py -1 R1.fastq -2 R2.fastq -m 14
          --careful -o SPAdes
```

(where spades.py is the python script that runs the assembler; R1.fastq and R2.fastq are the paired-end read files; --careful is the parameter used to minimize mismatches and indels; -o is the output file prefix; and -m designates the memory (in GB) allotted for this task) (*see* **Note 16**).

On MIRA (Mimicking Intelligent Read Assembly):

(a) In the same folder, create a text file (referred to as a manifest file ending with the extension of .conf) that is customized for a *de novo* assembly of paired-end Illumina read files (*see* **Note 18**):

```
project = DeNovoAssembly
job = genome,denovo,accurate
parameters = -NW:cmrnl=no -GE:not=4
readgroup = SomePairedEndIlluminaReads
data =R1.fastq R2.fastq
technology = solexa
template_size = autorefine
segment_placement = ---> <---
```

(b) In the same working directory, run MIRA assembly to completion using the command:

```
mira Manifest.conf
```

(where mira calls for MIRA assembler; Manifest.conf is the name of the manifest file you created in **step 3** and should be located in your working directory) (*see* **Note 16**).

3.10.3 Troubleshooting for Failed Assemblies and Validation for Successful Assemblies

Genome assemblies may not always be successful or even run to completion due to various factors. Below are some directions for troubleshooting a failed assembly as well as validating an assembled genome.

1. Genome assembly is a very memory-intensive process and may be too much for a desktop computer to handle. If switching to a better computing station is not possible, try reducing the input file sizes as follows: (1) shortening the names of the reads can reduce file size significantly (*see* **Note 20**); (2) repeat the QC step (*see* Subheading 3.10.1) to remove any secondary contaminant(s); (3) raw reads can be filtered from base-call qualities given advanced knowledge on FASTQ file structures. Note that any manipulation of raw data should be saved into a new file to avoid corrupting files (e.g., incomplete read information, disrupted paired reads).
2. Poor sample sources and poor library preparation often generate numerous short, noncontinuous contigs. These contigs should be checked for termination by aberrant primer tags [44].

3. A preliminary poxviral genome-sized sequence (~130–360 kb) should be carefully examined. Tanoti, a BLAST-guided reference-based short read aligner, can be used to map raw reads to the assembled poxvirus genome; the SAM file that is created can be visualized in the Tablet program [45] to examine coverage across suspicious junctions, or make manual base calls to adjust/correct the preliminary sequence into the final genome. The assembled genome can also be checked with JDotter and BBB. A dotplot displays regions of indels (disjointed diagonal lines), repeats, and rearrangements.

3.11 Phylogenetic Tree

Phylogenetic trees aim to represent the evolutionary relationships between homologous DNA and protein sequences. However, there are several situations that may reduce the accuracy of the trees: (1) alignment of sequences that are paralogous rather than orthologous, (2) recombination events that substitute other homologous sequences, (3) inclusion of regions containing multiple indels that reduce the accuracy of MSAs, and (4) inclusion of sequences with very different nucleotide composition. For poxviruses, the use of approximately 75 kb from the most conserved central core of the genome works well, as does the concatenation of the proteins encoded by this region. Smaller alignments are much quicker to calculate and we usually use the RNA polymerase subunit RPO147 or a concatenation of about seven proteins for quick trials. If highly diverse viruses are required in a tree, then it is better to use protein sequences, whereas DNA sequences are more discriminatory if viruses are very similar. Likewise, if a figure aims to compare orthopoxviruses, nothing is gained by inclusion of G+C-rich genomes; in fact by adding such diverse sequences that require many indels to create an alignment, the quality of the MSA is likely to be reduced. Ultimately, the user must select a meaningful dataset to optimize alignment speed and accuracy.

RAxML [19] is a fast, high-performance command-line program that can be used to generate a maximum-likelihood (ML) phylogenetic tree for chordopoxviruses. The ML analysis is a computationally intensive process, but also one of the most stringent phylogenetic methods. It builds every possible tree for the input MSA, and then employs a substitution model to calculate the probability of each phylogeny branch; the tree with the highest likelihood of occurrence is returned. Note that RAxML treats gaps as missing data, and the user should examine the alignments around gaps because alignment algorithms frequently make errors in these regions.

1. Install RAxML according to computer specs, here https://cme.h-its.org/exelixis/software.html.
2. Identify an appropriate substitution model for this particular protein set:

```
raxmlHPC-PTHREADS-AVX2 -p 12345 -m PROTGAMMA-
AUTO -s MSA.fasta -n AUTO
```

(PTHREADS-AVX2 was determined specifically based on the processor type of the computing station and will differ between computers; -p is a random number for troubleshooting purposes; -m PROTGAMMAAUTO determines an appropriate substitution model under the gamma model of rate heterogeneity; -s is the name of the MSA file in FASTA or PHYLIP format; -n specifies the name of the output file).

3. Run the maximum-likelihood phylogenetic analysis with the ide ntified model from **step 2**:

```
raxmlHPC-PTHREADS-AVX2 -f a -p 12345 -s MSA.
fasta -x 54321 -# 1000 -m PROTGAMMALG -n out-
put -T 4
```

(specifying "a" under the –f option turns on rapid bootstrap analysis; -p and –x are two different random numbers for troubleshooting purposes; -# specifies the number of bootstrap replications to be made; -m specifies the best model identified by **step 2** (e.g.: LG); -n specifies the name of the output file; -T specifies the number of thread to be used).

The results are written in Newick format (with bootstrap labels) and the bipartitionsBranchLabels file prior to can be visualized in the MEGA tool [46]. Although MEGA can display trees with a graphical user interface (GUI), it runs a maximum-likelihood analysis significantly slower than RaxML.

A rectangular phylogenetic tree is the common tree seen with an inferred or arbitrary root (a node that dictates the earliest common ancestor) often created by the inclusion of a distantly related sequence called the "outgroup." A radiation phylogenetic tree is used when the root (the common ancestor) is not inferred. The latter is the preferred format for chordopoxvirus phylogenetic tree, as it does not suggest an ancestral root that dictates the direction of evolution. It should be kept in mind that a phylogenetic tree represents an *average* of the genetic differences regardless of the localization of these substitutions. A gene could have several domains subjected to different mutational constraints. To fully analyze these differences and their implications (synonymous vs. non-synonymous changes, or functionally similar aa substitution), one needs to examine the actual MSA.

3.12 Staying on Top of Your Research: Data Notification Alerts

With the development of new sequencing technologies, which have drastically reduced the cost of DNA sequencing, the NCBI database has since accumulated more than 400 million whole-genome sequences (WGS); this is 2800× more than 15 years ago.

The VOCs database has grown from 30 poxvirus sequences in 2004 to 114 genomes in 2012, to now 350+ genomes in 2017. However, many of the predicted genes have no function assigned to them. Currently, there are gt;500 hypothetical protein families yet to be characterized in VOCs. Similarly, 1 in 3 NCBI protein sequences has no assigned function, and 1 out of 10 is annotated as conserved hypothetical protein [47]. It is important to note that the function of a poxvirus protein is probably as likely to be discovered by researchers working on an ortholog in a distantly related organism (not poxviruses) as a group working on poxviruses. Therefore, in addition to "keeping up with the literature," it is key to be aware of new additions in the databases. Fortunately, several notification systems have been developed to provide researchers with a variety of "alerts."

With regard to sequence data, NCBI can be configured to send emails when new sequences match a user-defined query search term (https://ncbiinsights.ncbi.nlm.nih.gov/2013/11/14/setting-up-automatic-ncbi-searches-and-new-record-alerts/). For example, when searching the NCBI nucleotide database, the following query retrieves all complete poxvirus genomes: ((poxviridae[Organism]) AND complete[Title]) AND genome[Title]. Users can then enter this into the NCBI search box, and press the *Create Alert* button to set up email alerts on this particular search term. The query can be modified or restricted to specifically look for only chordopoxvirus protein sequences via search term chordopoxvirinae[Organism] against the NCBI protein database. PDBAlerts follow the same principle with 3D structures in the Protein Data Bank (PDB).

Similarly, there are several ways to be notified about new literature. First, most scientific journals email users the Table of Contents from new issues. Second, Rich Site Summary (RSS) feeds provide updates from a variety of sources (web blogs, journal contents, PubMed searches) to an RSS Reader such as Feedly (https://feedly.com/). Third, the social media tool Twitter can be set up to broadcast when new articles matching specific keywords are released; for example, we have established a twitter account (@pox_papers) to tweet when the keyword "poxvirus" appears in PubMed papers.

Finally, the Internet brings many researchers together via online research topic themed blogs and discussion forums. Bioinformaticians seem to be an especially helpful/collaborative group of researchers and the many bioinformatics blogs attest to their willingness to share knowledge and expertise. If you have a bioinformatics problem, it is well worth reviewing topics at the Biostars Forum (https://www.biostars.org), and if you cannot find the answer—post a question yourself!

4 Notes

1. A comprehensive 2017 review on virus databases [48].
2. VBRC (https://4virology.net/) was formerly part of the Poxvirus Bioinformatics Resource Center (PBRC; http://www.poxvirus.org) and the Viral Bioinformatics Resource Center (VBRC; http://www.biovirus.org).
3. Current desktop computers have more than sufficient "horsepower" to run basic bioinformatics tools; it is more useful to buy extra RAM and a second computer screen (for displaying data and results). Most of the software described in this chapter is available in either (a) a simple WWW form (e.g., BLAST at NCBI) interface that sends requests to a remote computer and displays the results in a WWW browser, (b) simple command-line programs, or (c) a Java client-server format in which the local client program is far more sophisticated than a WWW interface and only connects to the remote server to download information from a database or to off-load computationally intensive calculations. Java is a multi-platform, platform-independent, object-oriented programming language.
4. SQL stands for *structured query language* database.
5. Downloading Java Web Start is straightforward and takes only a few minutes. When Java Web Start downloads one of the Java clients of VOCs, VGO, or JDotter to the user's local machine for the first time, it displays a warning window by default, which informs the user that permitting this program to be installed on his or her computer is potentially dangerous. This is because the software is like most other programs on your computer and can write to your hard drive. The default warning message also includes information about the origin of the software and developers to allow the user to determine if the software comes from a trustworthy site. We recommend that you accept the software. The VBRC site is secure behind a firewall to prevent potential hackers from tampering with these files. Our group uses this software daily in exactly the same way as external users; therefore, we would quickly detect any problems with the software.
6. An example of the results from a user-defined input file is a simple format text file defining position, length, and color of boxes to be drawn on the VGO sequence map. We frequently use it to graphically display results of otherwise obtuse text files generated by promoter prediction programs.
7. Small ORFs less than 65 codons are typically not annotated as it is rare that these encode functional proteins due to their small sizes. However, the likelihood of putative genes to be real increases if there are conserved orthologous sequences in other species.

8. Use control or command (Apple) key for multiple selections.
9. Scrolling is faster with fewer items displayed on the screen. To maximize speed, do not display start/stop codons unless needed and zoom in before scrolling.
10. The annotated virus genes are shown in dark blue above (transcribed to the right) and below (transcribed to the left) the scale bar.
11. The TTTTTNT (T_5NT) sequence is a transcriptional stop signal for EARLY poxvirus genes.
12. The genome files will be downloaded either as .fasta files or .fna files. FASTA file format is the standard file type for any sequence reads, whereas FNA file specifies for nucleic acid. For downstream processing, FNA files should be changed to FASTA format simply by renaming the file to have a .fasta extension.
13. Sequence files like FASTQ read files, SAM (Sequence Alignment/Map), and BAM (Binary Alignment/Map) files can range between 100 MB and >10 GB. These files are often too large for a text editor to handle, and should be opened in the CL window or, alternatively, a GUI program called BAMseek.
14. Read more about SAM/BAM file formats and the bitwise flag stats here: https://samtools.github.io/hts-specs/SAMv1.pdf.
15. Commands related to the same protocol can be streamlined into one pipeline by adding two ampersands (&&) between each command. This way, the program will run ONLY if the previous command ran successfully. Apple computer users can additionally save the connected commands as a bash script (with .sh file extension) in the directory with all the required files, and execute it with command "./name_of_bash_file.sh" in the terminal.
16. To keep a log on the command-line processes, add ">&log.txt" to the end of a command that saves the standard output (any generated text during the process that gets outputted into the CL window) into a text file called log.txt. In another tab, the log file can be tracked in real time by using the command "tail –f log.txt."
17. For frequent usage of command-line programs, users should familiarize themselves with methods that set different folders to the user's system PATH, and thereby call bioinformatics program from the CL window without having to (1) put read files into the tool folder, or (2) specify complete file path to programs each time.
18. Inclusion of the "*-NW:cmrnl=no*" parameter overrides common halt in assembly due to read name size exceeding 40 characters. Specifying "*-GE:not=4*" asks the computer to run

the assembly process in parallel (in multiple threads of 4 if permitted computer specs) to improve efficiency.

19. Use "sed"—a substitution command (in the CL window)—or use the "rename_prefix=" parameter in MIRA.
20. If working with paired-end files, R1 and R2 read names need to be distinguishable to avoid an error that may occur in the downstream assembly. In the working directory, run the sample script below (rename.pl) in the CL window by entering command "perl rename.pl > R1.fastq"; this script looks for strings with a common prefix, for example, "@ NS500766:3:HHC7FBGX2:"; it then appends a 1 to the end of each search result. In real practice, different prefixes should be checked by running the "head R1.fastq" command, and the script should be modified and repeated for R2 reads accordingly:

```
#!/usr/bin/perl

open (FILEHANDLE," R1.fastq");
while ($line=<FILEHANDLE>) {
     if ($line =~ m/^\@NS500766:3:HHC7FBGX2:/) {
          $line =~ s/\n$/\/1\n/;
          print $line;
     }
     else {
          print $line;
         }
}
```

Acknowledgments

The authors wish to thank the many programmers, researchers, and students who have contributed to the Virus Bioinformatics Resource software. This work has been supported by funds from the Natural Sciences Engineering Research Council of Canada. Drs. C. Upton, R. M. L. Buller, and. E. J. Lefkowitz were the original developers of the Poxvirus Bioinformatics Resource Center.

References

1. Goebel SJ, Johnson GP, Perkus ME, Davis SW, Winslow JP, Paoletti E (1990) The complete DNA sequence of vaccinia virus. Virology 179:247–266
2. Bennett M, Tu S-L, Upton C, McArtor C, Gillett A, Laird T et al (2017) Complete genomic characterisation of two novel poxviruses (WKPV and EKPV) from western and eastern grey kangaroos. Virus Res 242:106–121
3. Laird MR, Langille MGI, Brinkman FSL (2015) GenomeD3Plot: a library for rich, interactive visualizations of genomic data in web applications. Bioinformatics 31:3348–3349
4. Upton C, Slack S, Hunter AL, Ehlers A, Roper RL (2003) Poxvirus orthologous clusters: toward defining the minimum essential poxvirus genome. J Virol 77:7590–7600

5. Upton C, Hogg D, Perrin D, Boone M, Harris NL (2000) Viral genome organizer: a system for analyzing complete viral genomes. Virus Res 70:55–64
6. Sonnhammer E, Durbin R (1995) A dot-matrix program with dynamic threshold control suited for genomic DNA and protein sequence analysis (Reprinted from Gene Combis, vol 167, pg GC1-GC10, 1996). Gene 167:GC1–GC10
7. Brodie R, Roper RL, Upton C (2004) JDotter: a Java interface to multiple dotplots generated by dotter. Bioinformatics 20:279–281
8. Sievers F, Higgins DG (2014) Clustal Omega, accurate alignment of very large numbers of sequences. Methods Mol Biol 1079:105–116
9. Katoh K, Misawa K, Kuma K, Miyata T (2002) MAFFT: a novel method for rapid multiple sequence alignment based on fast Fourier transform. Nucleic Acids Res 30:3059–3066
10. Edgar RC (2004) MUSCLE: a multiple sequence alignment method with reduced time and space complexity. BMC Bioinformatics 5:113
11. Hillary W, Lin S-H, Upton C (2011) Base-By-Base version 2: single nucleotide-level analysis of whole viral genome alignments. Microb Inform Exp 1:2
12. Tcherepanov V, Ehlers A, Upton C (2006) Genome Annotation Transfer Utility (GATU): rapid annotation of viral genomes using a closely related reference genome. BMC Genomics 7:150
13. Soding J, Biegert A, Lupas AN (2005) The HHpred interactive server for protein homology detection and structure prediction. Nucleic Acids Res 33:W244–W248
14. Chevreux B (2007) MIRA: an automated genome and EST assembler
15. Bankevich A, Nurk S, Antipov D, Gurevich AA, Dvorkin M, Kulikov AS et al (2012) SPAdes: a new genome assembly algorithm and its applications to single-cell sequencing. J Comput Biol 19:455–477
16. Li H, Durbin R (2009) Fast and accurate short read alignment with Burrows-Wheeler transform. Bioinformatics 25:1754–1760
17. Li H, Handsaker B, Wysoker A, Fennell T, Ruan J, Homer N et al (2009) The sequence alignment/map format and SAMtools. Bioinformatics 25:2078–2079
18. Breese MR, Liu Y (2013) NGSUtils: a software suite for analyzing and manipulating next-generation sequencing datasets. Bioinformatics 29:494–496
19. Stamatakis A (2006) RAxML-VI-HPC: Maximum likelihood-based phylogenetic analyses with thousands of taxa and mixed models. Bioinformatics 22:2688–2690
20. Altschul SF, Madden TL, Schaffer AA, Zhang J, Zhang Z, Miller W et al (1997) Gapped BLAST and PSI-BLAST: a new generation of protein database search programs. Nucleic Acids Res 25:3389–3402
21. Madden T (2013) The BLAST sequence analysis tool.
22. Satheshkumar PS, Moss B (2009) Characterization of a newly identified 35-amino-acid component of the vaccinia virus entry/fusion complex conserved in all chordopoxviruses. J Virol 83:12822–12832
23. Satheshkumar PS, Moss B (2012) Sequence-divergent chordopoxvirus homologs of the O3 protein maintain functional interactions with components of the vaccinia virus entry-fusion complex. J Virol 86:1696–1705
24. Da Silva M, Upton C (2005) Host-derived pathogenicity islands in poxviruses. Virol J:2, 30
25. Upton C (2000) Screening predicted coding regions in poxvirus genomes. Virus Genes 20:159–164
26. Da Silva M, Upton C (2005) Using purine skews to predict genes in AT-rich poxviruses. BMC Genomics 6:22
27. Boratyn GM, Schaeffer AA, Agarwala R, Altschul SF, Lipman DJ, Madden TL (2012) Domain enhanced lookup time accelerated BLAST. Biol Direct 7:12
28. Papadopoulos JS, Agarwala R (2007) COBALT: constraint-based alignment tool for multiple protein sequences. Bioinformatics 23:1073–1079
29. Kelley LA, Sternberg MJE (2009) Protein structure prediction on the Web: a case study using the Phyre server. Nat Protoc 4:363–371
30. Kim DE, Chivian D, Baker D (2004) Protein structure prediction and analysis using the Robetta server. Nucleic Acids Res 32:W526–W531
31. Zhang Y (2008) I-TASSER server for protein 3D structure prediction. BMC Bioinformatics 9:40
32. O'Dea MA, Tu S-L, Pang S, De Ridder T, Jackson B, Upton C (2016) Genomic characterization of a novel poxvirus from a flying fox: evidence for a new genus? J Gen Virol 97:2363–2375
33. Pettersen EF, Goddard TD, Huang CC, Couch GS, Greenblatt DM, Meng EC et al (2004) UCSF chimera–A visualization system for exploratory research and analysis. J Comput Chem 25:1605–1612
34. Bairoch A (1993) The prosite dictionary of sites and patterns in proteins, its current status. Nucleic Acids Res 21:3097–3103

35. de Castro E, Sigrist CJA, Gattiker A, Bulliard V, Langendijk-Genevaux PS, Gasteiger E et al (2006) ScanProsite: detection of PROSITE signature matches and ProRule-associated functional and structural residues in proteins. Nucleic Acids Res 34:W362–W365
36. Marchler-Bauer A, Derbyshire MK, Gonzales NR, Lu S, Chitsaz F, Geer LY et al (2015) CDD: NCBI's conserved domain database. Nucleic Acids Res 43:D222–D226
37. Notredame C, Higgins DG, Heringa J (2000) T-Coffee: a novel method for fast and accurate multiple sequence alignment. J Mol Biol 302:205–217
38. Subramanian AR, Kaufmann M, Morgenstern B (2008) DIALIGN-TX: greedy and progressive approaches for segment-based multiple sequence alignment. Algorithms Mol Biol 3:6
39. Rissman AI, Mau B, Biehl BS, Darling AE, Glasner JD, Perna NT (2009) Reordering contigs of draft genomes using the Mauve Aligner. Bioinformatics 25:2071–2073
40. Hoen AG, Gardner SN, Moore JH (2013) Identification of SNPs associated with variola virus virulence. BioData Min 6:3
41. Smithson C, Purdy A, Verster AJ, Upton C (2014) Prediction of Steps in the Evolution of Variola Virus Host Range. PLoS One 9:e91520
42. Flygare S, Simmon K, Miller C, Qiao Y, Kennedy B, Di Sera T et al (2016) Taxonomer: an interactive metagenomics analysis portal for universal pathogen detection and host mRNA expression profiling. Genome Biol 17:111
43. Juenemann S, Prior K, Albersmeier A, Albaum S, Kalinowski J, Goesmann A et al (2014) GABenchToB: a genome assembly benchmark tuned on bacteria and benchtop sequencers. PLoS One 9:e107014
44. Smithson C, Imbery J, Upton C (2017) Re-assembly and analysis of an ancient variola virus genome. Viruses 9:E253
45. Milne I, Bayer M, Stephen G, Cardle L, Marshall D (2016) Tablet: visualizing next-generation sequence assemblies and mappings. Methods Mol Biol 1374:253–268
46. Tamura K, Stecher G, Peterson D, Filipski A, Kumar S (2013) MEGA6: molecular evolutionary genetics analysis version 6.0. Mol Biol Evol 30:2725–2729
47. Sivashankari S, Shanmughavel P (2006) Functional annotation of hypothetical proteins—a review. Bioinformation 1:335–338
48. McLeod K, Upton C (2017) Virus databases. Reference Module in Biomedical Sciences. Elsevier

Chapter 3

Simple, Rapid Preparation of Poxvirus DNA for PCR Cloning and Analysis

Rachel L. Roper

Abstract

This chapter describes the simple, rapid, and inexpensive preparation of template DNA from poxvirus-infected cells, plaques, or crude virus stocks for PCR amplification. This technique is reliable and robust and only requires centrifugation, detergent, and protease treatment. The resulting DNA template preparation is suitable for PCR amplification for screening viruses, cloning, transfection, and DNA sequencing.

Key words DNA purification, PCR, Virus preparation, Virus screening, Cloning, DNA sequencing, Vaccinia, Modified vaccinia Ankara (MVA), Poxvirus

1 Introduction

Poxviruses cause disease in humans and animals worldwide. Smallpox is estimated to have killed ~500 million people during the twentieth century, but it was eradicated from nature by 1980 through a worldwide vaccination program headed by the World Health Organization [1]. Vaccination of civilian populations in the USA has ceased due to the poor safety record of the live vaccinia virus vaccine, but lab workers, first responders, and US military personnel continue to be vaccinated due to biowarfare/bioterrorism concerns. Most extant human-infecting poxviruses are zoonotic and transmitted mostly via rodents, including monkeypox, which caused an outbreak in the US Midwest in 2003 with more than 80 cases reported [2, 3]. Monkeypox virus is endemic in Africa and has a case fatality rate of 10% [2, 4]. Cowpox and Orf viruses also infect humans, but these infections are rarely fatal [5]. Molluscum contagiosum virus is only known to infect humans, is common worldwide, and is emerging as a sexually transmitted disease [6, 7]. In addition, new poxviruses are identified each year in animal populations, and several zoonotic poxviruses appear to be emerging worldwide in humans [8], such as Cantagalo in South America [9, 10], Tanapox in Africa (and also detected in Europe

Jason Mercer (ed.), *Vaccinia Virus: Methods and Protocols*, Methods in Molecular Biology, vol. 2023, https://doi.org/10.1007/978-1-4939-9593-6_3, © Springer Science+Business Media, LLC, part of Springer Nature 2019

and the USA among travelers) [11, 12], and buffalopox in India [13]. Thus, continued research and development of diagnostics and safe and effective poxvirus vaccines and treatments remain desirable.

Poxviruses are also often used as recombinant vaccine vectors for infectious diseases and cancer therapies [14–16]. Myxoma virus, which normally infects rabbits, has emerged as a potential oncolytic treatment for human cancer [17]. The only HIV vaccine with demonstrated efficacy in humans thus far employs a canary-pox vector, and this entered NIH-sponsored Phase 2b/3 trials in 2016 [18]. Poxviruses such as vaccinia virus (VACV) are suitable as vectors because they are easily grown to high titer in a wide variety of animals and cell types, can accommodate insertions of large pieces of DNA into their genomes, are very stable even when dried, and induce robust B- and T-cell immunity [19]. VACV is the most commonly used poxvirus vaccine vector; however its use is limited by its potential virulence, especially in immunocompromised hosts. The modified vaccinia Ankara (VACV-MVA) strain is much more attenuated, but its replicative capacity and immunogenicity may be limiting [19–23]. Raccoonpox virus (RCNV) is a naturally occurring attenuated North American poxvirus that is of interest evolutionarily, as a vaccine vector platform, and for its oncolytic therapy potential [24, 25]. We have recently reported that RCNV is less virulent and much safer than VACV in immunocompromised or pregnant mouse models [25–27]. Different poxviruses may find utility in different clinical strategies.

With a wide variety of poxvirus research progressing, it is often desirable to rapidly screen for recombinant viruses or amplify genes for cloning from unpurified poxvirus (crude stock preparations, plaques, or infected cells) by the polymerase chain reaction (PCR). This chapter describes simple, rapid, robust, inexpensive, and reproducible protocols for the preparation of template DNA from vaccinia virus-infected cells for PCR, without the need for extensive DNA purification. There are many chemical procedures and commercial columns available for DNA purification; however, some of the chemicals commonly used are hazardous, and the column kits are expensive. One of the main advantages of PCR is that it can amplify specific sequences from unpurified DNA. It has been estimated that each vaccinia virus-infected cell contains approximately 10,000 copies of the poxvirus genome, so ample template DNA is present in a single cell [28]. However, simply lysing infected cells does not consistently yield a product that can be used as a PCR template, as cells contain known, and unknown, inhibitors of the PCR amplification. The protocols described here provide methods to remove, denature, or degrade cellular inhibitors of PCR in three simple steps in less than 1 h. The first is the removal of cellular debris by centrifugation after cell lysis. It is possible to successfully amplify DNA from the supernatants of centrifuged lysed infected cells, but it is not possible to amplify DNA from a whole crude preparation of virus-infected cells or cell pellets. Second, PCR-compatible detergents (e.g., NP-40 (or IGEPAL), Tween 20, TritonX-100) are required [29]. The detergents may act to release bound proteins from the template DNA, denature DNases or proteins that bind dNTPs or cations, or disrupt the virus particles (although large amounts of cytoplasmic viral DNA is likely present). Third, proteinase K is used to degrade inhibitory proteins, probably those bound tightly to DNA. Finally, the proteinase K must be heat inactivated so that it does not degrade the Taq (or other DNA polymerases) required for PCR amplification. PCR products generated by this technique may be used for DNA cloning, transfection assays, DNA sequencing, as well as screening and analyzing recombinant virus genomes [30–34]. This technique has been employed successfully with vaccinia, MVA, rabbitpox, cowpox, myxoma, and ectromelia viruses, suggesting that it is widely applicable to poxviruses of different genera [30–36].

Two protocols are described, one is for template preparation for PCR amplification from a crude virus stock, infected cells lysed by freeze-thawing (the type that is commonly found in a poxvirus laboratory). The preparation of crude vaccinia virus stocks has been described at length elsewhere (*see* **Notes 1** and **2**). A second protocol describes template preparation directly from an infected tissue culture well via direct aspiration of cells and medium (or agarose) from an infected area of cells. This allows for rapid screening of recombinant viruses during their amplification. This technique has been used successfully on virus that was grown in cell culture for as little as 8 h. Single plaques picked from under liquid medium or in agarose plugs have also been PCR amplified successfully by this method. When virus is prepared from fresh infected cells rather than a crude stock, the detergent and proteinase K are added first to disrupt the cell and release viral DNA and the centrifugation step is done afterwards to remove the cellular debris. Another advantage of this technique is that a small area of an infected well can be tested without terminating the growth of the virus in the well.

As an example application of this technique, we describe the use of this protocol in the construction, screening, and analysis of a recombinant MVA virus containing the gene for a cancer-associated protein, mesothelin [37]. The parental MVA and recombinant MVA with the mesothelin gene cloned into the genome, with and without the immunosuppressive A35 gene, MVA Meso and MVA Meso ΔA35 [35, 38] were analyzed using primers specific for the mesothelin gene or amplifying across the insertion site (Fig. 1) so that the different sized insertion fragments were detected in the plasmid (pLW44) and the recombinant viruses. No mesothelin gene insertion was detected in the parental MVA virus as expected. This technique offers a rapid and easy method for analyzing pox viral genomes.

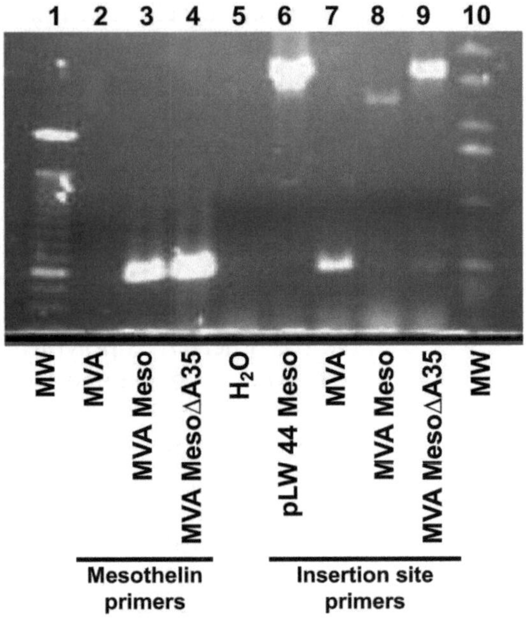

Fig. 1 Example of DNA amplification from crude virus preparation for analysis of recombinant virus genomes. Crude virus stocks of wild-type parental MVA, recombinant MVA with the mesothelin gene cloned into the genome (MVA Meso), and MVA Meso without the A35 gene (MVA MesoΔA35) were prepared from infected BHK cells. Template was prepared for PCR as described in Subheading 3.1. PCR amplification was performed using Taq under standard reaction conditions using primers specific for the mesothelin gene (lanes 2–4) or primers that amplify across the insertion site (lanes 5–9). Note that different sized insertion fragments were detected in the plasmid (pLW44 Meso), and the recombinant viruses (MVA Meso and MVA MesoΔA35). No mesothelin gene insertion was detected in the parental MVA virus as expected (lanes 2 and 7). Negative control H_2O contained no template DNA (lane 5). PCR products were analyzed on 0.8% agarose gel and imaged on a Gel Doc

2 Materials

Make sure that all reagents are sterile, pure, and PCR grade since this material will ultimately be used as template in a PCR reaction. Cell culture medium does not need special purity.

1. Detergent solution: 2× PCR buffer with 0.9% NP40 (or IGEPAL) 0.9% Tween, 20 mM Tris–HCl, pH 8.3, 3 mM $MgCl_2$; 100 mM KCl (stored at −20 °C).
2. 5 mg/mL Proteinase K (stored at −20 °C).
3. 37 °C or 45 °C water bath.
4. 94 °C Heating block.

3 Methods

3.1 Amplification from a Crude Virus Preparation (Fig. 1)

Remember that the end product will be used as PCR template. Autoclave all plastics and use sterile solutions and techniques to avoid contamination.

1. Centrifuge the crude virus preparation in a microfuge (14,000 × *g*) for 10 s to pellet any cellular debris. The amount of crude virus preparation in the tube is not important as long as there is adequate volume to remove 10 μL of supernatant (*see* **Note 2**).
2. Remove 10 μL of the supernatant and dispense into a fresh microfuge tube.
3. Add 10 μL of detergent solution (*see* **Notes 3–5**).
4. Add 1.2 μL of proteinase K solution and mix.
5. Incubate for 30 min to 1 h at 37 °C, or 30 min at 45 °C (*see* **Note 6**).
6. Heat inactivate the proteinase K for 10 min at 94 °C (*see* **Note** 7).
7. Centrifuge in a microfuge for 5 s to collect the reaction mixture into the bottom of the tube.
8. Use 10 μL of this preparation in a 100 μL PCR reaction following standard PCR protocols (*see* **Notes 8** and **9**).

3.2 Amplification from Virus from an Infected Well

Remember that the end product will be used as PCR template. Autoclave all plastics and use sterile solutions and techniques to avoid contamination.

1. Scrape a micropipette tip through an infected area of cells while aspirating 10 μL of medium and cells and dispense into a microfuge tube. Alternatively, pick a plaque in an agarose plug and place into 30 μL of media or 10 mM Tris pH 9.
2. Vortex and remove 10 μL to a microfuge tube, reserving the remaining sample for virus amplification.
3. Add 10 μL of detergent solution to tube (*see* **Notes 3–5**).
4. Add 1.2 μL of proteinase K solution.
5. Incubate for 30 min to 1 h at 37 °C, or 30 min at 45 °C (*see* **Note 6**).
6. Heat inactivate the proteinase K for 10 min at 94 °C (*see* **Note** 7).
7. Centrifuge the microfuge tube (14,000 × *g*) for 10 s to remove cellular debris.
8. Pipette out 10 μL of this supernatant to use in a 100 μL PCR reaction following standard PCR protocols (*see* **Notes 8** and **9**).

4 Notes

1. This technique has been used on several mammalian cell lines commonly used for growth of orthopoxviruses, e.g., vaccinia or MVA, including HeLa, BS-C-1, and BHK. While numerous cell lines have not been screened, there has been no cell line tested in which this technique does not work.
2. For the purposes of this report, a crude virus stock is defined as virus-infected cells that have been concentrated approximately ten times relative to the volume in which the cells were grown during the virus amplification (e.g., 20 mL of culture grown in one T75 flask makes 2 mL of crude stock, or one well of a 6-well plate makes 100 μL of crude virus preparation), and frozen and thawed three times to break open the cells. Ideally, every cell has been infected. The crude virus preparation may be resuspended in medium (e.g., minimum essential media), and the presence of 10% fetal bovine serum does not interfere with these protocols. Do not sonicate your crude virus preparation prior to performing this PCR-prep protocol because sonication breaks DNA.
3. The detergent solution can be made with either NP-40 (which is now sometimes difficult to purchase) or IGEPAL. If an alternate detergent is desired, a solution of 2% Triton X-100 in 2× PCR buffer was found to be as effective as the combination of 0.9% NP40/IGEPAL and 0.9% Tween 20 in most, but not all, cases. Special care should be taken to avoid the introduction of SDS, or other known inhibitors of the PCR reaction or polymerases.
4. Commercial PCR buffers may be used to make up the detergent solution. For this protocol, the detergents are dissolved in 2× concentration PCR buffer so that when the detergent is mixed with the infected cell preparation the final concentration of the PCR buffer is 1×, as this will be introduced directly into the PCR reaction. The user should take this into account when preparing the PCR reaction so as not to add too much 10× PCR buffer.
5. For analyzing multiple reactions simultaneously, it is convenient to make a master mix of detergents and proteinase K in the proper proportions immediately prior to use, aliquot the mix into microfuge tubes, and then add virus samples. However, remember to be careful with your techniques. The virus material you are preparing will provide the template in a PCR reaction. Contamination by any means should be strictly avoided.

6. The length of the protease digestion does not seem to be crucial. Times and temperatures known to work have been provided as guidelines.
7. Normally the detergent and proteinase K solution is incubated in a 45 °C water bath, and the heat inactivation step is incubated in a 94 °C heating block; however either method of heating should suffice as long as the required temperature is maintained. Be careful not to allow water to contaminate the lip of the microfuge tube in a water bath, and make certain that there is good contact between the microfuge tube and the walls of the heating block. Also make certain that your 94 °C heat inactivation of proteinase K proceeds for the full 10 min. Lower incubation temperatures or shorter times may result in the remaining active proteinase K digesting the PCR polymerase enzyme and inhibiting amplification.
8. Quantity of template preparation: 10 μL of virus DNA template prepared by this method should be used in a 100 μL PCR reaction, or 5 μL in a 50 μL PCR reaction. 1 μL has also been found to be sufficient template in many cases.
9. Samples prepared by this method work best when used for PCR within a few days (stored at 4 °C). To test your primers and PCR reaction conditions, you can use your primers on purified viral DNA or an appropriate plasmid as positive control templates. If there is no PCR product, try using *less* template mixture per PCR reaction. Often the use of more template mixture increases inhibitory contaminants with a concomitant reduction in PCR product. The most common problem experienced with this protocol is the **omission of steps**. If you omit any step, the procedure will NOT work. Each step has been tested and is required for this protocol.

References

1. Mahalingam S, Damon IK, Lidbury BA (2004) 25 years since the eradication of smallpox: why poxvirus research is still relevant. Trends Immunol 25:636–639
2. Chen N, Li G, Liszewski MK, Atkinson JP, Jahrling PB, Feng Z et al (2005) Virulence differences between monkeypox virus isolates from West Africa and the Congo basin. Biochemistry 340:46–63
3. McCollum AM, Damon IK (2013) Human Monkeypox. Oxford University Press. Clin Infect Dis cit703
4. Lederman ER, Reynolds MG, Karem K, Braden Z, Learned-Orozco LA, Wassa-Wassa D et al (2007) Prevalence of antibodies against orthopoxviruses among residents of Likouala region, Republic of Congo: evidence for monkeypox virus exposure. Am J Trop Med Hyg 77:1150–1156
5. Lewis-Jones S (2004) Zoonotic poxvirus infections in humans. Curr Opin Infect Dis 17:81–89
6. Molino AC, Fleischer AB, Feldman SR (2004) Patient demographics and utilization of health care services for molluscum contagiosum. Pediatr Dermatol 21:628–632. Blackwell Science Inc
7. Senkevich TG, Koonin EV, Bugert JJ, Darai G, Moss B (1997) The genome of molluscum contagiosum virus: analysis and comparison with other poxviruses. Biochemistry 233:19–42

8. Shchelkunov SN (2013) An increasing danger of zoonotic orthopoxvirus infections. PLoS Pathog 9:e1003756. (G F Rall, Ed.) Public Library of Science
9. Damaso CR, Esposito JJ, Condit RC, Moussatché N (2000) An emergent poxvirus from humans and cattle in Rio de Janeiro State: Cantagalo virus may derive from Brazilian smallpox vaccine. Biochemistry 277:439–449
10. Oliveira DB, Assis FL, Ferriera PCP, Bonjardim CA, de Souza Trindade G, Kroon EG et al (2013) Group 1 vaccinia virus zoonotic outbreak in Maranhao State, Brazil. Am J Trop Med Hyg 89:1142–1145
11. Dhar AD, Werchniak AE, Li Y, Brennick JB, Goldsmith CS, Kline R et al (2004) Tanapox infection in a college student. N Engl J Med 350:361–366
12. Stich A, Meyer H, Köhler B, Fleischer K (2002) Tanapox: first report in a European traveller and identification by PCR. Trans R Soc Trop Med Hyg 96:178–179
13. Kolhapure RM, Deolankar RP, Tupe CD, Raut CG, Basu A, Dama BM et al (1997) Investigation of buffalopox outbreaks in Maharashtra State during 1992-1996. Indian J Med Res 106:441–446
14. Campbell CT, Gulley JL, Oyelaran O, Hodge JW, Schlom J, Gildersleeve JC (2013) Serum antibodies to blood group A predict survival on PROSTVAC-VF. Clin Cancer Res 19:1290–1299. American Association for Cancer Research
15. Hui EP, Taylor GS, Jia H, Ma BB, Chan SL, Ho R et al (2013) Phase I trial of recombinant modified vaccinia Ankara encoding Epstein-Barr viral tumor antigens in nasopharyngeal carcinoma patients. Cancer Res 73:1676–1688
16. Gómez CE, Nájera JL, Krupa M, Esteban M (2008) The poxvirus vectors MVA and NYVAC as gene delivery systems for vaccination against infectious diseases and cancer. Curr Gene Ther 8:97–120
17. Rahal A, Musher B (2017) Oncolytic viral therapy for pancreatic cancer. J Surg Oncol 116(1):94–103
18. Rerks-Ngarm S, Pitisuttithum P, Nitayaphan S, Kaewkungwal J, Chiu J, Paris R et al (2009) Vaccination with ALVAC and AIDSVAX to prevent HIV-1 infection in Thailand. N Engl J Med 361:2209–2220
19. Tscharke DC, Karupiah G, Zhou J, Palmore T, Irvine KR, Haeryfar SM et al (2005) Identification of poxvirus CD8+ T cell determinants to enable rational design and characterization of smallpox vaccines. J Exp Med 201:95–104
20. Belyakov IM, Earl P, Dzutsev A, Kuznetsov VA, Lemon M, Wyatt LS et al (2003) Shared modes of protection against poxvirus infection by attenuated and conventional smallpox vaccine viruses. Proc Natl Acad Sci U S A 100:9458–9463
21. Earl PL, Americo JL, Wyatt LS, Eller LA, Whitbeck JC, Cohen GH et al (2004) Immunogenicity of a highly attenuated MVA smallpox vaccine and protection against monkeypox. Nature 428:182–185
22. Graham JH, Graham VA, Bewley KR, Tree JA, Dennis M, Taylor I et al (2013) Assessment of the protective effect of Imvamune and Acam2000 Vaccines against aerosolized monkeypox virus in cynomolgus macaques. J Virol 87:7805–7815
23. Guzman E, Cubillos-Zapata C, Cottingham MG, Gilbert SC, Prentice H, Charleston B et al (2012) Modified vaccinia virus Ankara-based vaccine vectors induce apoptosis in dendritic cells draining from the skin via both the extrinsic and intrinsic caspase pathways, preventing efficient antigen presentation. J Virol 86:5452–5466
24. Evgin L, Vaha-Koskela M, Rintoul J, Falls T, Le Boeuf F, Barrett JW et al (2010) Potent oncolytic activity of raccoonpox virus in the absence of natural pathogenicity. Mol Ther 18:896–902
25. Jones GJB, Boles C, Roper RL (2014) Raccoonpox virus safety in immunocompromised and pregnant mouse models. Vaccine 32:3977–3981
26. Fleischauer C, Upton C, Victoria J, Jones GJ, Roper RL et al (2015) Genome sequence and comparative virulence of raccoonpox virus: the first North American poxvirus sequence. J Gen Virol 96:2806–2821
27. Roper RL (2017) Poxvirus Safety analysis in the pregnant mouse model, vaccinia, and raccoonpox viruses. Methods Mol Biol 1581:121–129
28. Joklik WK, Becker Y (1964) The replication and coating of vaccinia DNA. J Mol Biol 10:452–474
29. Erlich HA (ed) (1989) PCR technology: principles and applications for DNA amplification. M Stockton Press, New York
30. Sung TC, Roper RL, Zhang Y, Rudge SA, Temel R, Hammond SM et al (1997) Mutagenesis of phospholipase D defines a superfamily including a trans-Golgi viral protein required for poxvirus pathogenicity. EMBO J 16:4519–4530
31. Roper RL, Moss B (1999) Envelope formation is blocked by mutation of a sequence related to the HKD phospholipid metabolism motif in

the vaccinia virus F13L protein. J Virol 73:1108–1117

32. Roper RL, Wolffe EJ, Weisberg A, Moss B (1998) The envelope protein encoded by the A33R gene is required for formation of actin-containing microvilli and efficient cell to cell spread of vaccinia virus. J Virol 72:4192–4204
33. Wolffe EJ, Katz E, Weisberg A, Moss B (1997) The A34R glycoprotein gene is required for induction of specialized actin-containing microvilli and efficient cell-to-cell transmission of vaccinia virus. J Virol 71:3904–3915
34. Roper RL, Payne LG, Moss B (1996) Extracellular vaccinia virus envelope glycoprotein encoded by the A33R gene. J Virol 70:3753–3762
35. Rehm KE, Roper RL (2011) Deletion of the A35 gene from Modified Vaccinia Virus Ankara Increases Immunogenicity and Isotype Switching. Vaccine 29:3276–3283
36. Roper RL (2004) Rapid preparation of vaccinia virus DNA template for analysis and cloning by PCR. Methods Mol Biol 269:113–118
37. Zervos E, Agle S, Freistaedter AG, Jones GJ, Roper RL (2016) Murine mesothelin: characterization, expression, and analysis of growth and tumorigenic effects in a murine model of pancreatic cancer. J Exp Clin Cancer Res 35:39
38. Rehm KE, Jones GJB, Tripp AA, Metcalf MW, Roper RL (2010) The poxvirus A35 protein is an immunoregulator. J Virol 84(1):418–425

Chapter 4

Construction and Isolation of Recombinant Vaccinia Virus Expressing Fluorescent Proteins

N. Bishara Marzook and Timothy P. Newsome

Abstract

Vaccinia virus recombinants that express fluorescent proteins have a variety of applications such as the identification of infected cells, efficient screening for genetically modified strains, and molecular characterization of virus replication and spread. The detection of fluorescent proteins and viral–fluorescent fusion proteins by fluorescence microscopy is noninvasive and can be used to describe protein localization in live cells and track the intracellular movement of virus particles. This chapter describes a number of approaches for the construction of plasmids and subsequent generation and isolation of fluorescent recombinant viruses.

Key words Fluorescent proteins, Fluorescence microscopy, Recombinant vaccinia virus

1 Introduction

A number of characteristics of vaccinia virus (VACV) make it amenable to genetic modification, particularly for the generation of viruses that express fluorescent proteins. The most popular methods for generating recombinant VACV take advantage of the high levels of homologous recombination in the host cytoplasm that is a feature of VACV replication [1, 2]. Homologous recombination can be used to promote the insertion of exogenous DNA (transfected into infected cells) into replicating viral DNA by the careful design of plasmids carrying regions of homology to the viral genome [3]. With a large, double-stranded DNA genome, VACV has the capacity to stably accept sizeable (up to 25 kb) insertions of exogenous DNA without substantial attenuation [4, 5].

The first application of recombinant fluorescent VACV was a virus that expresses green fluorescent protein (GFP) under a strong viral promoter allowing the visualization of infected cells as early as 1 h postinfection (hpi) [6, 7]. It was soon recognized that wrapped virus (WV), one of the two mature infectious forms produced during virus replication, could be labeled by fusing GFP to WV-specific

Jason Mercer (ed.), *Vaccinia Virus: Methods and Protocols*, Methods in Molecular Biology, vol. 2023,
https://doi.org/10.1007/978-1-4939-9593-6_4, © Springer Science+Business Media, LLC, part of Springer Nature 2019

viral membrane proteins [8–10]. Analysis of these viruses enabled the tracking of virus particles in live, infected cells, and led to one of the best understood models of intracellular virus motility [11, 12]. The large particle size of VACV and lack of a rigid icosahedral or helical symmetry in the capsid architecture are factors that may contribute to the success of this approach. It has become apparent that many VACV genes can be replaced to express viral–fluorescent fusion proteins with minimal effects on infectivity and replication, which serve as valuable research reagents into many aspects of the host–pathogen interface [13–15]. Recent innovations in this technology include imaging virus particle architecture with super-resolution microscopy [16–18], visualization of VACV recombination in live cells [19], and developing reporters for all stages of viral gene expression [20].

Strategies that have been successfully used in the field to generate fluorescent VACV lines are diverse and, to some extent, dictated by the application. We outline a number of methodologies for designing the construction of fluorescent viruses and applying optimal selection and screening strategies. We discuss the choice of fluorescent proteins and the consideration of multichannel imaging, the range of promoters available, and the selection of viral genes as targets for protein fusions. We compare the brightness of plaques formed by fluorescent viruses that use either endogenous or engineered promoters of different strengths to drive the expression of fluorescent proteins and viral–fluorescent fusion proteins.

2 Materials

1. Standard cell culture consumables and pipettes.
2. 100 mm × 21 mm cell culture dishes, 6-well and 24-well cell culture plates.
3. Cell scrapers.
4. Humidified incubator, 37 °C, 5% CO_2.
5. Cell lines: BS-C-1 (ATCC CCL-26) cells, HeLa (ATCC CCL-2) cells (*see* **Note 1**).
6. Complete 2× Eagle's minimum essential medium supplemented with 4 mM glutamine, 0.2 μg/mL penicillin, 0.2 μg/mL streptomycin, and 20% FBS (2×MEM).
7. Complete Eagle's minimum essential medium (high glucose) supplemented with 2 mM glutamine, 0.1 μg/mL penicillin, 0.1 μg/mL streptomycin, and 10% FBS for cell growth (MEM) and 0% FBS for infection (serum-free medium or SFM).
8. Phosphate-buffered saline (PBS).

9. 2% Low-melting-point (LMP) agarose in water, autoclaved.
10. Carboxymethylcellulose (CMC).
11. GPT selection reagent: 500× Mycophenolic acid (12.5 mg/mL), 100× aminopterin solution (25 mg/mL xanthine, 1.5 mg/mL hypoxanthine, 0.2 mg/mL aminopterin, 1 mg/mL thymidine in 0.08 M NaOH).
12. Transfection reagent such as Lipofectamine 2000 (Invitrogen).
13. Liquid nitrogen.
14. Crystal violet (0.5% (w/v) in a 20% methanol solution).

3 Methods

The frequency of recovery of recombinant viruses created using standard homologous recombination varies according to a number of factors: size of regions of homology, size of insertion, linear or circular plasmids used for transfection, efficiency of transformation, and site of insertion [21, 22]. Under optimal conditions, recombinant viruses might be generated at a frequency of 10^{-4} to 10^{-3} [3] (*see* **Note 2**). To facilitate the recovery of fluorescent viruses using standard recombination, screening and selection strategies are integrated into the approach.

Screening facilitates the identification of recombinant viruses by the expression of an easily detectable marker such as β-galactosidase [23, 24] or fluorescence [6, 7]. Screening does not increase the frequency with which the desired recombinant virus is recovered but aids their identification from the parental line used for their generation. In some cases, the fluorescence of plaques formed by viruses expressing the desired fluorescent protein or viral–fluorescent fusion protein can be used to successfully screen for recombinant viruses. This will depend on the brightness and stability of the particular fluorescent protein used and the strength of the promoter (either synthetic or endogenous). In instances where the desired virus is insufficiently bright, a second open reading frame expressing a fluorescent protein can be expressed at intermediate stages in the recovery of the recombinant virus (using transient-dominant selection, TDS; *see* Subheading 3.3.3) [25].

Selection is used for the recovery of recombinant viruses through the preferential growth of the recombinant virus over the parental strain in the presence of biochemical selection. The *E. coli* xanthine-guanine phosphoribosyl transferase gene (*gpt*) gene enables the utilization of the nucleotide precursors xanthine and hypoxanthine in the presence of mycophenolic acid (MPA), an inhibitor of purine synthesis [26]. Recombinant viruses that express *gpt* are able to form plaques under MPA selection [3]. *gpt* selection can be used to select for insertion events throughout the VACV genome unlike selection protocols based on ΔTK viruses that are only suitable for cloning into the thymidine kinase locus [27]. With appropriately designed recombination vectors, TDS can be used such that the *gpt* transgene is excised at the final stages of recovery of a recombinant virus [25] (*see* **Note 3** for other options).

Rescue of attenuated growth associated with specific virus mutant strains allows for the efficient selection of rescued recombinant viruses that will display an altered plaque morphology and outgrow the parental strain, greatly aiding their isolation and subsequent purification [3, 28]. Although a powerful strategy for isolating fluorescent recombinant viruses, it is only applicable for a locus which is associated with an attenuated growth phenotype and the availability of the mutant strain and rescue construct. For example, loss of B5 expression results in a small plaque phenotype that is restored by expression of B5-GFP and has been used to generate the pB5R-GFP fluorescent recombinant virus [10].

The approaches described here are optimized for generating recombinant VACV-Western Reserve but are suitable for other VACV strains (Lister, MVA, Copenhagen) and other orthopoxviruses (cowpox and ectromelia virus) [29–32].

3.1 Plasmid Construction

Since recombinant VACV generation relies on homologous recombination, the plasmid vectors must be carefully designed, depending on the virus background being used, and where the fluorescent gene is to be inserted, both of which will affect the plaque selection strategy subsequently used (*see* Subheading 3.5). All plasmids have the pBS SKII backbone, containing a multiple cloning site (MCS), an ampicillin resistance gene, and T7 and T3 promoters. The various configurations of alignment of homologous sequences, fluorescence genes, and selection markers (if any) are outlined (*see* Figs. 1 and 2). Some general considerations apply to generating a strategy to create a recombinant VACV that apply to many of these strategies.

3.1.1 N-Terminal or C-Terminal Tagging

To generate a recombinant VACV that expresses a fusion protein from the endogenous locus, a configuration that attaches the fluorescent protein to the N-terminus or C-terminus of the viral target must be selected. The goal is to interfere with the function of the endogenous gene and its neighboring genes as little as possible. This is important if the localization of the fluorescent–viral fusion protein is to reflect the localization of the untagged protein. It is also critical that the fluorescent–viral fusion protein rescues the function of the endogenous protein, particularly if it has an important role, or is essential in virus replication. Attaching an approximately 20 kDa protein may interfere with a viral protein's ability to interface with binding partners or assemble into capsid structures. Correct folding may be compromised if the linkage site is embedded deep within the three-dimensional structure of the protein. In

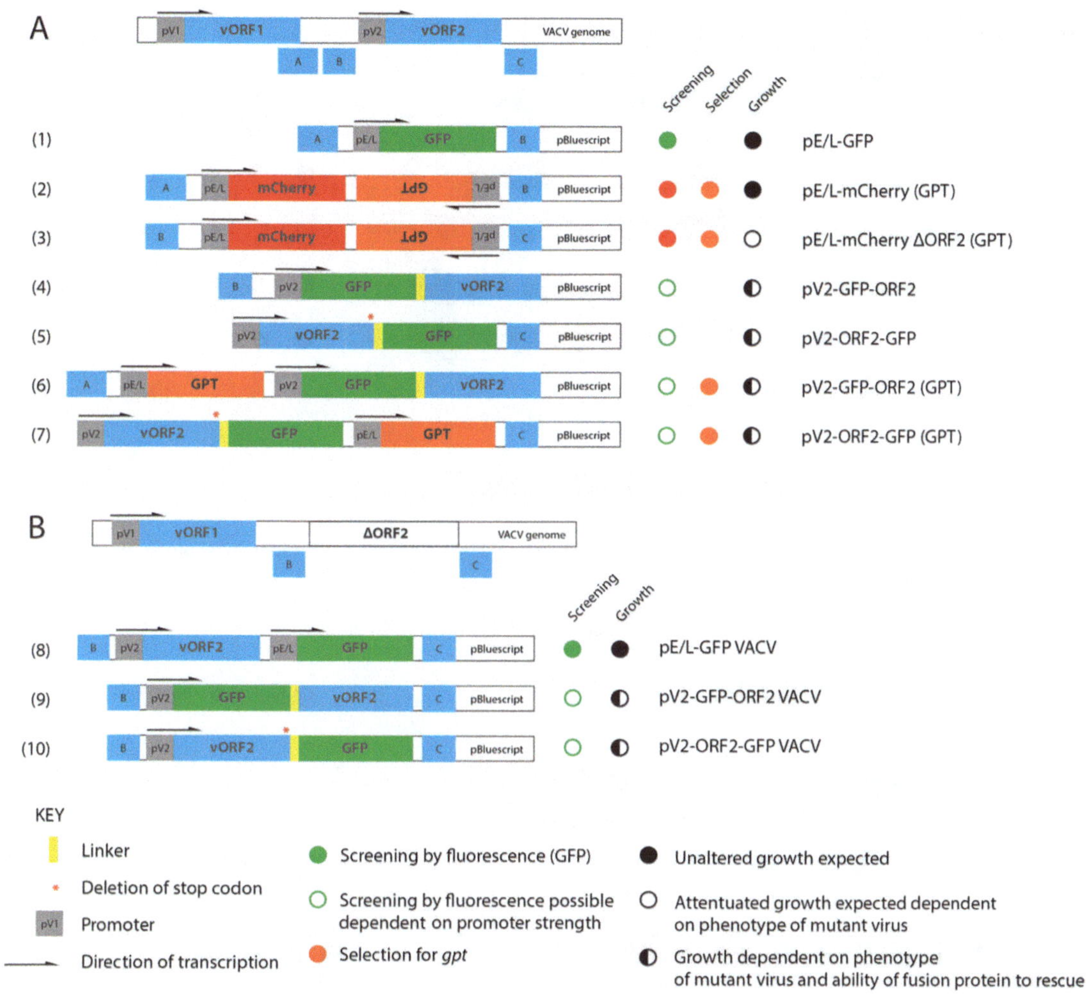

Fig. 1 (**a**) Fluorescent recombinant VACVs can be generated from a parental strain (unmodified) by the screening of plaques for fluorescence (1, 4, 5) or in tandem with biochemical selection for *gpt* expression (2, 3, 6, 7). Screening for fluorescence is reliable when a strong, synthetic promoter is used (1–3) or when the endogenous promoter is highly active (4, 5, 6, 7; e.g., A3L, F13L). If the endogenous promoter strength is weak (F1L) or unknown, then the capacity to select for recombinants is desirable. (**b**) Fluorescent recombinant VACV can be generated by rescue of a mutant strain that displays an attenuated growth and small plaque phenotype. Rescue of the growth defect can be used to select for the insertion of a transgene (8) or to replace the endogenous gene with an in-frame fusion to a fluorescent open reading frame. The ability of a viral–fluorescent fusion protein to restore gene function is critical for strategies 4, 5, 6, 7, 9, 10

most cases the structure of a protein target will be unknown but common pitfalls can be avoided by searching for features such as transmembrane domains. For example, online bioinformatics analysis programs such as InterPro (http://www.ebi.ac.uk/interpro/) identifies a transmembrane domain at the C-terminus of F1 (residues 204–221), a prediction borne out by structure–function analysis that revealed a mitochondrial targeting sequence [33]. The location of its transmembrane domain guided the strategy for tag-

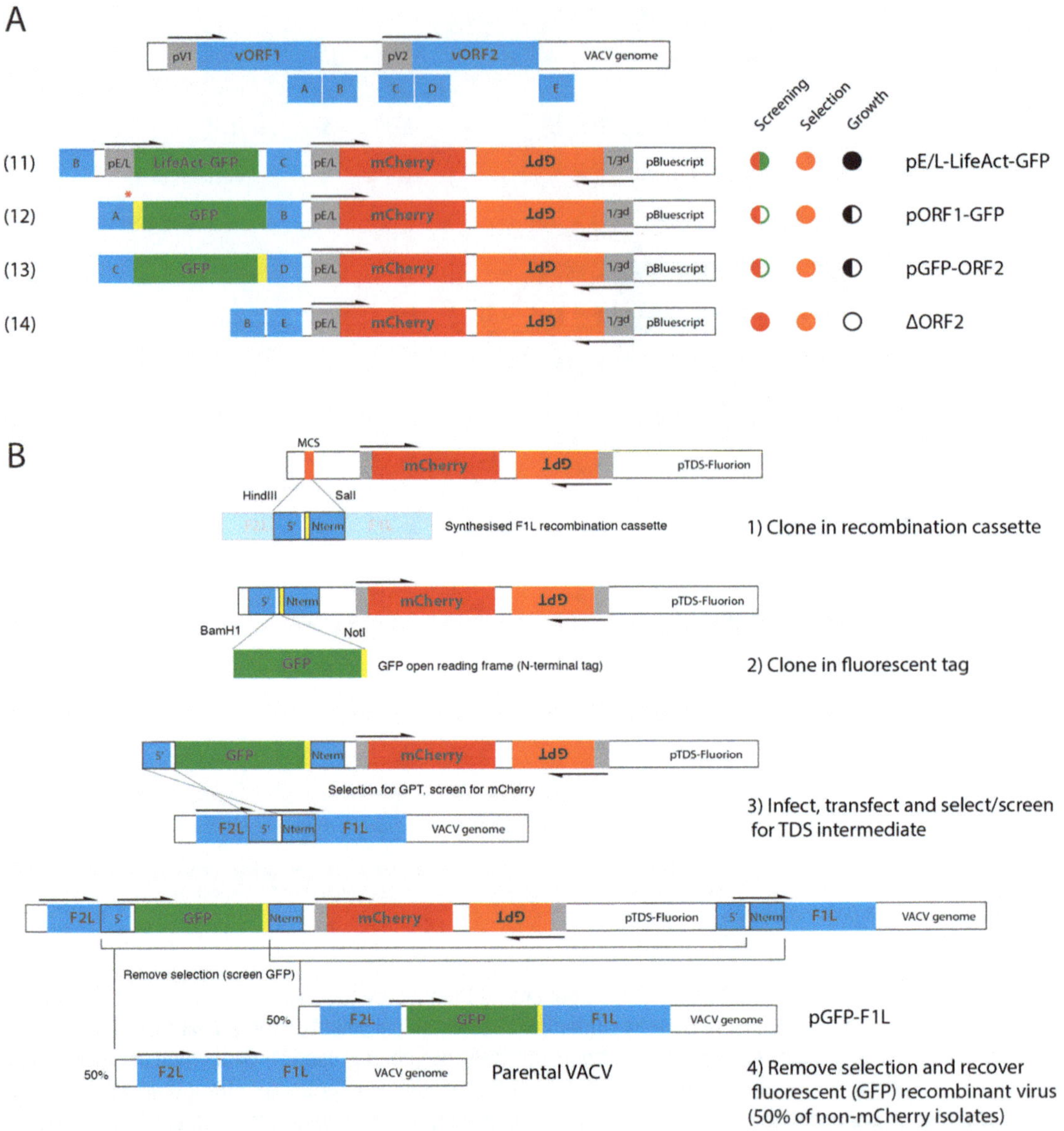

Fig. 2 (**a**) Strong screening (mCherry) and selection (*gpt*) can be combined with a TDS approach to recover recombinant viruses that express fluorescent transgenes (11), delete viral genes (14), or replace the endogenous gene with an in-frame fusion to a fluorescent open reading frame (12, 13). One advantage of TDS is that the recombinant viruses generated have the screening and selection genes excised. (**b**) Example of the TDS-Fluorion vector to generate an N-terminally GFP-tagged F1L recombinant virus (pGFP-F1, strategy 13). Generating the recombination plasmid requires two cloning steps: firstly, the left and right homology arms are synthesized and cloned into the TDS-Fluorion vector (1); secondly, the desired fluorescent protein is added in-frame (2). TDS intermediates that express mCherry and *gpt* are screened/selected for (3). These are the products of a single recombination event at either the 5′ or the N-terminal regions (*see* **Note 10**). Upon the removal of selection, excision of the pTDS-Fluorion sequence from the viral genome owing to recombination at the duplicated sequences will result in approximately equal frequencies, in reversion to the parental strain (recombination at the 5′ region) or resolution to pGFP-F1 (recombination at the N-terminal region)

ging F1 at the N-terminus (Fig. 2), and generating pB5R-GFP [10] and pA36-YFP [34]. In the absence of identifiable structural features, it may be advisable to attempt tagging both ends individually and then determine viability (*see* **Note 4**). In cases where the modular domain of a protein is well characterized, integrating GFP between domains may also be possible while maintaining protein function [35].

A further consideration is that the insertion of an ORF encoding a fluorescent protein into the VACV genome has the potential to disrupt promoter elements from neighboring genes and internal ORFs. The VACV genome is highly compact with viral ORFs present on both strands and often only small intergenic regions. There are instances of overlapping ORFs and cases where the promoter of one gene lies in the ORF of another. Early and late promoter units often reside in close proximity to the start codon of VACV genes and can be identified by consensus sequences [36, 37]. In the example of F1L, the F2L ORF terminates only 11 nucleotides upstream of the F1L start codon (Fig. 3). In this case, a fusion at the C-terminus of F2L is likely to interfere with F1L expression. For the tagging of the C-terminus of F13L, we duplicated 18 nucleotides to ensure that F12L expression was not disrupted (Fig. 4). Although a duplicated sequence does risk genomic instability, in our hands this virus has maintained the GFP insert stably and its small size is likely below that required for efficient recombination [21].

3.1.2 Designing a Linker

A linker sequence encoding a short sequence of amino acids is usually added at the fusion between a viral protein and fluorescent tag. The linker will act to maintain activity of the tagged protein by providing a separation distance between the two functional modules (the fluorescent protein and viral protein). Our standard linker is five amino acids (N-terminal tag: GGRSG; C-terminal tag: GSAAA; Figs. 3 and 4) that incorporates a NotI restriction enzyme site for ease of cloning and modularity. Although we have had success with this linker, maintaining protein folding and function can require longer, structured linkers [38, 39].

3.1.3 Selecting a Fluorescent Protein

Enhanced GFP (EGFP) is the most common fluorescent protein used to tag VACV proteins due to its brightness, photostability, and speed of folding [40]. For dual imaging, GFP can be simultaneously imaged with high spectral separation in combination with mRFP or mCherry (a derivative of mRFP with superior spectral qualities) [41]. Triple-channel imaging was successfully achieved with recombinant viruses using Venus (a YFP variant with improved brightness, chemical stability, and folding) [42], mCherry, and Tag-BFP [20]. Surprisingly, it was recently found that tagging F13 to mCherry resulted in a virion wrapping defect not seen with GFP fusions, despite the similarity in size and folding of these two pro-

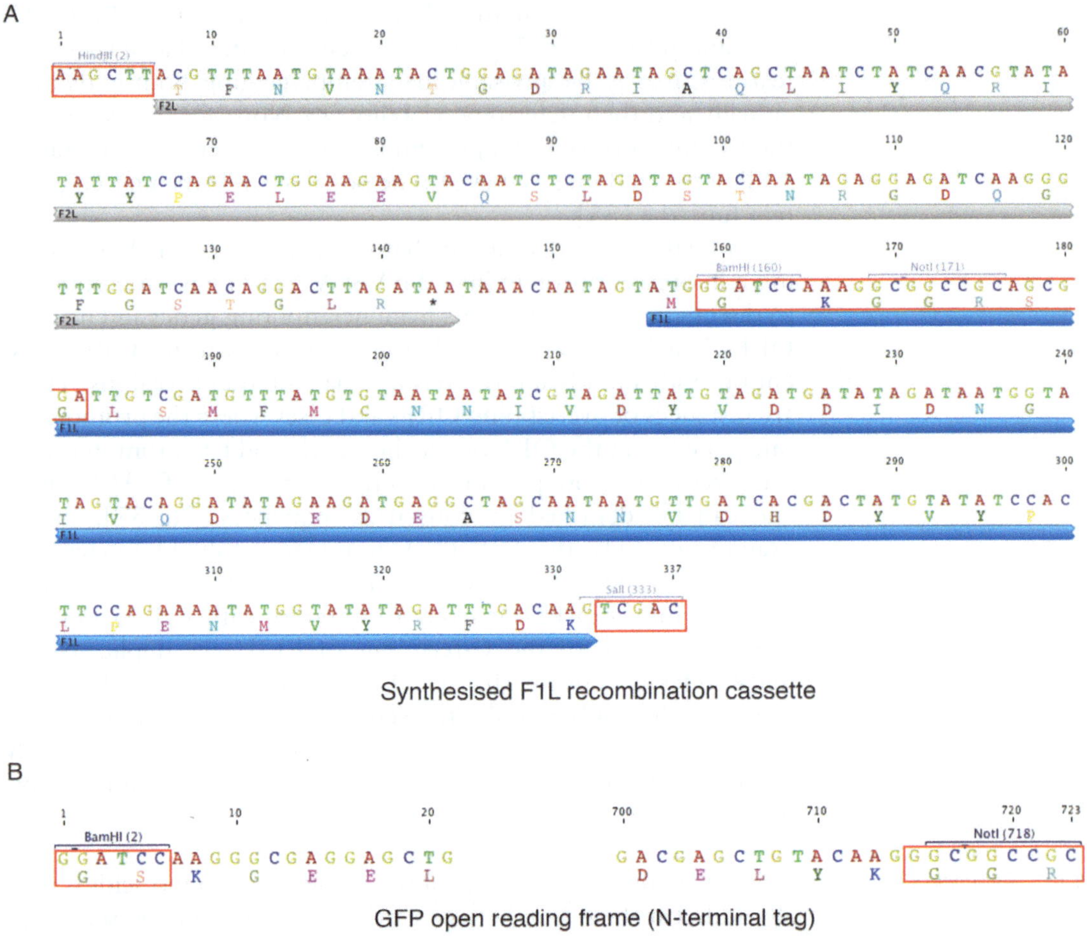

Fig. 3 Construction of recombination plasmid to generate pGFP-F1L fluorescent recombinant virus. (**a**) A genomic region surrounding the start codon of the F1L ORF is ordered as synthesized DNA, which includes the left arm (31,175–31,024, accession AY243312.1) and right arm (31,023–30,874). The sequence is modified to include 5′ and 3′ cloning sites (HindIII and SalI) that allow the sequence to be cloned into the TDS-Fluorion vector. If necessary, HindIII and SalI sites are removed from the internal sequence without modifying any ORFs (such as substitutions at the third position of codons, not necessary in this case). A linker/cloning site is added immediately after the ATG start codon of F1L allowing the in-frame cloning of a fluorescent tag ORF. The resulting fusion ORF will add a G residue in position 2 and a linker consisting of GGRSG between the fluorescent tag and the N-terminus of F1. (**b**) The GFP ORF can be cloned as an N-terminal fusion using the sites BamHI and NotI. Sequences modified from the genomic sequence or tag ORF are boxed in red

teins [43]. Hence some caution needs to be exercised in the choice of new fluorescent proteins as improved variants such as mScarlet [44] and superfolder GFP [45] are made available.

3.2 Expression of an Exogenous Fluorescence Transgene

Driving strong expression of a fluorescent protein from the viral genome can be used to label infected cells, visualize cellular compartments in infected cells, and facilitate recovery of mutant viruses.

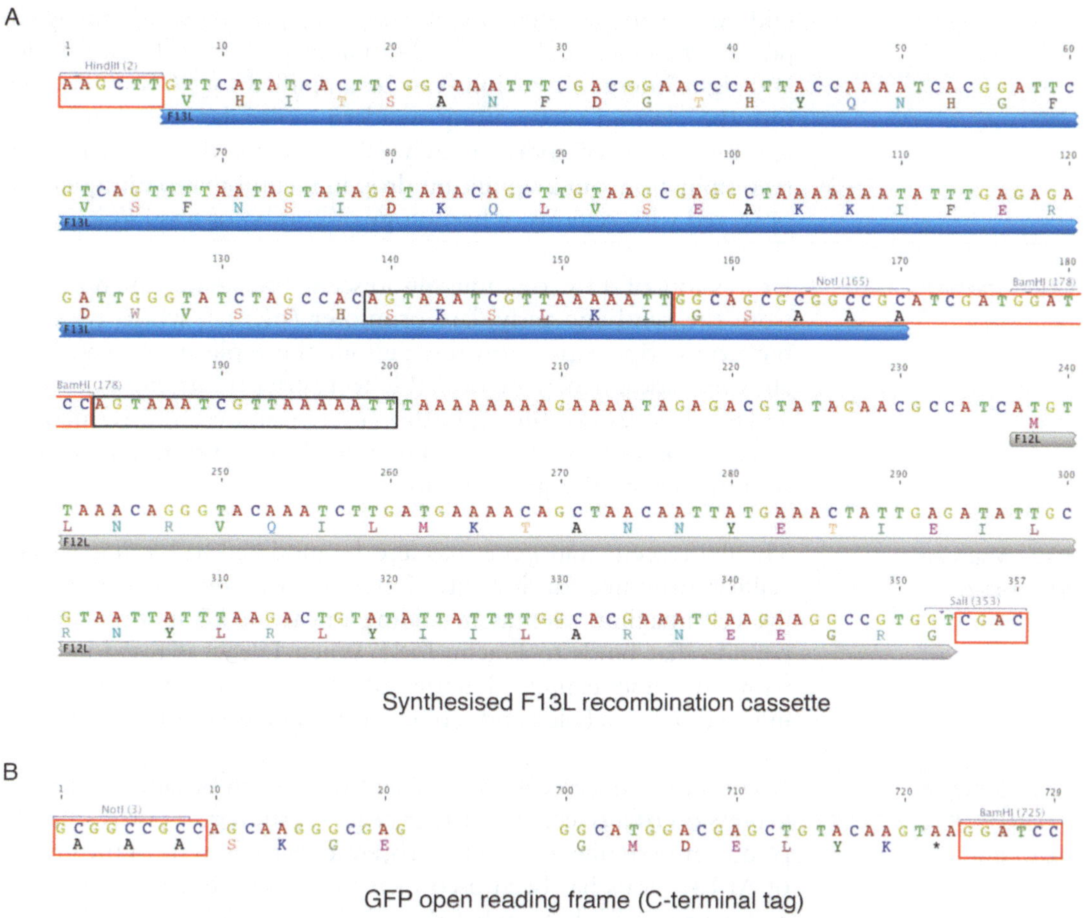

Fig. 4 Construction of recombination plasmid to generate pF13L-GFP fluorescent recombinant virus. (**a**) A genomic region surrounding the stop codon of the F13L ORF is ordered as synthesized DNA, which includes the left arm (40,983–40,834) and right arm (40,851–40,681). The sequence is modified to include 5′ and 3′ cloning sites (HindIII and SalI) that allow the sequence to be cloned into the TDS-Fluorion vector. The F13L stop codon is deleted and a linker/cloning site is added allowing the in-frame cloning of a fluorescent tag ORF. The resulting fusion ORF will add a linker consisting of GSAAA between the C-terminus of F13 and the fluorescent tag. A small sequence (40,851–40,834, boxed in black) is duplicated to ensure that no promoter elements of F12L are disrupted following the modification of the viral genome at the 3′ end of F13L. (**b**) The GFP ORF can be cloned as a C-terminal fusion using the sites NotI and BamHI. A stop codon is added immediately preceding the BamHI site. Sequences modified from the genomic sequence or tag ORF are boxed in red

3.2.1 Selecting an Insertion Site

Genes encoding for fluorescent genes like EGFP or mRFP can be incorporated into the VACV genome in a neutral genomic region that will accept inserts without phenotypic consequences (Fig. 1a, strategy 1). Sites that have been used for this purpose include 122,815*122,817 between the genes A11R and A12L [29], 82,854*82,555 between J4L and J5R, and 30343*30344 between K7R and F1L [20] (numbers refer to VACV Western Reserve accession AY243312.1; * marks insertion site). In these cases, the insertion site lies at the intergenic region where two genes have their 3′

ends that run together, minimizing the possibility of disrupting promoter elements. Regions of homology corresponding to the left and right intergenic regions can be amplified by PCR and cloned into the plasmid vector using standard molecular techniques. Left and right arms of approximately 300 bp are sufficient to mediate recombination and a significant drop in recombination frequency is observed with regions of homology of 200 bp or less [21].

3.2.2 Selecting a Promoter

Expression of an exogenous fluorescent transgene can be driven from the synthetic early/late promoter (pE/L), which produces high transcript expression throughout the replication cycle [46]. In some cases, it may be desirable to restrict transgene expression to specific stages of the replication cycle using early, intermediate, or late promoters [20] or modulate the levels of transgene expression by varying the spacer region [47].

3.2.3 Modifying the Transgene

The fluorescent transgene may also be modified to localize to subcellular structures during infection and allow their visualization in live-cell imaging, such as by adding Lifeact, a 17-amino acid-long peptide that binds to F-actin [48]. When Lifeact is fused in-frame with GFP and expressed under pE/L, the actin cytoskeleton of infected cells is labeled and can be imaged at high resolution (Fig. 5).

3.2.4 Strategies to Efficiently Recover Recombinant Virus

Recovery of fluorescent recombinant viruses can be aided by the co-expression of a selectable marker (Fig. 1a, strategy 2) or rescue of a plaque growth phenotype (Fig. 1b, strategy 8). Rescue of the growth of ΔF13L virus has been used to recover recombinant viruses [28] and we successfully targeted the A36R locus with a recombination cassette that rescued ΔA36R and inserted pE/L Lifeact-GFP (Fig. 5) [49]. In both cases (F13L and A36R), the insertion was targeted to the 5′ end of the gene following the stop codon.

The strong expression of a fluorescent protein can be used to facilitate the recovery of viruses with specific genes disrupted, usually in combination with a selectable marker such as *gpt*. In these cases, the left and right arms are designed such that a double-recombination event will remove a specific sequence that will include a part of the ORF (Fig. 1a, strategy 3). Again, care must be taken that promoter elements of neighboring genes are not disrupted, which would confound mutant analysis. If removal of the selection and screening markers is desirable, then adopting a TDS approach will allow the excision of the markers following the removal of the selection media (Fig. 2a, strategy 14).

3.3 Fluorescent Tagging of Viral Proteins

The most commonly used lab strain of VACV, Western Reserve (accession AY243312) encodes 223 proteins (www.viprbrc.org) involved in DNA transcription and replication, virulence factors that modulate host responses, as well as various structural proteins that constitute the virion itself. Fluorescently tagging a viral protein

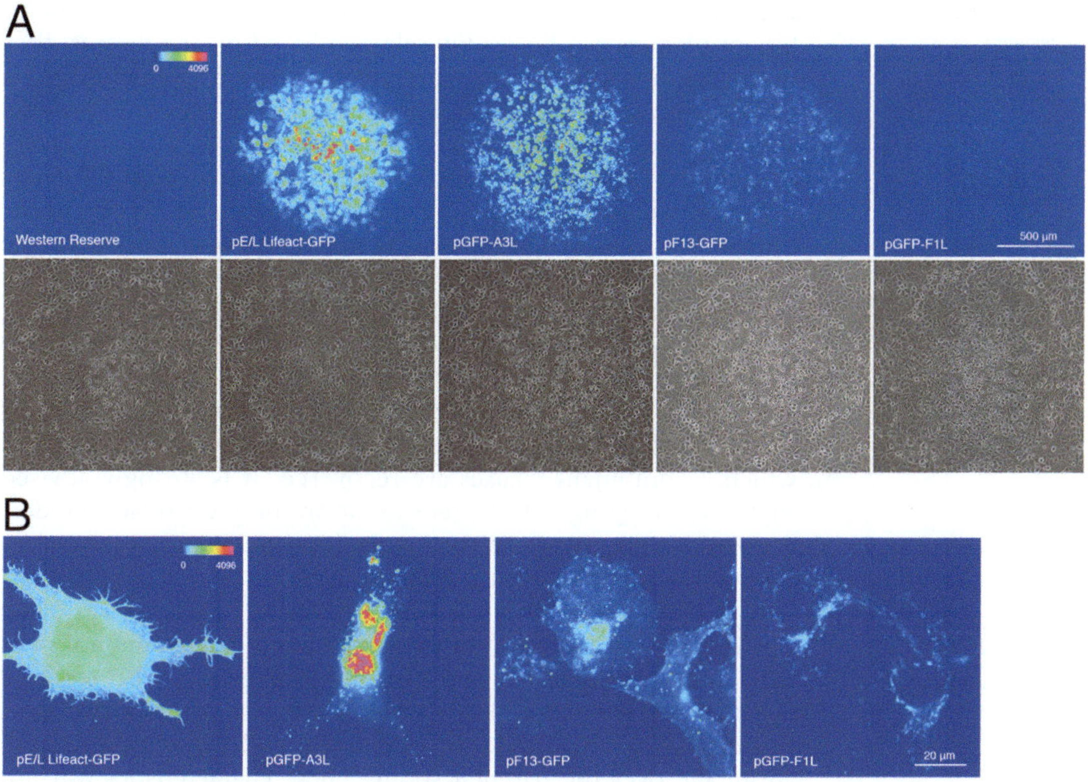

Fig. 5 (**a**) Micrographs of plaques formed by the indicated viruses 3 dpi on monolayers of BS-C-1 cells overlaid with CMC-MEM. (**b**) Micrographs of cells infected with the indicated viruses and fixed (4.7% paraformaldehyde for 10 min). All images were captured using the same settings (**a**: 50-ms exposure with 10× objective; **b**: 20-ms exposure with 60× oil immersion objective) on a Nikon Eclipse T*i*-E inverted microscope with an ANDOR Zyla sCMOS camera and a 470 nm LED light source (100%). Pixel saturation is indicated by a rainbow scale (0–4096)

can be used to track individual virus particles or to identify viral protein localization and/or function. The advantage of incorporating an in-frame fusion ORF at the endogenous locus is that its expression is under the control of the native promoter and the encoded protein does not have to compete with the endogenous protein. If loss of the gene under investigation is associated with a mutant phenotype, then the ability of the fusion ORF to rescue that phenotype can readily be assayed. For essential genes, fluorescently tagged recombinant viruses will only be able to be recovered if the expressed fusion ORF rescues gene function. The techniques and plasmids employed for the generation of these viruses are shared with those described in Subheading 3.2.1 except instead of targeting the insertion to a neutral genomic region, plasmid vectors must be carefully designed such that fluorescent genes are inserted in-frame at the endogenous locus. The plasmid vector design will differ depending on the parental virus being used, and the method of virus selection employed.

3.3.1 By Fluorescence Screening

If the viral protein being tagged is abundantly expressed, recombinant plaques can be readily selected based on the fluorescence of the fusion protein (Fig. 1a, strategies 4 and 5). The brightness of the fluorescent protein and how the plaques are visualized will also determine how successful this approach is. If plaques are identified by eye with an epifluorescence microscope, GFP and mCherry will be easiest to screen for, due to the sensitivity of the human eye. Fluorescent proteins with emission spectra at the shorter wavelength range (Cerulean, Tag-BFP) or weakly expressed fusion proteins may be effectively screened by utilizing sensitive CCD cameras in combination with an epifluorescence microscope. Recombinant VACV pYFP-A3 [34], pGFP-E2 [50], and pRFP-A3 [51] were generated in this way. As screening solely for fluorescence does not increase the frequency with which recombinant viruses are recovered, it is strongly advised that when tagging VACV genes that are only expressed moderately or weakly, or where the expression level is unknown, selection or rescue of a growth defect is used.

3.3.2 By Plaque Phenotype and Selection

The recovery of a fluorescent virus is greatly aided if a mutant virus with a growth defect is available for the gene to be targeted (Fig. 1b, strategies 9 and 10). The fluorescent recombinant virus will display an altered plaque phenotype and will have a growth advantage over the parental strain. These characteristics will not only facilitate the identification of recombinant virus but also greatly assist its subsequent purification. The limitation of this approach is that it is only applicable when the mutant strain is available and the strain has a growth defect. This strategy was successfully used to generate pGFP-F12L [50], pB5R-GFP [10], pA36R^{YdF}-YFP [34], pA36R-YFP, and pF13L-GFP [8]. In principle, the expression of a selection marker can be used to enhance the recovery of a fluorescent recombinant virus but this is rarely used (Fig. 1a, strategies 6 and 7).

3.3.3 By TDS with gpt Selection and Fluorescence Screening

We have developed a methodology that combines strong screening (mCherry) and selection (*gpt*) with a modular TDS vector that allows for the streamlined generation of recombination vectors and the efficient isolation of recombinant fluorescent viruses [52] (Fig. 2b). The use of TDS results in a fluorescent virus that does not retain screening or selection markers and can therefore be further modified to express other fluorescent gene fusions or incorporate loss-of-function modifications. This approach takes advantage of the ever-diminishing costs of DNA synthesis. Left and right arms of homology are designed around the insertion site and modifications, such as small duplications to avoid disrupting promoters of neighboring genes, or elimination of internal restriction sites, are introduced (Figs. 3 and 4) (*see* **Note 5**). The sizes of the left and right arms are selected based on the costs of DNA synthesis

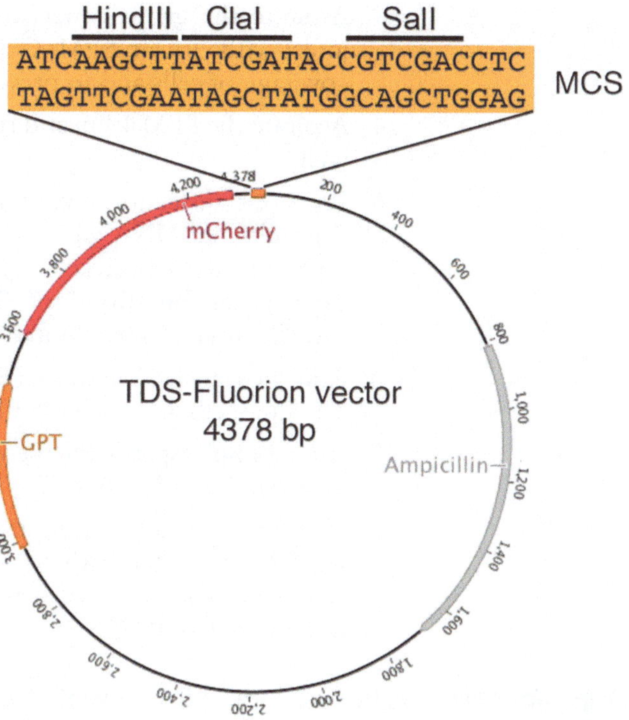

Fig. 6 Map of the TDS-Fluorion vector used to construct recombination plasmids for the recovery of fluorescent viruses using TDS with mCherry screening and *gpt* selection (*see* **Note 11**). Synthesized DNA fragments are cloned into the MCS using HindIII and SalI restriction enzymes

and the optimal length for recombination frequency. For the construction of pGFPF1L and pF13L-GFP, we synthesized left and right arms of approximately 150 bp each. Following cloning of the synthesized DNA into the TDS-Fluorion vector (Fig. 6), a modular fluorescent tag is added. Hence the recombination vector is constructed in two simple cloning steps.

3.4 Infection/Transfection

Once the desired recombination strategy has been selected and the recombination vector has been constructed, plasmid DNA is transfected into infected cells to enable recombination to occur in the host cytoplasm (*see* **Note 6**). The gene of interest and strategy selected will dictate the parental strain used in this step.

1. Seed a 6-well plate with BS-C-1 or HeLa cells 24 h prior to infection, such that they will be around 70% confluent on the day of infection. Incubate cells in MEM in a CO_2 incubator at 37 °C (*see* **Note 7**).
2. Prepare a dilution of parental virus stock in SFM (500 μL per well of a 6-well plate) such that the MOI < 1.

3. Aspirate the growth medium, wash cells in warm PBS, and infect with virus in SFM. Gently rock the plate to ensure even coverage of cells and to prevent drying.
4. Aspirate the SFM 1 hpi and rescue cells in 2 mL MEM (2 mL/well).
5. Prepare transfection mix in 1.5 mL tubes. For each well, add 3 μg plasmid DNA and 1 μL of Lipofectamine 2000 to two separate tubes (or according to manufacturing instructions), each containing 100 μL of SFM. After 5 min, add the diluted Lipofectamine to the diluted DNA, mix, and stand for 20 min.
6. Add the transfection mix to each well dropwise not more than 1 h after cells are rescued in MEM. Incubate at 37 °C.
7. After 24 hpi, aspirate the media and scrape cells in SFM. Collect in sterile 1.5 mL tubes.
8. Perform three alternating freeze-thaw rounds in liquid nitrogen/37 °C water bath to lyse cells and release intracellular virus particles. This lysate can be used for plaque assays directly or stored at −80 °C.

3.5 Plaque Selection and Purification

The basic principle of a viral plaque assay is to perform serial dilutions of a virus stock, such that at a given dilution individual virus plaques, initiated by a single plaque-forming unit, are discernible in a cell monolayer. After Subheading 3.4, **step 8**, the cell lysate contains a mix of parental and recombinant viruses. Therefore, a plaque assay of this lysate needs to be performed to recover plaques formed by recombinant viruses. The basic method for carrying out a plaque assay is provided below, followed by specific techniques related to the desired method of plaque selection. We strongly advise that readers refer to "Construction and isolation of recombinant vaccinia using genetic markers" in a previous volume of this book, which contains detailed advice on the methods used here [3].

1. Seed a 6-well plate with BS-C-1 in MEM such that they are ~90% confluent on the day of infection.
2. Seed a 12-well plate of HeLa in MEM such that they are ~90% confluent when plaques have been isolated (in 3–5 days).
3. Prepare tenfold serial dilutions of the cell-virus lysate in 500 μL of SFM (from Subheading 3.4).
4. Aspirate the MEM, add each dilution to a well (ensure that the dilutions go up to 10^{-6} or 10^{-7}), and leave for 1 h in the CO_2 incubator at 37 °C. The plate can be gently rocked at 30 min to ensure even distribution of virus particles. Prepare the agarose overlay by melting 2% LMP agarose in the microwave and cooling in a 45 °C water bath. After 1 hpi, aspirate the SFM and wash cells twice with PBS.

5. Mix equal volumes of warm (not hot) 2% LMP agarose and 2× MEM and carefully pour 2 mL into each well before it sets. Leave to cool for 5 min before incubating at 37 °C in a CO_2 incubator for 3–5 days postinfection (dpi). If using the GPT selection method, refer to Subheading 3.5.3.
6. After a minimum of 3 dpi, view plaques under an inverted microscope and select the well with the dilution in which plaques are at their highest concentration while still being well spaced and distributed.
7. See Subheadings 3.5.1, 3.5.2, and 3.5.3 for selecting most appropriate plaque selection procedure. Once at least ten plaques have been selected, they can be picked using a P1000 pipette tip. Pierce the agarose directly on top of your selected plaque and push down to the bottom of the dish to retrieve a plug of agarose and detach infected cells. Deposit the plug into 500 μL of SFM and pipette up and down a few times to expel all cells into the media.
8. Freeze-thaw all the picked samples in liquid nitrogen/37 °C water bath three times.
9. Use the samples to infect HeLa cells in a 24-well plate. As above, media is aspirated and cells are washed in PBS before 500 μL of medium containing a single picked plaque is added. Cells are incubated for 48 h to allow for amplification of the virus derived from the plaque.
10. Scrape the cells in each of the wells in the 24-well plate. It is only necessary to proceed with two isolates that have successfully amplified and exhibit fluorescence.
11. Use these two isolates to repeat **steps 1–10** at least 2–3 times, or until you have obtained a pure stock of recombinant virus (all plaques are fluorescent).

3.5.1 By Fluorescence Screening

Use an epifluorescence microscope to observe the selected well and use the appropriate light source and filter block to identify fluorescent plaques. Recombinant viruses that express high levels of mCherry or GFP fusion proteins should be readily visible by eye. Weakly fluorescent plaques or Cerulean/Tag-BFP-expressing plaques may need to be identified through a CCD camera or a higher magnification to increase sensitivity (*see* **Note 8**).

3.5.2 By Rescue of a Growth Defect

If the recombinant virus is recovered by virtue of a growth advantage over the parental strain, such as by rescue of a growth defect, select plaques that display a plaque morphology resembling VACV Western Reserve. To identify recombinant (rescued) plaques, it is important to generate control plaques (mutant parental virus, VACV Western Reserve) in parallel for comparison. Screening for plaque size should be conducted simultaneously with screening for fluorescence (Subheading 3.5.1).

3.5.3 By TDS and Metabolic Selection/Fluorescence Screening

GPT selection reagents are added to the medium to suppress the growth of viruses that do not express *gpt*. In the TDS-Fluorion vector, *gpt* is expressed simultaneously with mCherry. To isolate recombinant viruses using TDS, recombinant viruses derived from single recombination events at either of the homology arms, which incorporate the entire plasmid, are selected and screened for.

Under *gpt* selection, intermediate isolates are purified to clonality (*gpt*-resistant, mCherry-positive) following plaque selection and amplification (*see* **Note 9**). Perform **steps 1–11** (Subheading 3.5) with the modification that GPT selection reagents are added to the growth media 24 h prior to performing the plaque assay or amplifying the picked plaques. GPT selection reagent is also included in the agarose overlay when amplifying plaques and the recovery medium when picked plaques are amplified in HeLa cells. These modified **steps 1–11** (Subheading 3.5) are repeated until the intermediate (*gpt*-resistant, mCherry-positive) is pure. Confirm expression of mCherry (as in Subheading 3.5.1).

Once pure, the intermediate is amplified in the absence of *gpt* selection, **steps 9–10** (Subheading 3.5). This will allow excision of the TDS-Fluorion vector from the viral genome by a recombination event between either of the duplicated sequences. Approximately half of the mCherry-negative isolates will express the desired fluorescent fusion protein. Perform **steps 1–11** (Subheading 3.5) as described and screen for fluorescence of the desired fusion protein and loss of strong mCherry expression (as in Subheading 3.5.1).

3.6 Virus Amplification

Once sufficient rounds of plaque purification are complete, the virus stock needs to be amplified for further use. Please refer to [53].

3.7 Recombinant Virus Verification and Analysis

Recombinant viruses need to be verified by molecular means to ensure that the genome carries the predicted modifications. Viral genomes should be prepared [54] and used as templates for PCR that amplify the novel fusion ORF and boundaries where recombination has taken place. These PCR products should then be sequenced.

The replication dynamics of the recombinant virus generated must also be assessed. This can be done by performing a plaque assay of both parental and recombinant viruses.

1. Seed a 6-well plate with BS-C-1 cells in MEM such that they are ~90% on the day of infection.
2. Prepare tenfold serial dilutions of the pure amplified virus in 500 μL of SFM.
3. Mix equal volumes of CMC and 2×MEM supplemented with 20% FBS and pour 2 mL into each well. Incubate at 37 °C in a CO_2 incubator for 3–5 days (depending on expected replication rates of parental and/or recombinant VACV strains).

4. Aspirate the CMC-MEM mix from each well and wash three times with PBS.
5. Fix and stain cells with crystal violet for 10 min, followed by at least three rounds of washing in PBS.
6. Image the plaques using a scanner (a high-resolution gel scanner will do).
7. Compare plaque sizes of parental virus to that of the recombinants using an image analysis tool such as ImageJ and quantification followed by statistical analysis (GraphPad Prism).

Finally, the recombinant fluorescent virus is ready for analysis by immunofluorescence to observe colocalization with cellular or viral markers, live-cell microscopy, and a range of other assays.

4 Notes

1. We use HeLa cells as in our hands they display superior transfection frequencies to BS-C-1 cells. HeLa are used in the steps where cells are transfected and infected. BS-C-1 cells (or CV-1; ATCC CCL-70) form superior monolayers and are used for plaque assays.
2. Recent research has demonstrated that these efficiencies can be greatly enhanced using CRISPR-Cas9 genome editing [55].
3. A small GFP–bsd fusion protein functions as both a fluorescent screening marker and for selection of plaque growth in the presence of the protein synthesis blocker, blasticidin [29].
4. In some cases, tagging a gene at either end may not disrupt functionality. We have successfully tagged F12 at both the N-terminus and C-terminus ([50] and unpublished observations) without significant attenuation.
5. A number of vendors offer custom DNA synthesis (www.genscript.com/; sg.idtdna.com/). If the recombination cassette is ordered as a linear fragment, ensure that several nucleotides are added at each end such that the restriction enzymes are able to digest the fragment efficiently.
6. Although we have had success with transfecting plasmid DNA, there is evidence that linearizing DNA before transfection may enhance recombination frequency [21].
7. We recommend that these steps are conducted in duplicate and at least one recombinant virus is selected from each replicate. This ensures that the two isolated strains are derived from independent events. This is important as occasionally aberrant recombination events or background mutations will be recovered in selected recombinant viruses.

8. Recombinant viruses exhibiting weak fluorescence may be difficult to identify as plaques in 6-well plates. If no fluorescent viruses are recovered at this stage, amplified virus from plaques can be tested by infecting HeLa cells seeded on coverslips. At 8 hpi, infected cells are fixed, mounted onto slides, and viewed by epifluorescence microscopy to determine successful expression of a fluorescent fusion protein.
9. The expression of the desired fluorescent–viral fusion protein may be detected in these intermediary isolated viruses depending on how the plasmid is integrated. Approximately 50% of events will give rise to an intermediate that expresses a full-length fusion protein. If the viral protein is tagged with mCherry then it is unlikely that this signal will be able to be resolved against the background of pE/L expression. This should not present any hindrance in identifying the fusion protein following the removal of *gpt* selection.
10. Whether the recombination event occurs at the left arm (5′) or right arm (N-term) will not affect the final recombinant virus; it will result in a slightly different TDS-intermediate that expresses either a truncated GFP-F1 (encoding the first 50 residues of F1, depicted above) or a full-length GFP-F1 fusion protein (not depicted).
11. As mCherry and *gpt* are expressed from the same promoter (pE/L) they are oriented to run in opposing directions such that the plasmid is stable and not prone to deletions.

References

1. Ball LA (1987) High-frequency homologous recombination in vaccinia virus DNA. J Virol 61:1788–1795
2. Ball LA (1995) Fidelity of homologous recombination in vaccinia virus DNA. Virology 209:688–691
3. Lorenzo MM, Galindo I, Blasco R (2004) Construction and isolation of recombinant vaccinia virus using genetic markers. Methods Mol Biol 269:15–30
4. Smith GL, Moss B (1983) Infectious poxvirus vectors have capacity for at least 25 000 base pairs of foreign DNA. Gene 25:21–28
5. Moss B (1996) Genetically engineered poxviruses for recombinant gene expression, vaccination, and safety. Proc Natl Acad Sci U S A 93:11341–11348
6. Lorenzo MM, Blasco R (1998) PCR-based method for the introduction of mutations in genes cloned and expressed in vaccinia virus. Biotechniques 24:308–313
7. Dominguez J, Lorenzo MM, Blasco R (1998) Green fluorescent protein expressed by a recombinant vaccinia virus permits early detection of infected cells by flow cytometry. J Immunol Methods 220:115–121
8. Geada MM, Galindo I, Lorenzo MM, Perdiguero B, Blasco R (2001) Movements of vaccinia virus intracellular enveloped virions with GFP tagged to the F13L envelope protein. J Gen Virol 82:2747–2760
9. Hollinshead M, Rodger G, Van Eijl H, Law M, Hollinshead R, Vaux DJ et al (2001) Vaccinia virus utilizes microtubules for movement to the cell surface. J Cell Biol 154:389–402
10. Ward BM, Moss B (2001) Visualization of intracellular movement of vaccinia virus virions containing a green fluorescent protein-B5R membrane protein chimera. J Virol 75:4802–4813
11. Leite F, Way M (2015) The role of signalling and the cytoskeleton during Vaccinia Virus egress. Virus Res 209:87–99

12. Ward BM (2004) Pox, dyes, and videotape: making movies of GFP-labeled vaccinia virus. Methods Mol Biol 269:205–218
13. Carter GC, Rodger G, Murphy BJ, Law M, Krauss O, Hollinshead M et al (2003) Vaccinia virus cores are transported on microtubules. J Gen Virol 84:2443–2458
14. Dobson BM, Procter DJ, Hollett NA, Flesch IE, Newsome TP, Tscharke DC (2014) Vaccinia virus F5 is required for normal plaque morphology in multiple cell lines but not replication in culture or virulence in mice. Virology 456–457:145–156
15. Schmidt FI, Bleck CK, Reh L, Novy K, Wollscheid B, Helenius A et al (2013) Vaccinia virus entry is followed by core activation and proteasome-mediated release of the immunomodulatory effector VH1 from lateral bodies. Cell Rep 4:464–476
16. Horsington J, Turnbull L, Whitchurch CB, Newsome TP (2012) Sub-viral imaging of vaccinia virus using super-resolution microscopy. J Virol Methods 186:132–136
17. Humphries AC, Dodding MP, Barry DJ, Collinson LM, Durkin CH, Way M (2012) Clathrin potentiates vaccinia-induced actin polymerization to facilitate viral spread. Cell Host Microbe 12:346–359
18. Gray RD, Beerli C, Pereira PM, Scherer KM, Samolej J, Bleck CK et al (2016) VirusMapper: open-source nanoscale mapping of viral architecture through super-resolution microscopy. Sci Rep 6:29,132
19. Paszkowski P, Noyce RS, Evans DH (2016) Live-cell imaging of vaccinia virus recombination. PLoS Pathog 12:e1005824
20. Dower K, Rubins KH, Hensley LE, Connor JH (2011) Development of Vaccinia reporter viruses for rapid, high content analysis of viral function at all stages of gene expression. Antiviral Res 91:72–80
21. Yao XD, Evans DH (2001) Effects of DNA structure and homology length on vaccinia virus recombination. J Virol 75:6923–6932
22. Coupar BE, Oke PG, Andrew ME (2000) Insertion sites for recombinant vaccinia virus construction: effects on expression of a foreign protein. J Gen Virol 81:431–439
23. Chakrabarti S, Brechling K, Moss B (1985) Vaccinia virus expression vector: coexpression of beta-galactosidase provides visual screening of recombinant virus plaques. Mol Cell Biol 5:3403–3409
24. Panicali D, Grzelecki A, Huang C (1986) Vaccinia virus vectors utilizing the beta-galactosidase assay for rapid selection of recombinant viruses and measurement of gene expression. Gene 47:193–199
25. Falkner FG, Moss B (1990) Transient dominant selection of recombinant vaccinia viruses. J Virol 64:3108–3111
26. Boyle DB, Coupar BE (1988) A dominant selectable marker for the construction of recombinant poxviruses. Gene 65:123–128
27. Byrd CM, Hruby DE (2004) Construction of recombinant vaccinia virus: cloning into the thymidine kinase locus. Methods Mol Biol 269:31–40
28. Blasco R, Moss B (1995) Selection of recombinant vaccinia viruses on the basis of plaque formation. Gene 158:157–162
29. Wong YC, Lin LC, Melo-Silva CR, Smith SA, Tscharke DC (2011) Engineering recombinant poxviruses using a compact GFP-blasticidin resistance fusion gene for selection. J Virol Methods 171:295–298
30. Roscoe F, Xu RH, Sigal LJ (2012) Characterization of ectromelia virus deficient in EVM036, the homolog of vaccinia virus F13L, and its application for rapid generation of recombinant viruses. J Virol 86:13501–13507
31. Alejo A, Saraiva M, Ruiz-Arguello MB, Viejo-Borbolla A, de Marco MF, Salguero FJ et al (2009) A method for the generation of ectromelia virus (ECTV) recombinants: in vivo analysis of ECTV vCD30 deletion mutants. PLoS One 4:e5175
32. Lorenzo MM, Sanchez-Puig JM, Blasco R (2017) Vaccinia virus and Cowpox virus are not susceptible to the interferon-induced antiviral protein MxA. PLoS One 12:e0181459
33. Stewart TL, Wasilenko ST, Barry M (2005) Vaccinia virus F1L protein is a tail-anchored protein that functions at the mitochondria to inhibit apoptosis. J Virol 79:1084–1098
34. Arakawa Y, Cordeiro JV, Schleich S, Newsome TP, Way M (2007) The release of vaccinia from infected cells requires RhoA-mDia modulation of cortical actin. Cell Host Microbe 1:227–240
35. Rodger G, Smith GL (2002) Replacing the SCR domains of vaccinia virus protein B5R with EGFP causes a reduction in plaque size and actin tail formation but enveloped virions are still transported to the cell surface. J Gen Virol 83:323–332
36. Davison AJ, Moss B (1989) Structure of vaccinia virus late promoters. J Mol Biol 210:771–784
37. Davison AJ, Moss B (1989) Structure of vaccinia virus early promoters. J Mol Biol 210:749–769

38. Chen X, Zaro JL, Shen WC (2013) Fusion protein linkers: property, design and functionality. Adv Drug Deliv Rev 65:1357–1369
39. Klein JS, Jiang S, Galimidi RP, Keeffe JR, Bjorkman PJ (2014) Design and characterization of structured protein linkers with differing flexibilities. Protein Eng Des Sel 27:325–330
40. Day RN, Davidson MW (2009) The fluorescent protein palette: tools for cellular imaging. Chem Soc Rev 38:2887–2921
41. Shaner NC, Campbell RE, Steinbach PA, Giepmans BN, Palmer AE, Tsien RY (2004) Improved monomeric red, orange and yellow fluorescent proteins derived from Discosoma sp. red fluorescent protein. Nat Biotechnol 22:1567–1572
42. Nagai T, Ibata K, Park ES, Kubota M, Mikoshiba K, Miyawaki A (2002) A variant of yellow fluorescent protein with fast and efficient maturation for cell-biological applications. Nat Biotechnol 20:87–90
43. Carpentier DCJ, Hollinshead MS, Ewles HA, Lee SA, Smith GL (2017) Tagging of the vaccinia virus protein F13 with mCherry causes aberrant virion morphogenesis. J Gen Virol
44. Bindels DS, Haarbosch L, van Weeren L, Postma M, Wiese KE, Mastop M et al (2017) mScarlet: a bright monomeric red fluorescent protein for cellular imaging. Nat Methods 14:53–56
45. Pedelacq JD, Cabantous S, Tran T, Terwilliger TC, Waldo GS (2006) Engineering and characterization of a superfolder green fluorescent protein. Nat Biotechnol 24:79–88
46. Chakrabarti S, Sisler JR, Moss B (1997) Compact, synthetic, vaccinia virus early/late promoter for protein expression. Biotechniques 23:1094–1097
47. Di Pilato M, Sanchez-Sampedro L, Mejias-Perez E, Sorzano CO, Esteban M (2015) Modification of promoter spacer length in vaccinia virus as a strategy to control the antigen expression. J Gen Virol 96:2360–2371
48. Riedl J, Crevenna AH, Kessenbrock K, Yu JH, Neukirchen D, Bista M et al (2008) Lifeact: a versatile marker to visualize F-actin. Nat Methods 5:605–607
49. Marzook NB, Latham SL, Lynn H, McKenzie C, Chaponnier C, Grau GE et al (2017) Divergent roles of beta- and gamma-actin isoforms during spread of vaccinia virus. Cytoskeleton (Hoboken) 74:170–183
50. Dodding M, Newsome TP, Collinson L, Edwards C, Way M (2009) An E2-F12 complex is required for IEV morphogenesis during vaccinia infection. Cell Microbiol 11:808–824
51. Weisswange I, Newsome TP, Schleich S, Way M (2009) The rate of N-WASP exchange limits the extent of Arp2/3 complex dependent actin-based motility. Nature 458:87–91
52. Marzook NB, Procter DJ, Lynn H, Yamamoto Y, Horsington J, Newsome TP (2014) Methodology for the efficient generation of fluorescently tagged vaccinia virus proteins. J Vis Exp:e51151
53. Kotwal GJ, Abrahams MR (2004) Growing poxviruses and determining virus titer. Methods Mol Biol 269:101–112
54. Roper RL (2004) Rapid preparation of vaccinia virus DNA template for analysis and cloning by PCR. Methods Mol Biol 269:113–118
55. Yuan M, Zhang W, Wang J, Al Yaghchi C, Ahmed J, Chard L et al (2015) Efficiently editing the vaccinia virus genome by using the CRISPR-Cas9 system. J Virol 89:5176–5179

Chapter 5

Generation of Vaccinia Virus Gene Deletion Mutants Using Complementing Cell Lines

Amber B. Rico, Annabel T. Olson, and Matthew S. Wiebe

Abstract

This protocol describes how to couple two techniques, the generation of complementing cells lines and production of viral deletion mutants, to rapidly construct novel tools for poxvirus analysis. Specifically, the production and utilization of a complementing cell line expressing a poxvirus gene of interest are critical for the generation of poxvirus mutants in which essential genes are disrupted. Complementing cells are also valuable for the characterization of vaccinia genes in the absence of infection. Here, we detail the process of isolating vaccinia virus deletion mutants. Deletion mutant generation involves homologous recombination between replicating viral DNA and transfected DNA followed by selection and screening on a complementing cell line that provides the deleted gene in trans. Finally, deletion is confirmed by polymerase chain reaction, sequencing, and functional assays if available.

Key words Vaccinia, Deletion mutant, Complementing cells, Fluorescence, Essential gene

1 Introduction

Vaccinia single-gene deletion mutants provide powerful and straightforward tools with which to identify the function of individual vaccinia proteins. Such mutants can be generated in infected cells by exploiting recombination known to occur between replicating virus genomes and transfected DNA fragments containing a genetic marker flanked by sequences homologous to the region to be deleted [1–5]. Although resulting recombinants are initially uncommon among the progeny virus population, deletion mutants can be isolated efficiently through selection and plaque purification. However, an important caveat of this approach is the requirement that the gene of interest not perform a vital function, as loss of essential genes abrogates vaccinia replication and prevents the growth and subsequent isolation of deletion mutants. One approach to circumvent the high fitness cost resulting from loss of an essential gene is to employ temperature-sensitive point mutants or drug-inducible transgenes [6–11]. Mutants constructed using these

Jason Mercer (ed.), *Vaccinia Virus: Methods and Protocols*, Methods in Molecular Biology, vol. 2023, https://doi.org/10.1007/978-1-4939-9593-6_5,

techniques have provided numerous insights into the function of viral genes over the years. However, the conditional lethality inherent with temperature-sensitive mutants and the leakiness of inducible promoters in some contexts [12–14] introduce the possibility that these mutants do not exhibit true null phenotypes and are instead an intermediate phenotype due to incomplete repression. To address these points, an alternative approach is to use a cell line to express the viral protein and thus complement the necessary function during isolation and expansion of the recombinant deletion virus. Various methods for the generation of deletion mutants of essential genes have been developed using complementing cell lines [12–21]. These approaches all rely on the generation of a complementing cell line that provides the essential gene to be deleted in trans. Beyond this, approaches can vary widely. Importantly, though these methods were developed to disrupt essential viral genes, the technique is valuable in the generation of deletion mutants which display any fitness cost to the virus. Furthermore, critical insights can be gained from using complementing cell lines to study vaccinia proteins in the absence of infection [22, 23].

Guidelines or attributes for selection of genes suitable for this method have been described since the initial complementation experiments performed by Sutter et al. [20] and are useful to reiterate below.

1. Loss-of-gene function is likely to prevent viral replication or come at a high fitness cost.
2. The gene is required early and in amounts achievable by cellular expression. Alternatively, intermediate and late genes are amenable to this method if the protein products are stable and not degraded rapidly following the inhibition of host gene expression.
3. The deleted gene must be nontoxic to the cell. In the case of toxic genes, inducible expression systems may potentially be used to isolate deletion mutants.

This list of gene attributes is neither exhaustive nor completely prohibitive but should serve as a general rubric for determining when complementing cell lines can be employed for the deletion of essential genes.

Herein we describe our strategy for the generation of complementing cell lines as well as vaccinia virus deletion mutants. These procedures will detail the steps to (1) establish a complementing cell line using lentiviral transduction to achieve stable gene expression, (2) infect the complementing cell line with vaccinia virus and transfect with homologous recombination plasmid to generate a deletion virus, (3) isolate and enrich for the recombinant deletion virus through screening of fluorescent plaques on the complementing cell line, and (4) confirm deletion of vaccinia gene by PCR.

2 Materials

1. Lentiviral transfer vector containing codon-optimized ORF of gene of interest (GOI)/gene to be deleted.
2. BamHI enzyme.
3. Lentivirus generation system (pHAGE-HYG-MCS (pTransfer), pVSVG, pREV, and pGag/Pol).
4. 293T cells.
5. Dulbecco's phosphate-buffered saline (DPBS) without calcium, magnesium.
6. Trypsin-EDTA: 0.05% Trypsin, 0.53 mM EDTA.
7. Cell growth medium: DMEM supplemented with 10% fetal bovine serum (FBS), 100 U/mL penicillin, and 100 μg/mL streptomycin.
8. 1 M Sodium butyrate, sterile-filtered.
9. 0.5 M HEPES, pH 7.4, sterile-filtered.
10. Cell growth medium supplemented with 5 mM sodium butyrate (*see* **Note 1**).
11. Cell growth medium supplemented with 10 mM HEPES (*see* **Note 2**).
12. Dulbecco's modified Eagle's medium (DMEM).
13. Lipofectamine 2000.
14. Transfection medium: DMEM supplemented with 10% FBS.
15. 10 mL Syringes.
16. 0.45 μm Sterile filters.
17. 10 mg/mL Polybrene.
18. CV-1 cells.
19. Vaccinia virus.
20. Homologous recombination plasmid (*see* Subheading 3.3).
21. Disposable sterile cell lifters.
22. Tabletop centrifuges equipped to hold 1.5 and 15 mL tubes.
23. 10 mM Tris, pH 9, sterile-filtered.
24. Cup horn sonicator.
25. Crystal violet fixative: 0.1% Crystal violet, 3.7% formaldehyde.
26. Inverted, fluorescent microscope with appropriate filters.
27. GeneJET Genomic DNA Purification Kit (Thermo Scientific).
28. Polymerase chain reaction (PCR) machine and reagents.
29. PCR primers.
30. Other supplies: Dishes, plates, pipettes, centrifuge tubes, pipette tips, CO_2 incubators, water bath.

3 Methods

3.1 Establishing Complementing Cell Lines

Before vaccinia gene deletion viruses can be made for essential genes, a cell line expressing the vaccinia gene to be deleted must be engineered to provide the gene in trans during the recombination and purification stages of this protocol. Multiple methods have been used to establish complementing cell lines including transient, stable, and inducible gene expression. Here, the stable method of gene expression is detailed using lentivirus transduction. Numerous plasmid systems have been designed to produce replication-incompetent lentiviruses. The current protocol utilizes the third-generation packaging system in which all necessary genes are provided on four distinct plasmids. Care should be taken when working with lentiviruses to prevent the generation of replication-competent virus [24]. In this regard, third-generation packaging systems are considered safer than second-generation systems. We recommend using CV-1 or BSC40 cells as your parental cell line to construct the complementing cell line as they are highly amenable to transduction and highly permissive to vaccinia infection.

1. Generate complementing gene construct by amplifying vaccinia gene to be deleted from codon-optimized construct using primers with BamHI restriction sites and cloning the product into the BamHI site of pHAGE-HYG-MCS or other lentivirus transfer vectors (*see* **Note 3**).
2. Approximately 24 h before lentivirus generation, seed 293T cells in 10 cm dishes (*see* **Note 4**). Incubate the cells in a CO_2 incubator at 37 °C for 20–24 h until the cells are 80–90% confluent.
3. Assemble transfection reactions. In a 1.5 mL centrifuge tube, dilute 25 μL Lipofectamine 2000 in 500 μL DMEM. In a separate 1.5 mL centrifuge tube, dilute 10 μg pHAGE-HYG-complementing gene or pHAGE-HYG-MCS, 1 μg pVSVG, 0.5 μg pREV, and 0.5 μg pGag/Pol in 500 μL DMEM. Incubate separately for 5 min at room temperature. It is assumed that plasmids are supercoiled and highly purified prior to transfection.
4. Combine Lipofectamine and DNA and incubate for 20 min at room temperature.
5. During 20-min incubation, aspirate cell growth medium from 10 cm dish of 293T cells and carefully add 9 mL transfection medium.
6. Add 1 mL of Lipofectamine:DNA complex (*see* **step 4** above) dropwise to 10 cm dish of 293T cells.
7. Approximately 24 h after transfection, in a viral cell culture hood aspirate medium from 10 cm dish of transfected 293T

cells and add 10 mL cell growth medium supplemented with sodium butyrate.

8. Approximately 8 h after addition of sodium butyrate-supplemented cell growth medium, aspirate medium from 10 cm dish of transfected 293T cells in a viral cell culture hood and gently wash cells once with prewarmed DPBS.
9. Add 5 mL cell growth medium supplemented with HEPES.
10. Approximately 24 h after addition of HEPES-supplemented cell growth medium, collect medium containing lentiviruses from 10 cm dishes of transfected 293T cells and filter purify using a 0.45 μM filter (*see* **Note 5**).
11. Add 10 mg/mL polybrene to filtered lentivirus to a final concentration of 8 μg/mL. Aliquot 1 mL volumes into 1.5 mL centrifuge tubes and use immediately or store at −80 °C until needed.
12. Split CV-1 cells into a 6-well plate (35 mm wells). Incubate the cells in a CO_2 incubator at 37 °C for 20–24 h until the cells are 80% confluent (*see* **Note 6**).
13. Aspirate cell growth medium from well and replace with 1 mL of lentivirus (*see* **step 11** above). Incubate for 16 h in a CO_2 incubator at 37 °C.
14. Aspirate lentivirus-containing medium and replace with 2 mL cell growth medium. Incubate for 24 h in a CO_2 incubator at 37 °C.
15. Split transduced CV-1 cells into two 10 cm plates per transduced 35 mm well of cells (*see* **Note 7**). Maintain one 10 cm plate in standard cell growth medium. Maintain one 10 cm plate in cell growth medium containing hygromycin (*see* **Note 8**). Incubate in a CO_2 incubator at 37 °C.
16. Allow several days for hygromycin selection to take place, splitting the cells as necessary (*see* **Note 9**). Continue passage of transduced CV-1 cell lines in hygromycin as required by cell growth.

3.2 Confirmation of Vaccinia Gene Expression in Complementing Cell Line

Following hygromycin selection, stable expression of the complementing protein in the CV-1 cell line should be confirmed by western blot analysis if possible. To evaluate protein expression, CV-1-complementing cells are harvested and assayed for the complementing vaccinia protein by western blot. Antibodies to the vaccinia protein itself may be used or in the case of vaccinia constructs that have been epitope tagged those antibodies may be used. Under ideal conditions, the complementing cell line should produce protein in amounts equivalent to, or higher than, levels that occur during wild-type infection. Therefore, if antibodies are available to the vaccinia protein of interest, lysate from infected cells should be analyzed in parallel to determine if expression from the complementing cell is equivalent to infected cells. If no antibodies are available, verification of mRNA expression is an option. Alternatively, a functional assay can be used to measure the activity of the viral protein (e.g., phosphorylation of a known substrate of a viral kinase), if one is known. Detailed description of these assays can be found in other sources.

3.3 Infection/Transfection Procedure

Homologous recombination will occur between replicating virus genomes and plasmid DNA introduced by transfection. For homologous recombination, a plasmid vector that possesses flanking sequences identical to the regions surrounding the gene to be deleted must be designed (*see* **Note 10**). Between the flanking sequences on the plasmid we recommend inserting the coding sequence for a fluorescent protein to be used during plaque purification as an indicator of recombinant virus (*see* **Note 11**). A variety of viral promoters may be used to express the marker gene [25, 26]. We have had success with the P11 late promoter originally described by Bertholet et al. and Wittek et al. [27, 28]. In this section, complementing cells are infected and then transfected with a linearized plasmid containing a genetic marker under a vaccinia promoter flanked by sequence homologous to that surrounding the gene to be deleted (Fig. 1).

1. Approximately 24 h before infection, seed 2 wells (*see* **Note 12**) of a 6-well plate with the CV-1-complementing cell line. Incubate the cells in a CO_2 incubator at 37 °C for 20–24 h until the cells are 100% confluent.
2. Thaw virus and sonicate twice for 15 s with 30 s of rest on ice between sonication cycles.
3. Prepare virus inoculum (500 μL/well) by diluting virus stock in DMEM to a multiplicity of infection (MOI) of 0.03 plaque-forming units per cell.
4. Aspirate cell growth medium from well and add virus inoculum. Place in a CO_2 incubator at 37 °C for 30 min rocking the dish every 10 min.
5. Aspirate virus inoculum from well and replace with 2 mL of cell growth medium. Incubate for 3 h in a CO_2 incubator at 37 °C.
6. Approximately 30 min before transfection, prepare transfection sample. In a 1.5 mL centrifuge tube, dilute 5 μL Lipofectamine 2000 in 150 μL DMEM. In a separate 1.5 mL centrifuge tube, dilute 1 μg linearized homologous recombination plasmid DNA in 150 μL DMEM. Incubate separately for 5 min at room temperature.

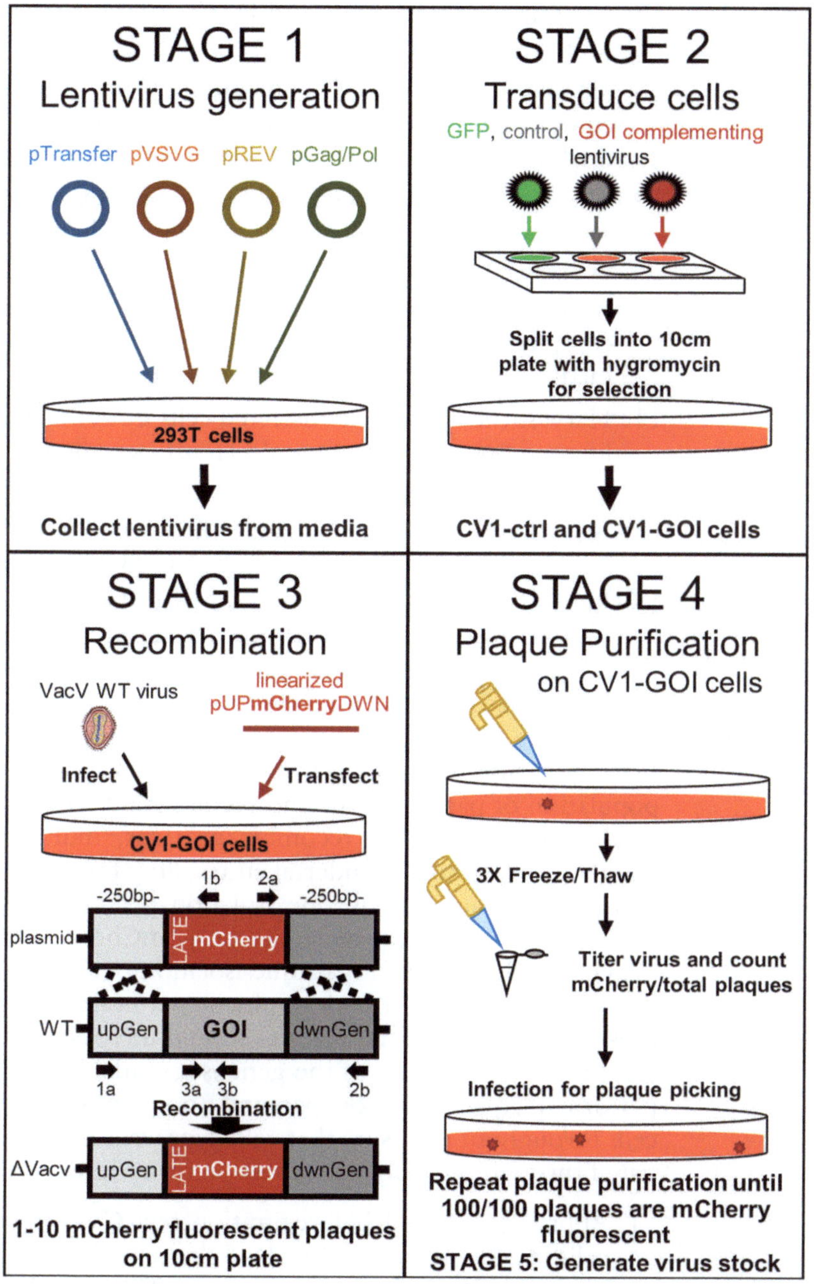

Fig. 1 Generation of a vaccinia virus deleted of an essential gene is completed in five stages. Stage 1 is to generate lentiviruses in 293T cells for stable expression of the GOI (gene of interest) in a cell line. Stage 2 uses the lentivirus to transduce CV1 cells for stable expression of GOI required for selection of vaccinia virus with deletion of essential GOI. Stage 3 recombination step in CV1-GOI-complementing cells occurs due to identical sequence flanking the mCherry gene to replace the GOI in the WT vaccinia virus genome. Stage 4 includes a number of plaque purifications on CV1-GOI complementing cells until 100/100 plaques observed are mCherry fluorescent representing a pure population. Stage 5 is the final step to generate a stock of the GOI knockout vaccinia virus

7. Combine Lipofectamine and DNA and incubate for 20 min at room temperature.
8. Aspirate cell growth medium from infected CV-1-complementing cell line culture and add 1.7 mL transfection medium.
9. Add 300 μL of Lipofectamine:DNA complex (*see* **step 7** above) to virus-infected CV-1-complementing cells. Incubate for 6 h in a CO_2 incubator at 37 °C.
10. Aspirate transfection medium from infected, transfected CV-1-complementing cell line culture and add 2 mL cell growth medium. Incubate for 48 h in a CO_2 incubator at 37 °C.
11. Harvest the infected, transfected cells from the well with a cell lifter and transfer into a sterile 15 mL centrifuge tube.
12. Spin infected and transfected cells at 4 °C for 10 min at 500 × *g*. Aspirate off supernatant. Suspend pellet in 200 μL 10 mM Tris pH 9 and transfer to a 1.5 mL centrifuge tube.
13. Release progeny virus by freeze-thaw lysis of the harvested cells by placing at −80 °C for 10 m and then thawing in a 37 °C water bath. Freeze-thaw three times. Progeny virus can be stored at −80 °C until titration and plaque purification.

3.4 Plaque Purification

Progeny virus (*see* **step 13** in Subheading 3.3) contains a mixed population of parental virus and recombinant virus (<0.1% of the total virus). To isolate recombinant virus, multiple rounds of plaque purification are undertaken to allow for complete removal of parental virus. The plaque isolation protocol described here involves fluorescent protein detection (mCherry), but there are many suitable markers for plaque isolation (*see* **Note 13**). As one goes through several rounds of plaque purification the ratio of fluorescent plaques to total plaques should increase until all plaques are fluorescent (Fig. 1). If the gene is essential it may be useful to perform plaque assays on parental (non-complementing) CV-1 cells throughout to ensure that wild-type virus has been fully eliminated through the plaque purification procedure.

1. Approximately 24 h before titration, seed CV-1-complementing cells in a 12-well plate. Incubate the cells in a CO_2 incubator at 37 °C for 20–24 h until the cells are 100% confluent.
2. Thaw progeny virus (*see* **step 13** in Subheading 3.3) and sonicate twice for 15 s with 30 s of rest on ice between sonications.
3. Make tenfold serial dilutions of progeny virus using dilutions from at least 10^{-5} to 10^{-8}.
4. Aspirate cell growth medium from 12-well plate and add 250 μL per well of respective serial dilutions. Place in a CO_2 incubator at 37 °C for 30 min rocking the dish every 10 min.

5. Aspirate virus inoculum from plate and replace with 1 mL of cell growth medium per well. Incubate for 48 h (*see* **Note 14**) in a CO_2 incubator at 37 °C.
6. Before fixing plate, it may be useful to observe mCherry expression compared to total plaque number (*see* **Note 15**). This will give you an idea of recombination frequency (*see* **Note 16**) and will serve as a guide for how many plaque-forming units to put down when beginning plaque isolation.
7. Stain cells by removing cell growth medium and adding 1 mL crystal violet fixative/well. Incubate for 30 min at room temperature.
8. Remove crystal violet fixative from wells, rinse, count plaques, and calculate titer of the virus in this crude preparation.
9. Approximately 24 h before beginning plaque isolation, seed several 15 cm plates with the CV-1-complementing cell line. Incubate the cells in a CO_2 incubator at 37 °C for 20–24 h until the cells are 100% confluent.
10. Sonicate virus twice for 15 s with 30 s of rest on ice between sonication cycles.
11. Dilute progeny virus (*see* **step 13** in Subheading 3.3) in DMEM (5 mL/dish) to approximately 5000 plaque-forming units per 15 cm dish.
12. Aspirate cell growth medium from dishes and add virus inoculum. Place in a CO_2 incubator at 37 °C for 30 min rocking the dish every 10 min.
13. Aspirate virus inoculum from the dish and replace with 25 mL of cell growth medium. Incubate for 48 h in a CO_2 incubator at 37 °C.
14. Using an inverted, fluorescent microscope, examine 15 cm dishes for fluorescence (*see* **Note 17**). Mark the position of fluorescent plaques (*see* **Note 18**) (Fig. 2).
15. To harvest plaques containing recombinant virus, aspirate medium and quickly scrape marked plaques with a sterile 100 μL pipette tip (*see* **Note 19**). Dip pipette tip into centrifuge tube containing 100 μL 10 mM Tris pH 9. Pipette up and down several times.
16. Release recombinant virus by freeze-thawing of the harvested plaque by placing at −80 °C for 10 m and then thawing in a 37 °C water bath. Freeze-thaw three times. Virus can be used immediately or frozen at −80 °C.
17. Approximately 24 h before the next round of plaque purification, seed several 10 cm plates with the CV-1-complementing cell line. Incubate the cells in a CO_2 incubator at 37 °C for 20–24 h until the cells are 100% confluent.

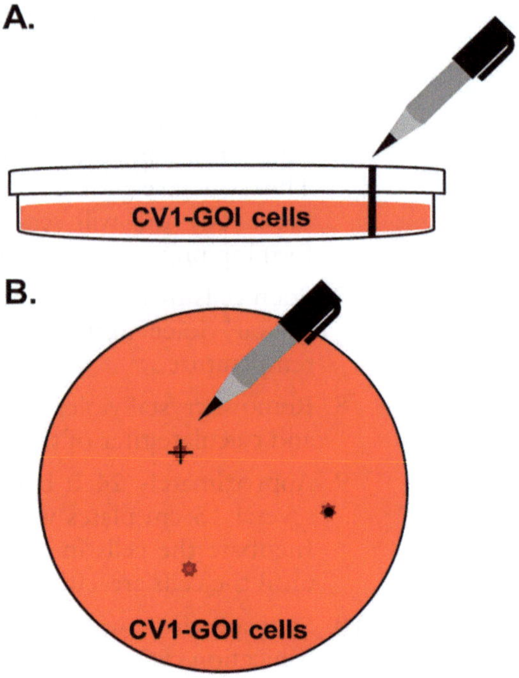

Fig. 2 Plaque picking technique. (**a**) Mark lid and plate to orient lid during plaque marking and plaque picking. (**b**) While looking under a fluorescent microscope, mark the lid with a cross or dot for reference of mCherry fluorescent plaques. Use these reference marks when picking plaques in the viral hood

18. Sonicate virus (*see* **step 16** above) twice for 15 s with 30 s of rest on ice between sonication cycles.
19. Aspirate cell growth medium from dishes and add 1, 5, or 10 μL of suspended virus inoculum diluted in 2 mL DMEM to each 10 cm plate. Place in a CO_2 incubator at 37 °C for 30 min rocking the dish every 10 min.
20. Aspirate virus inoculum from the dish and replace with 10 mL of cell growth medium. Incubate for 48 h in a CO_2 incubator at 37 °C.
21. Using an inverted, fluorescent microscope, examine 10 cm dishes for fluorescent plaques. Mark the position of fluorescent plaques and note the relative number of fluorescent and non-fluorescent plaques to monitor enrichment.
22. To harvest plaques containing recombinant virus, aspirate medium and quickly scrape marked plaques with a sterile 100 μL pipette tip. Dip pipette tip into centrifuge tube containing 100 μL 10 mM Tris pH 9. Pipette up and down several times.
23. Release recombinant virus by freeze-thawing of the harvested plaque by placing at −80 °C for 10 m and then thawing in a

37 °C water bath. Freeze-thaw three times. Virus can be used immediately or frozen at −80 °C.

24. Repeat **steps 16–22** three to five times or as needed to reach 100% fluorescent plaques (*see* **Note 20**).

3.5 Screening Resultant Deletion Virus

After 4–6 rounds of plaque purification, putative deletion viruses are expanded and analyzed by PCR using primers designed to amplify the DNA at and adjacent to the deletion site to confirm gene deletion (*see* **Note 21**) (Fig. 1, stage 3).

1. Approximately 24 h before small-scale expansion, seed CV-1-complementing cells in a 35 mm well. Incubate the cells in a CO_2 incubator at 37 °C for 20–24 h until the cells are 100% confluent.
2. Thaw recombinant virus from final plaque purification and sonicate twice for 15 s with 30 s of rest on ice between sonications.
3. Dilute 50 μL recombinant virus in 500 μL DMEM.
4. Aspirate cell growth medium from 35 mm well and add recombinant virus inoculum. Place in a CO_2 incubator at 37 °C for 30 min rocking the dish every 10 min.
5. Aspirate recombinant virus inoculum from well and replace with 2 mL of cell growth medium. Incubate for 48 h in a CO_2 incubator at 37 °C.
6. Harvest the infected cells from 35 mm well with a cell lifter and transfer into a sterile 15 mL centrifuge tube.
7. Wash plate once with 0.5 mL DPBS and add wash to the centrifuge tube.
8. Spin infected cells at 4 °C for 10 min at 500 × *g*. Aspirate off supernatant. Suspend pellet in 1 mL DPBS.
9. Spin infected cells at 4 °C for 10 min at 500 × *g*. Aspirate off supernatant. Suspend pellet in 1 mL 10 mM Tris pH 9 and transfer to a 1.5 mL centrifuge tube.
10. Release recombinant virus by freeze-thawing of the harvested cells by placing at −80 °C for 10 min and then thawing in a 37 °C water bath. Freeze-thaw three times. Recombinant virus can be stored at −80 °C until further expansion.
11. Purify DNA from 200 μL infected cell lysate (*see* **step 10** above) (*see* **Note 22**) using the GeneJET Genomic DNA Purification Kit following the manufacturer's recommendations.
12. Use 200 ng purified DNA as template for PCR reactions using vaccinia-specific primers that flank the deletion site (Fig. 1).
13. Resolve PCR products on an agarose gel.

14. After putative deletion viruses have been confirmed by PCR, recombinant viruses are amplified to make large-scale viral stocks of deletion virus [29].

Upon completion of these protocols, a robust system is in hand for further analysis of the viral gene both in the presence and absence of infection.

4 Notes

1. Sodium butyrate is a histone deacetylase that increases lentivirus titer by increasing the expression of viral genes.
2. Lentivirus production is sensitive to pH. Add HEPES to the medium to buffer the pH.
3. Gene synthesis services are offered by a variety of vendors which allow for gene modifications such as codon optimization and addition of epitope tag sequences. The authors have reliably used Operon and GenScript to produce constructs for the expression of tagged, codon-optimized vaccinia genes. Alternatively, DNA corresponding to the gene to be deleted can be amplified directly from purified vaccinia DNA by PCR. We recommend codon-optimized constructs as they result in increased expression of vaccinia proteins in our experience. Additionally, if using a gene synthesis service, we recommend having restriction sites added to both ends of the vaccinia gene so that the gene can be easily excised from the vector and ligated into the lentiviral transfer vector.
4. Each 10 cm dish of 293T cells produces adequate lentivirus to transduce five 35 mm wells (6-well plate) and is viable for 1 year. Scale lentivirus preparation accordingly being sure to include plates for an MCS-HYG control (transfer vector without viral GOI) with each preparation. In our experience, cells need to be transduced approximately every 6 weeks and therefore two 10 cm dishes per construct should be sufficient to create a year stock.
5. Appropriate biosafety procedure for lentiviral studies should be followed.
6. Include wells for an untransduced well and each lentivirus: the lentivirus-encapsidated pHAGE-HYG-complementing gene construct, the MCS-HYG control lentivirus, and a fluorescent-HYG control lentivirus (if available). The untransduced well will be used to establish when selection is complete. The pHAGE-HYG-complementing gene construct will be used to establish the complementing cell line. MCS-HYG control cells and fluorescent-HYG control cells will be used to establish the rate at which cells are transduced and selected in the absence of

a toxic gene. Together these controls will allow for the determination of whether the vaccinia gene is toxic to cells when expressed alone.

7. One 10 cm plate will be used to begin hygromycin selection. One 10 cm plate will be used to passage without hygromycin. The purpose of the untreated dish is to monitor the complementing cell line for potential toxicity independent of hygromycin selection. If no toxicity is noted, this plate can be omitted.
8. For each cell line transduced, a hygromycin kill curve should be performed before selection to determine a suitable hygromycin concentration. Typical hygromycin working concentrations for mammalian cells range from 100 to 500 μg/mL, we use 200 μg/mL for CV1 cells.
9. Selection should continue until all cells in a non-transduced dish have died from hygromycin in 3–4 days and until transduced cells have fully recovered and reached 100% confluency; at this point transduced cells are ready to be used. Recovery times are gene dependent but typically range from 1 to 2 weeks. In some instances, vaccinia genes may be toxic to the cell when expressed. In such cases, transient or inducible constructs may be utilized.
10. We usually use 200–300 bp of homologous sequence on each flank of the gene to be deleted.
11. The genetic marker, mCherry in this protocol, may be placed under the control of a variety of viral promoters.
12. One well will be used as a recombination negative control infected with virus. The other well will be both infected with virus and transfected with the vector for homologous recombination.
13. The isolation of recombinant vaccinia virus is facilitated by the incorporation of genetic markers. Markers can be divided into two types: those that allow for detection of recombinants and those that select for recombinants. Genetic markers allowing for detection of recombination include fluorescent molecules and β-galactosidase. Selectable genetic markers include those that provide drug resistance to recombinant viruses or allow for virus spread/replication in particular cell lines.
14. Monitor plaque size using a light microscope and fix plates when plaques have reached an appropriate size (>1 mm).
15. Typical titers at this stage are approximately 10^8 plaque-forming units per mL for parental wild-type virus and approximately 10^5 plaque-forming units per mL for recombinant mCherry-expressing virus. If no fluorescent plaques are observed, it is possible that the complementing cell line is

unable to complement for the deleted gene in the context of infection. In such a scenario, this method of gene deletion may be unsuitable. Revisit assays for confirming protein expression by cells. Alternatively, the recombination rate might be lower than previously anticipated. Scaling up to a larger plate of cells or increasing the amount of transfected homologous recombination plasmid could help address this issue.

16. Recombination frequency can be determined by comparing the ratio of total plaques visible under the light microscope to those visible under the fluorescent microscope.
17. Thousands of wild-type plaques will be visible under a light microscope, whereas only 1–10 fluorescent plaques will be visible under the fluorescent scope at this stage. Fluorescent plaques represent recombinant progeny virus. If no fluorescent plaques are present among 5000, increased amounts of virus (between 50,000 and 500,000 plaque-forming units) per plate can be plated. The plate with the fewest plaques that gives one fluorescent plaque should be used. The goal at this step is to have plaques as well separated as possible while still having fluorescent plaques.
18. To accurately mark the location of recombinant fluorescent plaques, make a line on the side of the plate to denote orientation of plate cover and bottom. Next, place a dot on the area of the plate cover that corresponds to the center of a fluorescent plaque. If possible, select fluorescent plaques that are well separated from WT virus plaques during plaque picking.
19. To isolate a plaque in the viral hood, place the plate cover under the plate bottom, align orientation mark on the side of the plate, and scrape marked plaque using a pipette tip. If possible 4–6 recombinant plaques should be picked.
20. Plaque purification should proceed until the number of total plaques visible under the light microscope equals those visible under the fluorescent microscope, for example, 100 fluorescent plaques observed for an equal number of total plaques. As purification proceeds it may be suitable to use smaller cell culture dishes.
21. A variety of PCR reactions may be performed to test whether the vaccinia gene has been deleted and replaced with the fluorescent marker. The authors regularly perform the following reactions testing for the absence of a PCR product: 3a+3b, 1a+3b, and 3a+2b. Additional reactions are performed looking for the presence of a PCR product: 1a+1b and 2a+2b. Lastly, a reaction is performed with which the product size is important, 1a+2b; for this reaction a product equivalent to the size of the fluorescent marker + ~500 flanking nucleotides is expected as opposed to a product the size of the deleted

gene + ~500 flanking nucleotides. If the deleted vaccinia gene and the fluorescent marker are of similar size, the PCR products may require restriction digest or sequencing to determine identity.

22. Store the rest of this small-scale expansion at −80 °C. It will be used to create a large stock of deletion virus following confirmation that the gene has been deleted.

References

1. Mackett M, Smith GL, Moss B (1982) Vaccinia virus: a selectable eukaryotic cloning and expression vector. Proc Natl Acad Sci 79(23):7415–7419
2. Blasco R, Moss B (1991) Extracellular vaccinia virus formation and cell-to-cell virus transmission are prevented by deletion of the gene encoding the 37,000-Dalton outer envelope protein. J Virol 65(11):5910–5920
3. Blasco R, Cole NB, Moss B (1991) Sequence analysis, expression, and deletion of a vaccinia virus gene encoding a homolog of profilin, a eukaryotic actin-binding protein. J Virol 65(9):4598–4608
4. Mackett M, Smith GL, Moss B (1984) General method for production and selection of infectious vaccinia virus recombinants expressing foreign genes. J Virol 49(3):857–864
5. Roper RL, Wolffe EJ, Weisberg A et al (1998) The envelope protein encoded by the A33R gene is required for formation of actin-containing microvilli and efficient cell-to-cell spread of vaccinia virus. J Virol 72(5): 4192–4204
6. Chernos VI, Belanov EF, Vasilieva NN (1978) Temperature-sensitive mutants of vaccinia virus. I. Isolation and preliminary characterization. Acta Virol 22(2):81–90
7. Condit RCMA (1981) Isolation and preliminary characterization of temperature-sensitive mutants of vaccinia virus. Virology 113(1):224–241
8. Rempel RE, Anderson MK, Evans E et al (1990) Temperature-sensitive vaccinia virus mutants identify a gene with an essential role in viral replication. J Virol 64(2):574–583
9. Evans ETP (1992) Characterization of vaccinia virus DNA replication mutants with lesions in the D5 gene. Chromosoma 102(1):S72–S82
10. da Fonseca FG, Wolffe EJ, Weisberg A, Moss B (2000) Effects of deletion or stringent repression of the H3L envelope gene on vaccinia virus replication. J Virol 74(16):7518–7528
11. Szajner P, Weisberg AS, Moss B (2004) Evidence for an essential catalytic role of the F10 protein kinase in vaccinia virus morphogenesis. J Virol 78(1):257–265
12. Meng X, Wu X, Yan B, Deng J, Xiang Y (2013) Analysis of the role of vaccinia virus H7 in virion membrane biogenesis with an H7-deletion mutant. J Virol 87(14):8247–8253. https://doi.org/10.1128/JVI.00845-13
13. Maruri-Avidal L, Weisberg AS, Bisht H, Moss B (2013) Analysis of viral membranes formed in cells infected by a vaccinia virus L2-deletion mutant suggests their origin from the endoplasmic reticulum. J Virol 87(3):1861–1871. https://doi.org/10.1128/JVI.02779-12
14. Boyle KA, Greseth MD, Traktman P (2015) Genetic confirmation that the H5 protein is required for vaccinia virus DNA replication. J Virol 89:6312–6327
15. Warren RD, Cotter CA, Moss B (2012) Reverse genetics analysis of poxvirus intermediate transcription factors. J Virol 86(17):9514–9519. https://doi.org/10.1128/JVI.06902-11
16. Kolli S, Meng X, Wu X, Shengjuler D, Cameron CE, Xiang Y (2015) Structure-function analysis of vaccinia virus H7 protein reveals a novel phosphoinositide binding fold essential for poxvirus replication. J Virol 89(4):2209–2219. https://doi.org/10.1128/JVI.03073-14
17. Hyun SI, Weisberg A, Moss B (2017) Deletion of the vaccinia virus I2 protein interrupts virion morphogenesis, leading to retention of the scaffold protein and mislocalization of membrane-associated entry proteins. J Virol 91(15):e00558–e00517. https://doi.org/10.1128/JVI.00558-17
18. Meng X, Rose L, Han Y, Deng J, Xiang Y (2017) Vaccinia virus A6 is a two-domain protein requiring a cognate N-terminal domain for full viral membrane assembly activity. J Virol 91(10):e02405–e02416. https://doi.org/10.1128/JVI.02405-16
19. Maruri-Avidal L, Weisberg AS, Moss B (2013) Direct formation of vaccinia virus membranes from the endoplasmic reticulum in the absence of the newly characterized L2-interacting protein A30.5. J Virol 87(22):12313–12326. https://doi.org/10.1128/JVI.02137-13

20. Sutter G, Ramsey-Ewing A, Rosales R, Moss B (1994) Stable expression of the vaccinia virus K1L gene in rabbit cells complements the host range defect of a vaccinia virus mutant. J Virol 68(7):4109–4116
21. Olson AT, Rico AB, Wang Z, Delhon G, Wiebe MS (2017) Deletion of the vaccinia virus B1 kinase reveals essential functions of this enzyme complemented partly by the homologous cellular kinase VRK2. J Virol 91(15):e00635–e00617. https://doi.org/10.1128/JVI.00635-17
22. Borrego B, Lorenzo MM, Blasco R (1999) Complementation of P37 (F13L gene) knock-out in vaccinia virus by a cell line expressing the gene constitutively. J Gen Virol 80(Pt. 2):425–432. https://doi.org/10.1099/0022-1317-80-2-425
23. Holzer GW, Falkner FG (1997) Construction of a vaccinia virus deficient in the essential DNA repair enzyme uracil DNA glycosylase by a complementing cell line. J Virol 71(7):4997–5002
24. Schambach A, Zychlinski D, Ehrnstroem B, Baum C (2013) Biosafety features of lentiviral vectors. Hum Gene Ther 24(2):132–142. https://doi.org/10.1089/hum.2012.229
25. Chakrabarti S, Sisler JR, Moss B (1997) Compact, synthetic, vaccinia virus early/late promoter for protein expression. Biotechniques 23(6):1094–1097
26. Davison AJ, Moss B (1990) New vaccinia virus recombination plasmids incorporating a synthetic late promoter for high level expression of foreign proteins. Nucleic Acids Res 18(14):4285–4286
27. Bertholet C, Drillien R, Wittek R (1985) One hundred base pairs of 5′ flanking sequence of a vaccinia virus late gene are sufficient to temporally regulate late transcription. Proc Natl Acad Sci U S A 82(7):2096–2100
28. Wittek R, Hanggi M, Hiller G (1984) Mapping of a gene coding for a major late structural polypeptide on the vaccinia virus genome. J Virol 49(2):371–378
29. Cotter CA, Earl PL, Wyatt LS, Moss B (2017) Preparation of cell cultures and vaccinia virus stocks. Curr Protoc Protein Sci 89:5.12.1–5.12.18. https://doi.org/10.1002/cpps.34

Chapter 6

Vaccinia Virus Genome Editing Using CRISPR

Carmela Di Gioia, Ming Yuan, and Yaohe Wang

Abstract

CRISPR/Cas9, an RNA-guided targeted genome editing system, can make precise, targeted modifications to the genome in living cells. Here we describe how this method can be used to efficiently edit the vaccinia virus genome enabling the insertion of transgene(s) specifically into a targeted site.

Key words CRISPR, Cas9, Genome, Editing, Homologous recombination, Vaccinia virus

1 Introduction

Vaccinia virus (VACV) is a large dsDNA virus belonging to the family *Poxviridae*. Like other members of this family, VACV replicates in the cytoplasm using its own enzymes for transcription and replication of DNA that is ultimately wrapped to produce various forms of the virions which are eventually released from the cell. Since its use in the eradication of smallpox, VACV has been developed as a vector for infectious disease vaccine and cancer immunotherapy, particularly to create tumor-targeted replicating oncolytic VACV [1–4]. The ability of this virus to specifically target tumor cells and induce antitumor effects relies on the deletion of certain viral genes together with the insertion of therapeutic genes. This is achieved through homologous recombination, a strategy based on the transfection of a repair donor vector carrying the gene of interest into pre-infected mammalian cells. Unfortunately, only 1–5% of the recombinant virus will contain the exogenous DNA, while the rest are mutant viruses with random insertion [5]. To increase the efficiency of DNA homologous recombination the CRISPR-Cas9 system has shown promising results being able to induce double-strand breaks in specific sites of the genome under the direction of a single-guide RNA (sgRNA) [6]. Cas9/sgRNA complex can recognize and cleave a site preceding a 5′-NGG-3′ PAM sequence in mammalian cells [7], as shown in Fig. 1, making the CRISPR-Cas9 system an efficient tool to generate mutant VACVs in CV-1 cells [8].

Jason Mercer (ed.), *Vaccinia Virus: Methods and Protocols*, Methods in Molecular Biology, vol. 2023,
https://doi.org/10.1007/978-1-4939-9593-6_6, © Springer Science+Business Media, LLC, part of Springer Nature 2019

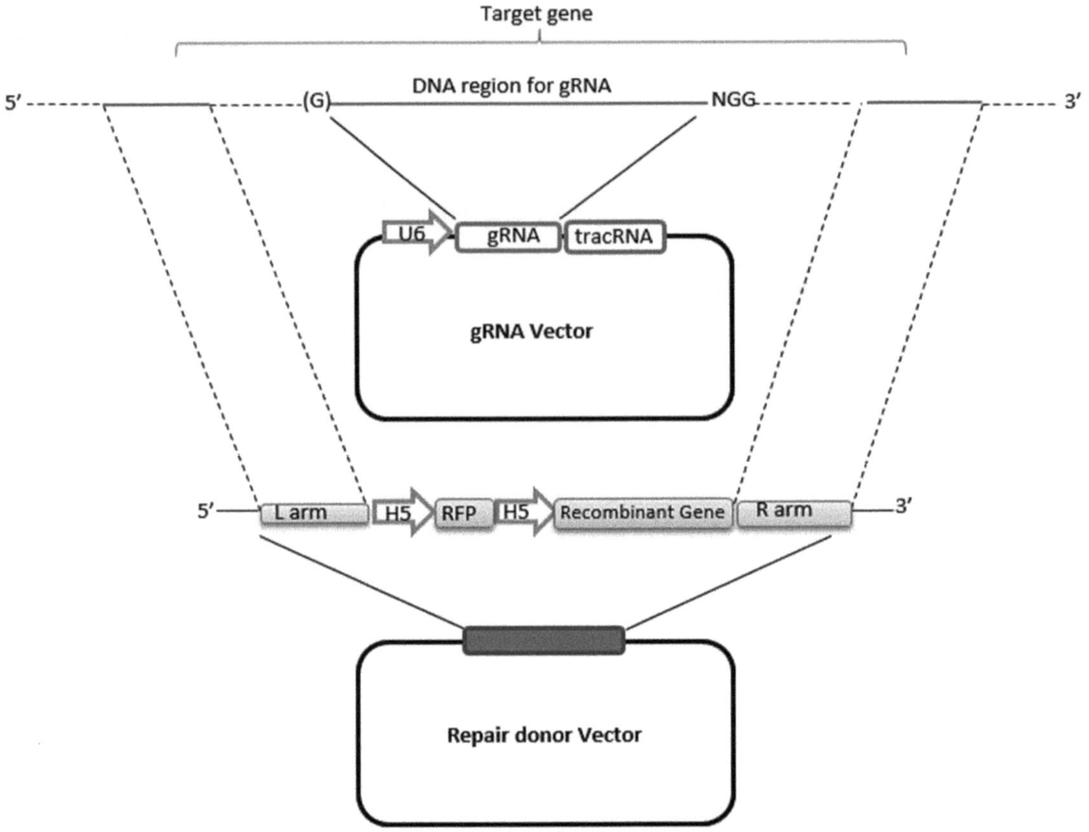

Fig. 1 Construction of gRNA vector and repair donor vector for creation of a mutant vaccinia virus. DNA sequence of a gRNA target region is cloned into a gRNA vector using U6 at the promoter. The repair donor vector is constructed containing left arm and right arm, being complementary to regions of the original vaccinia virus; the RFP and the recombinant gene, usually transcribed by two separate promoters (H5 promoter in this case)

2 Materials

2.1 Cloning of gRNA, Cas9 Construct, and Repair Donor Vector

1. Purified vaccinia virus genome.
2. Donor vector.
3. PCR reagents: PCR mastermix, Millipore water.
4. Agarose gel: Agarose powder, 1× Tris/borate/EDTA buffer, Millipore water, GelGreen.
5. Human codon-optimized Cas9 gene.
6. gRNA oligos.
7. 1× Annealing buffer: 10 mM Tris, pH 7.5–8.0, 50 mM NaCl, 1 mM EDTA
8. Mammalian expression cloning vectors.

9. Bacterial expression cloning vector carrying ampicillin resistance and LacZ gene.
10. DNA purification kit.
11. Ligation reagents: T4 Ligase, ligation buffer.
12. Competent bacterial cells.
13. 100 μg/mL Ampicillin agarose plates.
14. 100 μg/mL Ampicillin LB broth.
15. IPTG 0.1 mM, X-Gal 40 μg/mL.
16. Digestion reaction: Restriction enzymes, digestion buffer.

2.2 Transfection of Plasmids in CV1 Cells and Virial Production

1. Cell culture incubator: 37 °C with 5% CO_2.
2. CV1 cells.
3. Dulbecco's modified Eagle medium (DMEM).
4. Cell culture medium: DMEM 5% FBS, 100 U/mL penicillin, 100 U/mL streptomycin.
5. Backbone VACV (Lister strain).
6. PBS.
7. Trypsin.
8. Transfection reagents.
9. Cell scrapers.
10. Flasks, 6-well plates, cryovials.

3 Methods

3.1 Cloning of gRNA, Cas9 Construct, and Repair Donor Vector

1. Design a guide-RNA (gRNA) target sequence targeting the site of interest (*see* **Note 1**).
2. Use annealing buffer to set up a 25 μL reaction containing 50 μM of each strand of gRNA DNA in a PCR tube (*see* **Note 2**).
3. Anneal the oligonucleotides on a PCR machine. Heat to 95 °C, maintain the temperature for 2 min, cool to 25 °C over 45 min, and then cool to 4 °C and they are ready to use for ligation.
4. Centrifuge the PCR tube briefly to draw all moisture away from the lid.
5. Ligate 1 μL of annealed oligos with the gRNA cloning vector in a 10 μL ligation at 23 °C for 15–30 min in the presence of T4 Ligase and DNA ligation buffer.
6. The resulting ligation is then used to transform competent cells that are spread on an agar plate supplemented with ampicillin. A control-transformation is made using a ligation

mixture containing everything but the insert. After an overnight incubation the plate containing the desired insert is compared to the control plate that should show less colonies.

7. Pick colonies from the plate and grow in 3 mL of LB overnight supplemented with ampicillin.
8. Extract plasmids using plasmid Miniprep kit.
9. Confirm the presence of the insert using a standard digestion reaction with the appropriate restriction enzymes.
10. Sanger sequencing is used to confirm the correct sequence of the insert.
11. Construct the Cas9 expression vector as described in **steps 4–7** above (*see* **Note 3**).
12. Generate the repair donor vector: Amplify left and right arms of target by PCR using the VACV genome as template allowing up to 50 bp overlap with the target region; selected restriction enzyme sites are attached to the end of each PCR product for subsequent cloning (*see* **Note 4**).
13. Amplify the marker gene RFP (or GFP), therapeutic gene, and left and right arms with overhang A at 3′ end of the one strand of DNA by PCR (*see* **Note 5**). The H5 promoter sequence is incorporated into the marker and recombinant gene, respectively. Selected restriction enzyme sites are attached to the end of each PCR product to facilitate its downstream cloning into the DNA donor vector (*see* Fig. 1).
14. Check the size of the DNA fragments obtained through PCR on an agarose gel and purify the PCR product using a commercial gel DNA extraction kit following the manufacturer's instructions.
15. Ligate the purified PCR fragments into a commercial T-A cloning vector following the protocol of the manufacturer (*see* **Note 6**).
16. Grow positive colonies in 3 mL of ampicillin-LB broth overnight.
17. Extract plasmids using a Plasmid Miniprep kit according to the manufacturer's instructions.
18. Cut the purified plasmids with the specific enzymes for 1 h at the appropriate temperature. Check the digested plasmids to identify the plasmids with the right size of insert, and confirm the correct sequence of the insert by Sanger sequencing of the plasmid.
19. Upon confirmation of the correct sequence of the insert in the cloning plasmid, digest the insert from the cloning plasmid using chosen restriction enzymes and clone into the digested DNA donor vector as described in **steps 4–7** above.

20. Plasmids are amplified in transformed competent *E. coli* cells and purified using a plasmid miniprep kit according to the manufacturer's instructions.

3.2 Generate Mutant Vaccinia Virus in CV-1 Cells by Homologous Recombination

1. CV-1 cells are grown in cell culture medium and split at 80–90% confluence (*see* **Note 7**).
2. Seed 3×10^5 CV-1 cells in 2 mL of cell culture medium into one well of 6-well plate.
3. 24 h later, transfect 0.5 μg of Cas9 plasmid and 0.5 μg of gRNA plasmid into CV-1 cells using transfection kit.
4. Incubate cells for 24 h in tissue culture incubator.
5. Replace transfection medium with 2 mL of fresh cell culture medium.
6. Add 100 μL of vaccinia virus diluted in DMEM to 2×10^5 pfu/mL to the transfected CV-1 cells and mix well (*see* **Note 8**).
7. Transfect 1.0 μg repair donor vector 2 h after virus infection using a transfection kit.
8. Incubate plate in the cell culture incubator for 24 h.

3.3 Purification of Modified VACV

1. Detach the transfected CV-1 cells into the existing media with a cell scraper.
2. Collect cells in a cryovial and store at −80 °C for future use or proceed to **step 3** (*see* **Note 9**).
3. Seed fifteen 6-well plates containing 3×10^5 CV-1 cells in 2 mL of cell culture medium (*see* **Note 10**).
4. Thaw infected cell lysate from **step 2** in a 37 °C water bath for 3 min (*see* **Note 11**).
5. For each 6-well plate dilute 1 μL of infected cell lysate into 3 mL of DMEM and add 0.5 mL into each well of a 6-well plate (*see* **Note 12**).
6. Incubate later for 48 h in cell culture incubator.
7. Screen infected plates for RFP-positive plaques using the 10× objective under a fluorescence microscope (*see* Fig. 2).
8. Circle the location of the RFP-expressing plaques on the bottom of the plate using a marker pen under the guidance of a fluorescence microscope (*see* **Note 13**).
9. Prepare cryovial(s) containing 200 μL DMEM.
10. Aspirate all the medium from the well(s) containing the positive plaque(s).
11. Using a 200 μL tip, set the volume at 30 μL and take about 10 μL from the cryovial (*see* **Note 14**).
12. Holding the pipette push button, use the tip to scratch the area of the labeled plaque.

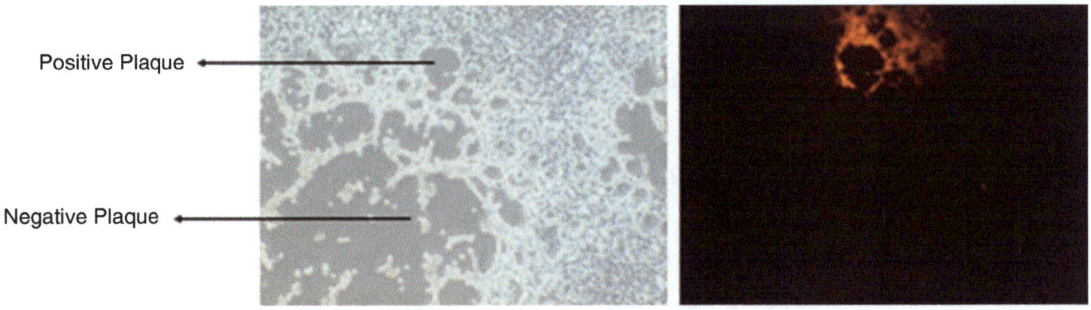

Fig. 2 Screening for RFP-expressing plaque

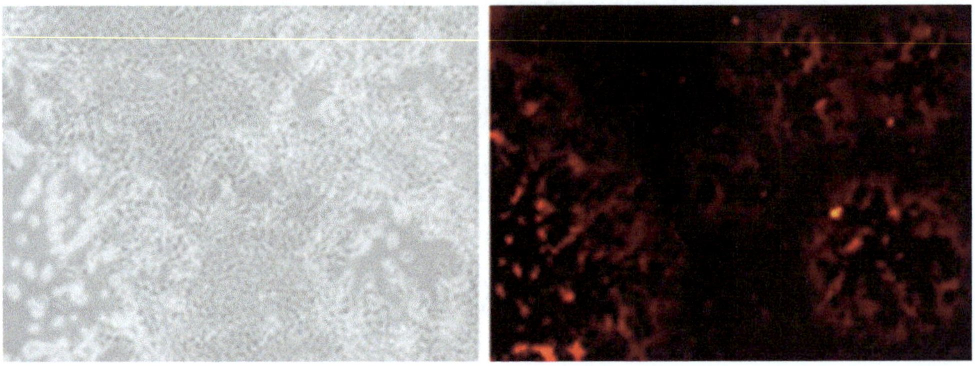

Fig. 3 Pure mutant VACV. When all infected cells show expression of RFP marker, the mutant VACV can be considered pure

13. Release the pipette push button and transfer the detached cells into the cryovial,
14. Repeat **steps 11–13** three times for each plaque and store the cryovial(s) at −80 °C for future use (*see* **Note 15**).
15. Seed one 6-well plate for each plaque harvested with each well containing 3×10^5 CV-1 cells in 2 mL of cell culture medium.
16. After 24 h, thaw the cryovial(s) from **step 14** in a 37 °C water bath for 3 min.
17. Add 0.5 μL of the infected cell lysate to one well of the 6-well plate.
18. Incubate plate in cell culture incubator for 48 h (*see* **Note 16**).
19. Select and harvest fluorescence-positive plaques as in **steps 7–14**.
20. Repeat **steps 15–19** until all the plaques formed from one positive plaque are RFP positive as shown in Fig. 3, and then proceed to **step 21**.
21. Harvest a RFP-positive plaque into a cryovial and store at −80 °C (*see* **Note 17**).

3.4 Verification of the Modification of Mutant VACV Using PCR

1. Seed one well of 6-well plate with 3×10^5 CV1 cells in 2 mL of cell culture medium.
2. Incubate cells for 24 h in cell culture incubator.
3. Thaw the RFP-positive plaque from Subheading 3.3, **step 21**, in a 37 °C water bath for 3 min.
4. Use 100 μL of the lysate to infect one well of the 6-well plate for 24 h (*see* **Note 18**).
5. Harvest the infected cells with a cell scraper and collect 1 mL in a 1.5 mL Eppendorf tube and the remaining 1 mL in a cryovial (*see* **Note 19**).
6. Centrifuge the infected cells collected in the 1.5 mL Eppendorf at 300 × *g* for 3 min.
7. Remove the supernatant and keep the cell pellet for immediate use in **step 8** or store the cell pellet at −80 °C for future use (*see* **Note 20**).
8. Extract the VACV DNA from the cell pellet harvested in **step 8** following the instructions of the commercial DNA extraction kit.
9. Perform 25 μL PCR reactions to amplify the recombinant gene, the DNA region spanning the target gene you aim to modify, and A52R as a control gene (*see* **Note 21**).
10. Resolve the PCR products using a 1% agarose gel.
11. If the DNA region spanning the target gene is negative and A52R is positive, then the mutant virus is correctly modified in the target gene.
12. Expand the corrected modified mutant virus in CV-1 cells using established protocols.

4 Notes

1. Tools for guide-RNA design such as http://crispr.dfci.harvard.edu/SSC/ would help you find the right sequence to use to induce a double-strand break (DSB) in the target site. The sequences surrounding the DSB are used to facilitate the pairing with the repair donor vector, which leads to the occurrence of homologous recombination. The abovementioned website is straightforward: easily insert the gene you plan to knock out and select two or three suggested guide-RNAs (pick the ones with higher score).
2. The annealing buffer composition (1×): 10 mM Tris, pH 7.5–8.0, 50 mM NaCl, 1 mM EDTA.
3. Any mammalian expression cloning vector can be used to insert the Cas9 gene lacking the nuclear localization signal.

4. The sequences of the left and right arm are about 500–600 bp.
5. RFP would give less background compared to GFP. The RFP gene insert can be designed to be chopped off using Cre-Lox system.
6. Using bacterial plasmids carrying LacZ and ampicillin resistance genes would allow you to recognize recombinant clones through the blue-white screening system.
7. Remove medium from the flask where cells are growing. Wash three times with PBS. Add trypsin-EDTA, enough to cover the flask's surface. Let the flask to be in the incubator at 37 °C, 5% CO_2, for 5 min. Resuspend cells in fresh new medium and count cells using a hemocytometer.
8. Dilution of VACV backbone is made to a concentration of 2×10^5 pfu/mL.
9. Remove 0.5 or 1 mL of the medium before scraping to fit the cell lysate in a 1.5 mL cryovial.
10. 15 or more plates can be seeded on the same day of repair donor vector transfection to proceed to Subheading 3.3, **step 2**, *immediately*.
11. Make sure to freeze vial from Subheading 3.3, **step 2**, before use, even if using on the same day of collection. The freezing-thawing cycle would allow to obtain a cell lysate containing released virus from burst cells.
12. The volume of cell lysate per plate needed to obtain optimal number of plaques might be different depending on the viral production. 1 μL/plate would usually allow you to obtain enough separated plaques. In case of over-infection proceed with increased dilution.
13. Six positive plaques should be enough to find the correct mutant virus containing the recombinant gene in the desired genome region.
14. 200 μL tips are preferably used to scratch the plaque.
15. There is no need to eject the medium after scratching the plaque because some of it wet the surface via capillarity.
16. Repeat the same process for each positive plaque found in **step 7**.
17. Remember to store the cell lysates of each step at −80 °C during plaque purification.
18. The time of infection varies from 24 to 48 h depending on the viral concentration in the cell lysate: 50% of cells should be rounded, easily detached, and RFP positive under fluorescence microscope. A rough titration step might be needed to evaluate the viral concentration: infect CV-1 cells in 6-well plate with different volume of virus.

19. Keep the cryovial at −80 while proceeding with the Eppendorf to verify the modification of the VACV.
20. In case you are not proceeding with the next step immediately, store the pellet at −80 °C.
21. PCR setting: First denature the DNA at 94 °C for 2 min, then 30 cycles of denaturation at 94 °C for 15 s, annealing at 52 °C for 15 s, and extension at 72 °C for 30 s. These settings might vary slightly depending on the primers used for your recombinant gene.

References

1. Cooney EL, Collier AC, Greenberg PD, Coombs RW, Zarling J, Arditti DE et al (1991) Safety of and immunological response to a recombinant vaccinia virus vaccine expressing HIV envelope glycoprotein. Lancet 337:567–572
2. Di Paola RS, Chen Y, Bubley GJ, Stein MN, Hahn NM, Carducci MA et al (2014) A national multicenter phase 2 study of prostate-specific antigen (PSA) pox virus vaccine with sequential androgen ablation therapy in patients with PSA progression: ECOG 9802. Eur Urol 68(3):372–373
3. Guo ZS, Bartlett DL (2004) Vaccinia as a vector for gene delivery. Expert Opin Biol Ther 4:901–917
4. Moss B (2013) Reflections on the early development of poxvirus vectors. Vaccine 31:4220–4222
5. Nakano E, Panicali D, Paoletti E (1982) Molecular genetics of vaccinia virus: demonstration of marker rescue. Proc Natl Acad Sci U S A 79:1593–1596
6. Cong L, Ran FA, Cox D, Lin S, Barretto R, Habib N et al (2013) Multiplex genome engineering using CRISPR/Cas systems. Science 339:819–823
7. Jinek M, Chylinski K, Fonfara I, Hauer M, Doudna JA, Charpentier E (2012) A programmable dual-RNA-guided DNA endonuclease in adaptive bacterial immunity. Science 337:816–821
8. Yuan M, Zhang W, Wang J, Al Yaghchi C, Ahmed J, Chard L et al (2015) Efficiently editing the vaccinia virus genome by using the CRISPR-Cas9 system. J Virol 89:5176–5179

Chapter 7

RNAi-Mediated Depletion of Poxvirus Proteins

Caroline Martin and Samuel Kilcher

Abstract

RNA interference (RNAi) allows for transient, targeted depletion of cellular or viral proteins. Previously, small interfering RNA (siRNA) screens targeting cellular factors successfully identified several host genes that are required for VACV infection, and other viruses such as HIV. In this chapter, we outline how RNAi can be adapted to unravel the functions of poxvirus genes, using a 96-well format. Additionally, we describe two different high-throughput methods (flow cytometry and automated microscopy) to assess infection levels of an engineered VACV that encodes a fluorescent reporter protein under an early and/or late viral gene promoter.

Key words siRNA, High-throughput screening, Poxvirus genes, Protein depletion, Transfection

1 Introduction

The central dogma of life states that genetic information is stored as DNA and translated to functional proteins, using RNA as an intermediate. However, not all RNAs found in plants, fungi, or animals encode proteins: short noncoding RNAs are an evolutionarily conserved regulator of gene expression and also play a key role in cell defense against intracellular pathogens (reviewed in [1]). This so-called RNA interference (RNAi) post-transcriptionally silences genes in a sequence-specific manner (reviewed in [2]). Double-stranded RNA (dsRNA) that is either encoded by the cell and expressed (microRNA), or introduced into the cell exogenously, is processed and cleaved into short single-stranded RNA (ssRNA) fragments by the cellular nuclease Dicer. In the case of exogenous RNA, these fragments are approximately 23 nucleotides long and are termed short interfering RNA (siRNA). The ssRNA is then loaded into the RNA-induced silencing complex (RISC) as a guide strand to its complementary target mRNA. Once the target mRNA is bound, it is either directly cleaved by the RISC subunit Argonaute or targeted for degradation in processing bodies (P-bodies). Therefore, RNAi enables the cell to regulate

Jason Mercer (ed.), *Vaccinia Virus: Methods and Protocols*, Methods in Molecular Biology, vol. 2023,
https://doi.org/10.1007/978-1-4939-9593-6_7,

endogenous protein levels, and protect itself from exogenous, pathogenic RNA expression.

Exploiting this cellular regulatory mechanism, RNAi was developed into a powerful tool that allows for depletion of any cellular or viral protein with a known coding mRNA sequence. As RNAi induces transient protein depletion, it allows for silencing of essential genes that cannot be studied by genetic deletion or identified from forward genetic screens. Cellular RNAi is triggered by synthetic siRNA sequences that are designed as complementary oligonucleotides (20–25 nucleotides) that specifically bind viral or cellular mRNA.

Previously, siRNA screens targeting cellular factors successfully identified host genes required for VACV infection [3, 4], and other viruses such as HIV [5, 6]. Adapting RNAi to unravel the function of VACV genes, we developed a custom-made siRNA library that targets 80 early viral genes with three individual siRNAs per gene [7]. Using this, we successfully identified the VACV uncoating factor [8]. In this chapter we describe how to perform a VACV-targeting RNAi screen using flow cytometry- and microscopy-based infection scoring assays. Analysis of screening datasets is not addressed as this is largely dictated by user-defined experimental setups and analysis parameters. Importantly, RNAi-mediated silencing of viral genes can be extended to other large DNA viruses such as herpes simplex virus and cytomegalovirus, as long as RNAseq data or transcriptomic data sets are available.

2 Materials

2.1 General

1. Phosphate-buffered saline (PBS).
2. Centrifuge with 96-well plate inserts and rotor.
3. Multichannel pipette.
4. Sterile laminar flow for biosafety level 2.

2.2 Cell Culture

1. Cells maintained at a low passage number (e.g., HeLa, BSC40, A549).
2. Cell culture medium: DMEM, 10% fetal bovine serum (FBS), 1× GlutaMAX, 1× penicillin/streptomycin, 1× nonessential amino acids for HeLa and A549s, add 1× sodium pyruvate for BSC40 cells.
3. For microscopy: Black 96-well plates, tissue culture treated (from Greiner-Bio One or Perkin and Elmer). For cytometry: Clear, 96-well, flat-bottom cell culture plate.

2.3 siRNA Library

1. siRNAs targeting viral early genes (duplex RNA strands (sense and antisense) with 3′ dTdT overhangs, designed and synthesized by Sigma-Aldrich) [8] (*see* **Notes 1** and **2**).

2. Sterile, nuclease-free water.
3. Aluminum tape cover foil (adhesive 96-well plate seal).

2.4 Reverse Transfection

1. Transfection medium: DMEM without supplements (DMEM (−)).
2. Lipofectamine RNAiMAX (Invitrogen).

2.5 Virus

1. Reporter virus (assay-dependent, e.g., luciferase- or GFP-expressing VACV) (*see* **Note 3**).
2. BSC40 cells for virus production and titration.
3. Infection medium: DMEM (−).

2.6 Analysis by Flow Cytometry

1. Trypsin/EDTA: 0.25% Trypsin, 0.02% EDTA.
2. Quenching buffer: 5% FBS in PBS.
3. Fixation solution (4% formaldehyde in PBS).
4. Cytometer (preferably with a 96-well plate loader).

2.7 Analysis by Immunofluorescence and Microscopy

1. Permeabilization-blocking buffer: 5% FBS, 0.1% Triton X-100 in PBS.
2. DNA stain: Hoechst 33342 (Invitrogen).
3. Primary antibody.
4. Secondary, fluorophore-coupled antibody.
5. Fixation solution: 4% Formaldehyde in PBS.

3 Methods

All procedures need to be carried out under sterile conditions at room temperature (RT) in a laminar flow hood, unless indicated otherwise. Cells are grown at 37 °C, 5% CO_2, in a humidified incubator. All cell culture reagents (1× PBS, trypsin, and media) are pre-warmed to 37 °C prior to use, unless specified otherwise. All volumes are for 1 well of a 96-well plate. Any indicated multiplicities of infection (MOIs) are based on VACV titers determined on BSC40 cells. The transfection protocol and final siRNA concentration were optimized for viral protein depletion efficiency and minimal cellular toxicity.

3.1 Cell Seeding

1. To achieve 90% confluency for transfection, split cells to 50% confluency 24 h prior to reverse transfection (*see* **Note 4**).

3.2 Reverse Transfection

1. Thaw diluted siRNA stock library (133 nM, 15 μL per well) on ice for 1 h (*see* **Note 5**). 133 nM Stock libraries should be prepared beforehand and can be stored at −80 °C (or −20 °C).

2. Spin the 96-well plates to collect the liquid at the bottom of the well (600 × *g*, 4 °C, 5 min), and then carefully remove aluminum plate seal.
3. Briefly vortex the RNAiMAX solution.
4. Prepare the RNAiMAX transfection reagent mixture: add 0.15 μL RNAiMAX to 15 μL cold DMEM (−) and incubate for 5 min (*see* **Notes 6** and 7).
5. Add 15 μL transfection reagent mixture to the 15 μL diluted siRNA (*see* **Note 8**).
6. Gently tap each side of the plate to mix the solutions (*see* **Note 9**).
7. After 45 min of incubation, host cells can be detached and prepared for seeding.
8. Wash the 15 cm cell culture plate once with 15 mL PBS.
9. Aspirate the PBS and add 5 mL trypsin/EDTA solution.
10. Incubate at 37 °C for 1–2 min and gently tap the sides of the dish to detach cells.
11. Add 10 mL cell culture medium to quench the trypsin digest and mix cell suspension by carefully pipetting up and down.
12. Determine the cell concentration.
13. Dilute the cell suspension in medium to a final concentration of 2500–20,000 cells/70 μL (*see* **Note 10**).
14. When the RNAiMAX/siRNA transfection mixture (30 μL) has been incubated for 1 h, add the 70 μL cell suspension (*see* **Note 11**).
15. Carefully tap the sides of the plate to ensure equal dispersion of the cells (*see* **Note 12**).
16. Place in incubator for 20 h at 37 °C (*see* **Note 13**).

3.3 Infection

1. Dilute virus in infection medium to the desired MOI (calculate for 100 μL per well) (*see* **Note 14**).
2. Aspirate siRNA transfection solution (*see* **Note 15**).
3. Add virus inoculum and place the plate in 37 °C incubator for 1 h.
4. Aspirate virus inoculum (*see* **Note 15**), add 150 μL cell culture medium, and place the plate in incubator. We define this time point as $t = 0$ h postinfection (hpi).

3.4 Analysis

After the desired time of infection, samples are ready for processing and analysis. The analysis method depends on the assay, the type of reporter virus used, the desired readout parameters, and the number of samples to be analyzed. In the following sections, we describe two different high-throughput methods (flow cytometry and

Table 1
Comparison of the advantages and disadvantages of flow cytometry versus automated microscopy for analysis of RNAi screens

Method	Advantages	Disadvantages
Flow cytometry	+ Simple analysis and small digital data volume + Fast sample preparation	– Impractical for plates when additional staining is required – No spatial/morphological information – Detaching the cells by proteolysis or ion chelation might perturb/cleave potential extracellular epitopes which might interfere with antibody staining
Microscopy	+ Spatial/morphological information easily accessible + Fast sample preparation and simple immunofluorescence staining possible + Potential cell toxicity easily quantifiable (cell detachment and cell loss) + Plates can be stored for several weeks prior to imaging	– Extensive and/or elaborate image analysis required – Large digital data volume

automated microscopy) both of which have advantages and disadvantages for assessing infection levels of an engineered VACV that encodes a fluorescent reporter protein under an early and/or late viral gene promoter (*see* Table 1).

3.4.1 Analysis by Flow Cytometry

1. Wash cells once with 200 μl PBS (*see* **Note 16**).
2. Add 80 μl trypsin/EDTA (*see* **Notes 17** and **18**) to detach cells and place in 37 °C incubator for 5 min (*see* **Note 19**).
3. Add 40 μL quenching buffer to each well.
4. Add 40 μL fixation solution to each well.
5. Incubate plates at RT in the dark for 15 min.
6. Samples are now ready to be analyzed by flow cytometry. Store plates in the dark at 4 °C (*see* **Note 20**).
7. If samples need to be stored for more than 2 days and/or the cytometer tubing is sensitive to PFA, transfer the samples to a 96-well plate with V-shaped wells. Centrifuge plates at 4 °C for 10 min at 600 × *g*, to collect cells at the bottom of the well.
8. Carefully aspirate supernatant and resuspend pellet in 120 μL quenching buffer (for resuspension volume *see* **Note 17**).

3.4.2 Analysis by Automated Microscopy

1. Wash cells twice with 200 μL PBS (*see* **Note 16**).
2. Add 60 μL fixation solution.

3. Incubate plates at RT in the dark for 15 min (*see* **Note 21**).
4. Wash twice with 200 μL PBS (*see* **Note 16**).
5. Add 80 μL permeabilization-blocking buffer.
6. Incubate plates for 30 min at RT in the dark.
7. Add 50 μL permeabilization-blocking buffer containing diluted primary antibody.
8. Incubate as suggested by the manufacturer (*see* **Note 22**).
9. Wash twice with 200 μL PBS.
10. Add 50 μL fluorophore-conjugated secondary antibody diluted in blocking buffer. If required, add a DNA stain to your antibody solution (e.g., DAPI or Hoechst) (*see* **Notes 22** and **23**).
11. Incubate plates for 1 h at RT in the dark.
12. Wash three times with 200 μL PBS.
13. Acquire images with an automated microscope or store plate in 150 μL PBS at 4 °C in the dark.

3.5 General Considerations

3.5.1 Dilution and Layout of siRNA Plates for Screening

1. For long-term storage, siRNAs should be kept at −20 or −80 °C at high concentration (10 mM). To prepare diluted siRNA plates for screening, thaw the stock siRNA library (10 mM) on ice for 1 h (*see* **Note 24**).
2. Spin the 96-well plates to collect the liquid at the bottom of the well (600 × *g*, 4 °C, 5 min), and then carefully remove aluminum plate seal.
3. Dilute the stock siRNA library in DMEM (−) to the desired final concentration (*see* **Note 25**).
4. We observed plate effects when screening in 96-well plates, which were mostly due to uneven warming of wells at the edge of the plate as compared to wells in the middle. To minimize and account for such location-dependent artifacts, we used the layout shown in Fig. 1. The outermost wells of the 96-well plate remain untreated. During the experiment, these wells are filled with medium to avoid a strong temperature gradient in the inner wells of the plate. To account for any other location-dependent effects, multiple wells containing either negative or positive controls are included at several positions on the plate (*see* **Note 26** and Fig. 1).

3.5.2 siRNA Knockdown Efficiency

SiRNA knockdown efficiency can be assayed by several different methods each of which has advantages and disadvantages (*see* Table 2):

1. **Western Blot:** Silence the gene of interest as described above. Depending on your antibody and target, you will have to upscale to larger wells to get sufficient protein. Harvest cells

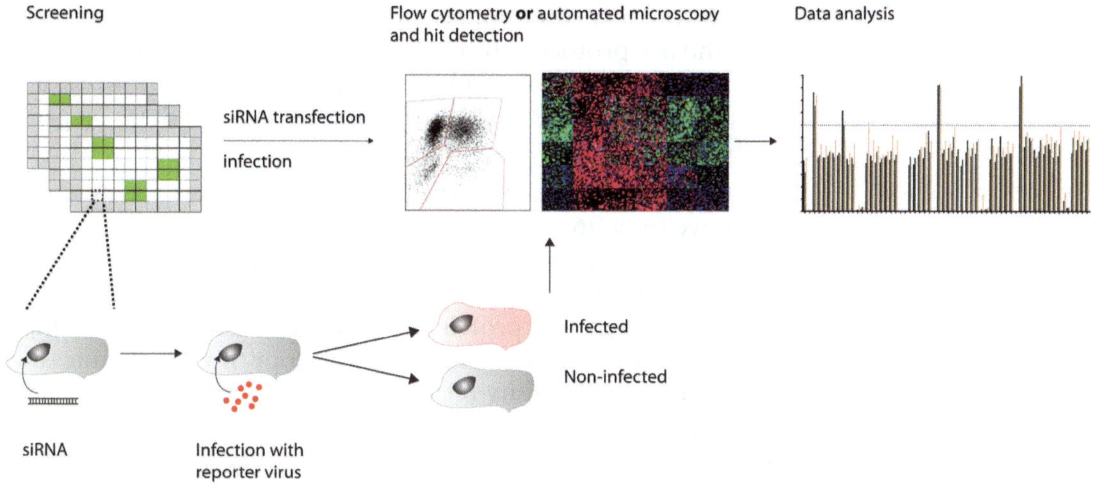

Fig. 1 Schematic workflow of a siRNA screen in 96-well plates. Cells are reverse transfected with siRNAs targeting the viral proteins of interest and incubated for the desired period of time. The transfected cells are then infected with a reporter virus. The duration of infection is assay dependent. After the desired time of infection, samples are ready for processing and analysis. The analysis method depends on the assay, the type of reporter virus used, the desired readout parameters, and the number of samples to be analyzed. We suggest two different high-throughput methods (flow cytometry and automated microscopy) to assess infection levels. Once the parameters of interest are measured, data can be analyzed

Table 2
Comparison of the advantages and disadvantages of immunofluorescence, western blot analysis, and late viral gene expression levels as readouts for siRNA knockdown efficiency

Method	Advantages	Disadvantages
Immunofluorescence	+ Direct detection of protein of interest	– Either protein needs to be fused to a detectable peptide sequence or a specific antibody needs to be available
	+ Can be done in a 96-well format, using a spectrofluorometer or infrared plate reader	– Unsuitable for proteins with very low abundance
Western blot analysis	+ Direct detection of protein of interest	– Either protein needs to be fused to a detectable peptide sequence or a specific antibody needs to be available – Larger sample volume required
Late viral promoter activity	+ No additional experiment required	– Reporter virus required

and process the sample for SDS-PAGE, following your standard protocol. Blot for viral protein levels in a siRNA knockdown sample and compare to the protein levels in a non-targeting siRNA control sample (e.g., AllStars Negative control siRNA).

2. **Immunofluorescence:** Silence the gene of interest as described above (in a 96-well plate for microscopy). Either infect with a virus that expresses a tagged version of the protein of interest or stain for the protein of interest as described in Subheading 3.4.2. Determine protein levels in the siRNA-treated sample by quantifying fluorescence microscopy images or measure fluorescence intensity relative to non-targeting control siRNA samples.
3. **Late viral gene expression levels**: Poxvirus gene expression follows a strictly regulated temporal cascade, in which the expression of certain early viral genes is strictly required for genome uncoating, replication, and expression of intermediate and late genes. Depletion of such early genes will result in abrogated late gene expression, which can be used to indirectly determine siRNA knockdown efficiency: To this end, siRNAs targeting, e.g., the viral uncoating factor D5 or the viral DNA polymerase E9 are transfected as described above. Cells are then infected with a virus that expresses a fluorescent reporter under a late viral promoter. When late viral gene expression is maximal in your control sample (in case of VACV this is 8 h p.i.), the samples are harvested and processed for analysis by flow cytometry as described in Subheading 3.4.1. The fluorescence intensity of D5 or E9 knockdown samples is compared to the non-targeting siRNA control. Lack of late gene expression indicates successful siRNA-mediated protein depletion.

3.5.3 Cellular Toxicity

SiRNAs targeting viral mRNAs are not expected to cause significant cellular toxicity. Thus, any eukaryotic cell toxicity prior to infection is a strong indicator of siRNA-mediated off-target effects. We recommend assessing toxicity using the following methods (*see* **Note 27**).

1. **Microscopy**: Reverse transfect the cells as described above (using microscopy 96-well plates). Incubate cells with siRNA for the time required but do not infect cells. Stain the cell nucleus as described in Subheading 3.4.2 of this chapter (antibody staining can be omitted). Image samples and quantify cell nuclei. Loss of nuclei indicates cell detachment and/or cell toxicity.
2. **Cell toxicity kits**: Reverse transfect the cells as described above (using microscopy 96-well plates). Incubate cells with

siRNA for the time required but do not infect cells. Use a commercial cell toxicity kit (e.g., Pierce LDH cytotoxicity assay kit) and follow the manufacturer's instructions.

4 Notes

1. To score as an early gene, the corresponding transcript had to be detected and classified as such in at least one of the VACV transcriptome studies [7, 9, 10].
2. Targeting intermediate or late viral transcripts should also be possible: Because the transcription of these genes occurs after viral DNA replication, transcript abundance will be significantly higher [11]. To compensate for increased mRNA abundance, siRNA concentrations may have to be adapted. In addition, it is important to keep in mind that several late VACV genes are produced from read-through transcripts. Therefore, targeting one transcript may affect the expression of multiple genes [10].
3. To score different stages of VACV infection, we used recombinant VACVs that express fluorescent reporter genes under the control of early, intermediate, or late viral promoters. Other reporter systems can also be used, for example recombinant VACV that expresses luciferase under the control of a viral promoter [12].
4. We find siRNA transfection to be the most efficient in cells of a low passage number (<15) that were split at sub-confluency in the days prior to transfection.
5. Optimally, the diluted stock library is prepared directly before each experiment. However, we found better reproducibility between repeats using aliquots of the same stock library dilution. Aliquots should be frozen and thawed only once. We suggest preparing 15 μL aliquots of the diluted stock library in 96-well plates. Store at −80 °C and use within 2 months.
6. To minimize pipetting error, prepare the RNAiMAX transfection mix in bulk for all wells and not individually.
7. We find siRNA transfection efficiency to decrease if the RNAiMAX is incubated for less than 5 min or more than 10 min.
8. To ensure optimal mixing of the solutions, tilt the 96-well plate toward you and directly add 15 μL transfection solution to the pre-pipetted siRNA solution.
9. Do not pipet up and down as this might create air bubbles.
10. The cell number to be seeded depends on the cell type, the desired confluency at the time of infection, and the siRNA

preincubation time (*see* **Note 13**). For HeLa cells, we suggest seeding 2500 cells for 48 h preincubation (sub-confluent at the time of infection), or 20,000 cells for 20 h preincubation (confluent at the time of infection).

11. To ensure optimal mixing between the solutions, tilt the 96-well plate toward you and directly add 70 ul cell suspension to the RNAiMAX/siRNA solution. Do not pipet up and down as this might create bubbles and clump the cells. Instead, tap the sides of the plate gently.
12. If you encounter uneven cell growth, it might help to let the plate sit at RT in the laminar flow for up to 45 min before placing in the incubator.
13. We find the optimum siRNA preincubation time to vary depending on the assay (as defined by maximum knockdown efficiency at minimum cell toxicity). Therefore, this parameter needs to be optimized for each assay. In most instances, the optimum time ranged from 20 to 48 h.
14. The MOI required depends on the assay. We found that higher MOI requires higher siRNA concentrations (e.g., a final siRNA concentration of 40 nM effectively silences viral early genes up to a MOI of 20). We suggest using the lowest possible MOI to ensure effective viral gene silencing without risking off-target effects.
15. Tilt the 96-well plate toward you and use a multi-aspirator at the bottom of the wells to prevent aspirating cells and to ensure complete removal of the transfection mixture. To prevent cross-contamination, wash aspirator tips between wells with different conditions by aspirating sterile ultrapure water or sterile PBS from a reagent reservoir.
16. Infected cells round up and are prone to detach from the plate. To prevent cell loss, add PBS carefully along the wall of the well. For aspiration *see* **Note 15.**
17. The volumes of trypsin/EDTA, 5% FBS stop solution, and 4% PFA fix solutions need to be adapted to the cytometer specifications (minimal volume required for analysis). Add the three solutions at a 2:1:1 ratio to obtain final concentrations of 1.25% FBS and 1% PFA.
18. In case trypsin and/or EDTA cannot be used, cells can alternatively be detached using a protease-free chelating solution: for a 10× citrate solution dissolve 1.35 M NaCl and 0.15 M $NaC_6H_7O_7$ in ultrapure water and sterile filter the solution through a 0.2 μm filter membrane. Please note that this detachment method is less efficient than trypsin digestion.
19. The incubation time required to fully detach cells depends on the cell density, the 96-well plate type used, and the cell type. Tap the plate gently to detach cells. To prevent clumping of

detached cells, keep trypsin incubation times as short as possible.

20. Plates should by analyzed as soon as possible as storage increases cell clumping. In addition, PFA may cause quenching of certain fluorophores over time.
21. If the antibody staining requires methanol fixation, use 60 μL cold (−20 °C) MeOH instead. Incubate at 4 °C in the dark for 10 min. As methanol permeabilizes eukaryotic cells, no further permeabilization steps are required.
22. Smaller volumes (e.g., 30 μL) can be used, if the plate is gently rocked during antibody incubation.
23. Alternatively, if no signal amplification is required, one can use a primary antibody that is directly coupled to a fluorophore and proceed directly to **step 9**.
24. The siRNA library (10 mM diluted in ddH_2O) can be stored at −20 °C as long as needed. Repeated freeze/thawing should be kept to a minimum. Once the siRNA is diluted in DMEM (−) it needs to be stored at −80 °C (*see* **Note 5**).
25. To minimize pipetting error, dilute larger volumes of each siRNA sample and then aliquot it directly into 96-well plates (*see* **Note 5**). Make sure to use a calibrated pipet for small siRNA volumes. For optimal mixing, pre-lay the DMEM (−) in a 96-well U-bottom plate, add the siRNA, and tap the plate carefully.
26. To test for any unspecific effect caused by the siRNA transfection itself, we use a scrambled siRNA sequence that does not complement any cellular or viral mRNA sequence (e.g., AllStars Negative Control siRNA, Qiagen). To control for siRNA transfection efficiency, a siRNA that silences a cellular housekeeping gene and induces cell death was used (e.g., AllStars Death Positive Control siRNA, Qiagen). Allow at least 48 h to observe cell death in the positive control. The control siRNA should be used at the same concentration as the siRNAs that target your viral transcript of choice.
27. Virus infection interferes with cellular gene expression and normal cellular metabolism. Therefore, we suggest using a toxicity assay and not a cell viability assay as a measurement of siRNA toxicity.

References

1. Agrawal N, Dasaradhi PVN, Mohmmed A, Malhotra P, Bhatnagar RK, Mukherjee SK (2003) RNA interference: biology, mechanism, and applications. Microbiol Mol Biol Rev 67(4):657–685
2. Rana TM (2007) Illuminating the silence: understanding the structure and function of small RNAs. Nat Rev Mol Cell Biol 8(1):nrm2085
3. Mercer J, Snijder B, Sacher R, Burkard C, Bleck CK, Stahlberg H et al (2012) RNAi screening reveals proteasome- and Cullin3-dependent stages in Vaccinia virus infection. Cell Rep 2(4):1036–1047

4. Sivan G, Martin SE, Myers TG, Buehler E, Szymczyk KH, Ormanoglu P et al (2013) Human genome-wide RNAi screen reveals a role for nuclear pore proteins in poxvirus morphogenesis. Proc Natl Acad Sci U S A 110(9):3519–3524
5. Nguyen DG, Wolff KC, Yin H, Caldwell JS, Kuhen KL (2006) 'UnPAKing' human immunodeficiency virus (HIV) replication: using small interfering RNA screening to identify novel cofactors and elucidate the role of group I PAKs in HIV infection. J Virol 80(1):130–137
6. Börner K, Hermle J, Sommer C, Brown NP, Knapp B, Glass B et al (2010) From experimental setup to bioinformatics: an RNAi screening platform to identify host factors involved in HIV-1 replication. Biotechnol J 5(1):39–49
7. Assarsson E, Greenbaum JA, Sundström M, Schaffer L, Hammond JA, Pasquetto V et al (2008) Kinetic analysis of a complete poxvirus transcriptome reveals an immediate-early class of genes. Proc Natl Acad Sci U S A 105(6):2140–2145
8. Kilcher S, Schmidt FI, Schneider C, Kopf M, Helenius A, Mercer J (2014) siRNA screen of early poxvirus genes identifies the AAA+ ATPase D5 as the virus genome-Uncoating factor. Cell Host Microbe 15(1):103–112
9. Rubins KH, Hensley LE, Bell GW, Wang C, Lefkowitz EJ, Brown PO et al (2008) Comparative analysis of viral gene expression programs during poxvirus infection: a transcriptional map of the Vaccinia and Monkeypox genomes. PLoS One 3(7):e2628
10. Yang Z, Bruno DP, Martens CA, Porcella SF, Moss B (2010) Simultaneous high-resolution analysis of vaccinia virus and host cell transcriptomes by deep RNA sequencing. Proc Natl Acad Sci 107(25):11513–11518
11. Sebring ED, Salzman NP (1967) Metabolic properties of early and late Vaccinia virus messenger ribonucleic acid. J Virol 1(3):550–558
12. Laliberte JP, Moss B (2014) A novel mode of poxvirus Superinfection exclusion that prevents fusion of the lipid bilayers of viral and cellular membranes. J Virol 88(17):9751–9768

Chapter 8

Assessing the Structure and Function of Vaccinia Virus Gene Products by Transient Complementation

Nouhou Ibrahim and Paula Traktman

Abstract

Poxviruses are large, complex dsDNA viruses that are highly unusual in replicating solely within the cytoplasm of the infected cell. The most infamous poxvirus was variola virus, the etiological agent of smallpox; today, poxviruses remain of biomedical significance, both as pathogens and as recombinant vaccines and oncolytic therapies. Vaccinia virus is the prototypic poxvirus for experimental analysis. The 195 kb dsDNA genome contains >200 genes that encode proteins involved in such processes as viral entry, gene expression, genome replication and maturation, virion assembly, virion egress, and immune evasion.

Molecular genetic analysis has been instrumental in the study of the structure and function of many viral gene products. Temperature-sensitive (*ts*) mutants have been especially useful in this endeavor; inducible recombinants and deletion mutants are now also important tools. Once a phenotype is observed following the repression, deletion, or inactivation of a particular gene product, the technique of transient complementation becomes central for further study.

Simply put, transient complementation involves the transient expression of a variety of alleles of a given viral gene within infected cells, and the evaluation of which of these alleles can "complement" or "rescue" the phenotype caused by the loss of the endogenous allele. This analysis leads to the identification of key domains, motifs, and sites of posttranslational modification. Subcellular localization and protein:protein interactions can also be evaluated in these studies. The development of a reliable toolbox of vectors encoding viral promoters of different temporal classes, and the use of a variety of epitope tags, has greatly enhanced the utility of this experimental approach for poxvirus research.

Key words Vaccinia virus, Transient complementation, Molecular genetics, Conditional lethal mutants, Structure/function

1 Introduction

Poxvirus overview. Vaccinia virus, the prototypic poxviruses, is a large, enveloped dsDNA virus with a 195 kb genome. The linear genome, which has covalently closed hairpin termini, encodes approximately 200 genes; this large genetic repertoire enables the virus to replicate solely in the cytoplasm of an infected cell. Three classes of temporally regulated genes are expressed during vaccinia life cycle: early genes, which are required for DNA replication, and

Jason Mercer (ed.), *Vaccinia Virus: Methods and Protocols*, Methods in Molecular Biology, vol. 2023,
https://doi.org/10.1007/978-1-4939-9593-6_8, © Springer Science+Business Media, LLC, part of Springer Nature 2019

intermediate and late genes, which are required for virion morphogenesis [1]. A single round of infection yields a large number of progeny virions, the great majority of which remain within the cytoplasm as infectious, mature virions (MV) that are surrounded by a single membrane. These virions are released only when the cell is disrupted. A minority of the virions acquire two additional membranes from the Golgi apparatus or late endosomes, traffic to the cell surface, and undergo exocytotic release. During release, one of the additional membranes is retained within the plasma membrane, and therefore the extracellular virions (EV) represent MV surrounded by an additional membrane referred to as the envelope. Whereas some EVs remain bound to the cell surface, the majority mediate proximal spread to adjacent cells and distal spread throughout cultures or infected hosts [2, 3].

In culture, the yield of virus from a round of infection ("single-step growth curve) is assessed by harvesting cells, releasing the intracellular and cell-associated virus, and performing titrations ("plaque assays") on fresh monolayers of BSC40 cells. The number of plaques that form ("plaque-forming units," or pfu) provides a quantification of the number of infectious virions. The plaques themselves form by viral spread from the initial infected cell to the surrounding cells, and this spread is mediated by EVs.

During the past 35 years, the field of poxvirus research has made enormous strides. The first full genomic sequence was reported in 1990 [4], and numerous studies have led to the characterization of early, intermediate, and late promoters [1, 5–7]. This knowledge of vaccinia gene expression and advances in recombinant DNA technology have enabled the construction of expression vectors that support the expression of viral gene products within infected cells. As described in this report, these vectors are fundamental to the structure/function analysis of viral proteins using transient complementation. In addition, these advances have also led to the development of vaccinia virus as an expression vector for exogenous cellular proteins [8, 9].

Poxvirus genetics and transient complementation overview. Fundamental research into the structure of poxviral proteins and their function during the infectious cycle has relied heavily on transient complementation. The fundamental goal of this assay is to assess the ability of a WT or variant viral protein to complement (rescue) the phenotype caused by the mutation of the endogenous allele [10–14]. Features of viral proteins such as structural domains and motifs, posttranslational modifications, and protein-protein and protein-DNA interactions have been identified and elucidated using this assay [10–14].

The toolbox for most transient complementation assays includes wild-type (WT) vaccinia virus, viral mutants of interest, and plasmid(s) enabling a collection of alleles to be expressed under the regulation of the desired temporal class of vaccinia promoters. Two collections of temperature-sensitive (*ts*) mutants are among the most valuable genetic tools: the Condit collection, which is the largest and most well characterized [15], and the Dales collection. The latter collection has recently been mapped and integrated with the Condit collection, facilitating its use [16, 17]. These *ts* mutants were the foundation for the forward genetic analysis of vaccinia virus. By comparing the permissive (31.5 °C) to the nonpermissive (39.7 °C) conditions, the impact of inactivation of a given gene product can be assessed [18–25]. *ts* mutants vary in their severity of phenotype, and "leaky" viruses can be difficult to study. However, *ts* viruses can be quite useful for the study of multifunctional proteins, since different mutant alleles may disrupt different activities of the protein [20, 26–28].

More recently, reverse genetic tools available to the field include inducible recombinants, in which genes are regulated by components of the *lac* or *tet* operons, making them inducible with IPTG or tetracycline/doxycycline, respectively [11, 13, 22, 29–32]. These recombinants have been instrumental in elucidating the role of a significant number of viral proteins of the intermediate and late temporal classes. One caveat to this system is that inducible recombinants are not readily constructed for early viral genes, because these genes are transcribed within the intact virion core. Repression of early genes would require the encapsidation of high levels of the lacI or tetR repressors, which does not appear to be feasible. A second caveat is that inducible recombinants may be somewhat leaky, which may be an issue for gene products that are needed at very low levels, such as enzymes.

The third group of viruses that have been used in transient complementation assays are deletion viruses, which are useful for genes whose loss gives a discernible phenotype but is not lethal for viral replication. To generate deletion viruses, the endogenous gene of interest is replaced with a reporter gene (e.g., GFP) or a selectable marker (e.g., NEO). Most recently, the development of complementing cell lines has enabled the construction and isolation of deletion viruses involving essential viral genes. The phenotype observed in the complete absence of the gene product of interest may be more severe and more readily studied that is possible using inducible recombinants or *ts* mutants [19, 30, 33–36]. For instance, analysis of the H7-deletion virus showed an absence of the short crescent membranes which were seen during infections with the H7-inducible virus; in this case, the somewhat leaky phenotype of the inducible recombinant and the absolute phenotype of the deletion mutant were complementary but not identical [30, 34]. Using the complementing cell line and the deletion virus, a variety of H7 alleles were assessed for their ability to support viral replication [37]. Similarly, functional analysis of the multifunctional H5 protein has benefited from the analysis of the different phenotypes seen using *ts* mutants and a deletion virus [20, 35, 38]. Although deletion viruses offer the most stringent genetic tool for assessment of protein function, the ability to perform temporal execution point studies with *ts* mutants can reveal different roles for a given protein at different stages of infection [13, 20, 28, 38]. In some cases, *ts* mutants are not available and an inducible recombinant is not feasible; using vΔI3, a null virus lacking the I3 ORF, our laboratory confirmed that the vaccinia virus single-stranded DNA-binding protein plays an essential role in viral replication [19].

Considerations for the establishment of a transient complementation assay. Because the Condit collection of *ts* mutants and most of the inducible and deletion viruses were generated in the Western Reserve (WR) strain of vaccinia virus [1, 17], this strain is used most often in genetic analyses. (The Dales collection of ts mutants was generated in the IHD strain, and in this case WT IHD should be used as a comparative.) A variety of plasmids that carry promoters of different temporal classes are used for transient expression in vaccinia virus [39]. In our laboratory, we have used pJS4, which carries two divergent synthetic early/late promoters [13, 14, 31], pUC1246, which carries a strong late promoter (from the ATI gene) [12, 31, 40], and pINT, which carries an intermediate promoter (from the G8 gene) [35]. The pInt plasmid has been used successfully for transient complementation analyses of early (H5, I3, D5) genes [10, 35, 41], and pUC1246 and pJS4 for late genes (F10, F18, A14, A17, I2) [11–14, 31].

Controls for each transient complementation assay include the empty plasmid and the plasmid containing the WT copy of the gene of interest. Cells must be infected with the mutant virus under both permissive and nonpermissive conditions to establish maximal and minimal levels of virus production. Parallel dishes should be infected under nonpermissive conditions and transfected with empty plasmid or the plasmid encoding the WT protein. Conditions must then be optimized to ensure the greatest difference in viral yield between these two latter conditions (>10-fold, optimally 100-fold). As mentioned below, conditions that can be varied include the multiplicity of infection (MOI), the expression plasmid used, the amount of DNA applied via transfection, the transfection reagent used, and the timing of transfection. Once optimal conditions are established, plasmids encoding a variety of alleles can be generated and tested. This repertoire can include alleles from different strains, truncated alleles, alleles with internal deletions, clustered charge-to-alanine mutants, phosphomimetic, phosphonull, etc.

When assessing the retention/loss of biological function, it is important to validate the stable expression and accumulation of the various mutant proteins. Therefore, transient complementation assays require either the availability of a good antibody specific to the protein of interest or the use of an epitope tag. Although epitope tags facilitate these assays, it is essential to choose an optimal tag that does not disrupt protein function. For each assay, the nature of epitope tag (i.e., size, charge) and its position on the protein of interest (e.g., N′- vs. C′-terminal) are important variables.

Epitope tags can disrupt protein localization, protein:protein interactions, or enzymatic activity, but we and others have successfully used epitope tags such as 3XFLAG, V5, and HA for which good commercial antibodies are available.

In the remainder of this chapter, we provide a general outline for the establishment and use of transient complementation to assess viral protein structure and function within vaccinia-infected cells (*see* Fig. 1). The name of this technique summarizes two

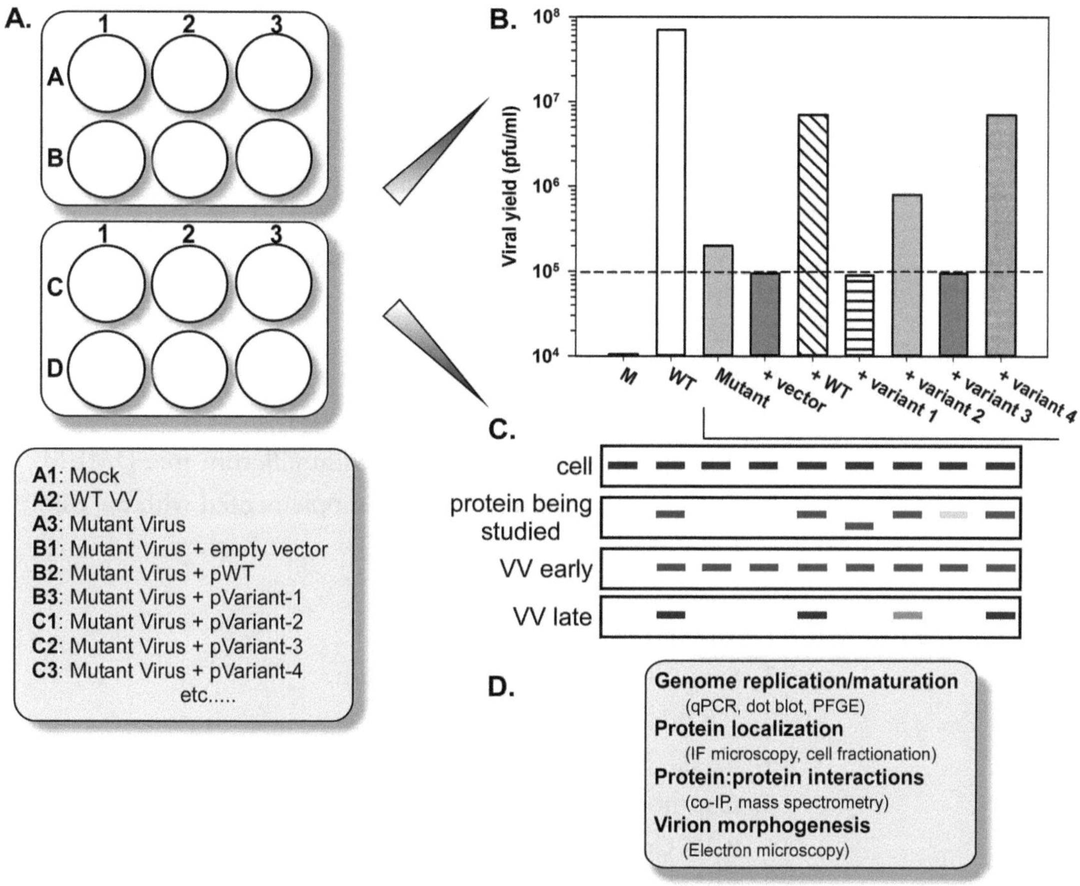

Fig. 1 Transient complementation: analysis of a viral mutant. (**a**) Confluent monolayers of BSC40 cells are grown in 6-well cluster dishes. Cells are left uninfected (mock), infected with WT vaccinia virus (VV) (+control) or with the mutant virus of interest under nonpermissive conditions (row A). Additional wells (rows B and C) are infected with the mutant virus and transfected with empty vector or plasmids encoding the WT protein of interest or a collection of variants. At 18 hpi, cells are harvested; plaque assays (**b**) are performed to determine viral yield (permissive conditions) and immunoblot analysis (**c**) to monitor the accumulation of the protein of interest as well as other early and late viral proteins. Further analyses, such as assessment of genome replication or virion morphogenesis, or of protein localization or protein:protein interactions, can also be performed (**d**). Data analysis: The plasmid-borne WT protein complements the mutant phenotype, increasing viral yield ~100-fold and restoring late protein accumulation. The same is true for variant 4. Variant 1 (minor truncation) does not complement; neither viral yield nor late protein accumulation is restored. Variant 2 shows significant but reduced complementation of viral yield; late proteins accumulate but to a lesser extent. Variant 3 does not accumulate and encodes an unstable protein

important features. First, the protein expressed from the transfected plasmid is being assessed for its ability to *complement* the mutant virus and restore a "WT-like" phenotype, such as the production of viral yield. Second, this plasmid-borne protein is only expressed *transiently* during the single round of infection/transfection: this is a functional assay and does not involve any durable change to the genome of the defective virus. Therefore, viruses produced during a transient complementation assay must be quantified under permissive conditions.

2 Materials

1. Vaccinia virus Western Reserve (WR) strain.
2. Temperature-sensitive mutant virus (*ts*).
3. Inducible recombinant virus: Isopropyl β-D-1-thiogalactopyranoside (IPTG) or tetracycline (TET)-dependent virus.
4. Deletion virus.
5. Supercoiled DNA plasmid preparations: pInt, pJS4, pUC1246.
6. BSC40 cells.
7. Adsorption and transfection medium: Serum-free DMEM.
8. Cell growth medium: DMEM supplemented with 5% FBS.
9. Phosphate-buffered saline (PBS).
10. Lipofectamine 2000.
11. Lipofectamine® LTX.
12. Six-well cluster tissue culture dish.
13. 1.5 mL Microcentrifuge tubes.
14. 37 °C, 5%CO_2 humidified incubator.
15. Cup-horn sonicator.
16. 0.1% Crystal violet in 3.7% formaldehyde.
17. Materials for immunoblot analysis.

3 Methods

The method outlined here is a general approach for transient complementation. Complementation of the mutant virus will occur once the infecting virus provides the necessary factors to drive the expression of the transfected viral gene, resulting in the transient "rescue" of the phenotype caused by the loss of the endogenous allele. A variety of cell lines can be used for transient complementation. Our standard cell line is the BSC40 cell line, an African green monkey kidney epithelial cell line that is highly permissible for vaccinia virus infection (*see* **Note 1**). For each transient complementation experiment three biological replicates should be performed to ensure that the data are reproducible and to assess whether changes in complementation activity reach statistical significance.

3.1 Infection

1. Seed BSC40 cells in 6-well cluster dish to be 90% confluent by the next day (*see* **Notes 2** and **3**). Incubate in a 5% CO_2-humidified incubator at 37 °C.
2. Preparation of virus inoculum: Approximately 18–24 h after seeding BSC40 cells, thaw WT virus and the chosen *ts* mutant virus on ice. Sonicate 2× for 15 s with rest on ice between sonication.
3. Dilution of virus stocks: Make a dilution of each virus to prepare an inoculum that represents a MOI of 3–5 pfu/cell in 500 μL serum-free DMEM/well. Remove cell growth medium from the cultures and add virus inoculum (*see* **Note 4**).
4. Incubate plates for 30 min in a 5% CO_2-humidified incubator, with rocking every 10 min. The temperature to be used depends upon the virus being studied, although most *ts* infections will be performed at 39.7 °C and most other infections at 37 °C (*see* **Note 5**).
5. Aspirate virus inoculum, add 1.0 mL of DMEM supplemented with 5% FBS, and incubate in a 5% CO_2-humidified incubator. The temperature to be used depends upon the experiment (*see* **Note 6**).

3.2 Preparation of DNA for Transfection

1. While the viral inoculum is adsorbing, transfection samples are set up in 1.5 mL centrifuge tubes.
2. Dilute 1 μg of plasmid DNA in 200 μL of serum-free DMEM in 1.5 mL tube and add 1 μL of Plus Reagent. Incubate for 5 min at room temperature (*see* **Note 7**).
3. Add 7.5 μL of Lipofectamine LTX Reagent in 200 μL of DNA:Plus Reagent solution. Incubate for 30 min at room temperature (*see* **Note 8**).
4. Add mixtures containing DNA:Lipofectamine complexes dropwise to the appropriate wells, swirling to mix (*see* **Note 9**).
5. Incubate cultures at 39.7 °C for 18 h (*see* **Note 10**).

3.3 Downstream Experiments

Upon completion of the infection/transfection assay, the harvested samples can be used in a variety of assays (Fig. 1). These assays include quantification of viral yield (plaque assays performed under permissive conditions), analysis of viral DNA accumulation (e.g., Southern dot blot, pulse-field gel electrophoresis), analysis of the accumulation, processing and posttranslational modification of viral proteins (immunoblot, ^{32}P-labeling), assessment of protein-protein interactions (co-immunoprecipitation), and analysis of virus morphogenesis (electron microscopy).

4 Notes

1. We recommend using the BSC40 cell line as it is widely used in poxvirus research. This cell line tolerates 31.5, 37, and 39.7 °C well. The cell line also forms flat monolayers and is optimal for plaque assay titrations. However, the use of other cell lines (CV-1, HeLa) can be attempted, although optimal conditions may be different.
2. We have optimized the format for 35 mm tissue culture dishes or 6-well cluster plates. Confluent monolayers (90% confluent, ~1 × 10^6 cells/well) are used for transient complementation assays. When follow-up studies require a greater number of cells, 6 cm dish of BSC40 cells (3 × 10^6 cells/dish) can also be used.
3. It is important to include the following non-transfected controls to establish baselines: uninfected, WT-infected, and *ts*-infected cells. Using a *ts* mutant *virus,* incubate one sample at 31.5 °C and another sample at 39.7 °C for positive and negative controls, respectively.
4. When using a different size plate or dish, the virus inoculum has to be adjusted. For 60 mm dishes, we recommend adsorption in 1 mL and culture in 4 mL. For 10 cm dishes, we recommend adsorption in 2 mL and culture in 10 mL.
5. When possible, experiments with *ts* mutant viruses should be performed at 39.7 °C, the nonpermissive temperature. However, when studying early proteins involved in DNA replication, it may be necessary to initiate infection at the permissive temperature in order to ensure that the infection proceeds far enough to allow complementation to occur [10]. When using an inducible or a deletion virus, the incubation temperature is 37 °C.
6. In some cases, shift-up or shift-down of the incubation temperature is used to assess the execution point of a *ts* mutant [20, 28, 35, 38]. As reported by DeMasi and Traktman (2000), shifting *ts*H5–4-infected BSC40 cells from 31.5 to 39.7 °C at 6 hpi was sufficient to block viral morphogenesis [20]. The H5 protein is essential for both DNA replication (early event) and virion morphogenesis (late event), but *ts*H5-4 only displays a defect in virion morphogenesis.
7. In addition to the controls outlined above in **Note 3**, it is important to include a dish that is infected under nonpermis-

sive conditions and transfected with an empty vector. Transfection often reduces viral yield, so it is important to compare experimental samples with this control.

8. We recommend optimizing transfection conditions by varying the amount of reagent, the amount of DNA, and the timing of transfection (immediately after the 30-min adsorption, or at 3 hpi, for example). We also recommend trying other transfection reagents if the initial results are not satisfactory. When testing several mutant alleles of the same gene (targeted mutations, truncations), assess the accumulation of the proteins, as some mutant proteins may be unstable and not accumulate.
9. In some circumstances, the transfection reagent may be toxic to the cells, so the medium should be removed, and cells refed with fresh growth medium at 5–7 h after transfection.
10. The total incubation time for this experiment is typically 18 h if viral yield is to be assessed. However, samples can be harvested at different times for different downstream assays. For example, 18–24-h incubation is needed for the assessment of viral yield, whereas 10 h is sufficient to assess DNA accumulation by southern dot blot and to assess protein accumulation by immunoblot analysis.

Acknowledgments

Some of the work described herein was supported by grants from the NIH to P.T. (R01 AI21758 and R01 AI107123). We appreciate the helpful comments of members of the Traktman lab. We acknowledge the many members of the poxvirus field whose laboratories have contributed to the development of tools for the genetic analysis of poxvirus research, especially Richard Condit, Bernard Moss, and Geoffrey Smith. We also apologize to the many scientists whose important work was not cited here.

References

1. Moss B (2013) Poxviridae. In: Knipe DM, Howley PM (eds) Fields virology, vol 2, 6th edn. Lippincott Williams & Wilkins, Philadelphia, PA, pp 2129–2159
2. Smith GL, Vanderplasschen A, Law M (2002) The formation and function of extracellular enveloped vaccinia virus. J Gen Virol 83(Pt 12):2915–2931
3. Condit RC, Moussatche N, Traktman P (2006) In a nutshell: structure and assembly of the vaccinia virion. Adv Virus Res 66:31–124
4. Goebel SJ, Johnson GP, Perkus ME, Davis SW, Winslow JP, Paoletti E (1990) The complete DNA sequence of vaccinia virus. Virology 179(1):247–263
5. Davidson AJ, Moss B (1989) Structure of vaccinia virus early promoters. J Mol Biol 210:749–769
6. Davison AJ, Moss B (1989) Structure of vaccinia virus late promoters. J Mol Biol 210(4):771–784
7. Yang Z, Reynolds SE, Martens CA, Bruno DP, Porcella SF, Moss B (2011) Expression profil-

ing of the intermediate and late stages of poxvirus replication. J Virol 85(19):9899–9908. https://doi.org/10.1128/JVI.05446-11
8. Carroll MW, Moss B (1997) Poxviruses as expression vectors. Curr Opin Biotechnol 8(5):573–577
9. Moss B, Flexner C (1989) Vaccinia virus expression vectors. Ann N Y Acad Sci 569:86–103
10. Boyle KA, Arps L, Traktman P (2007) Biochemical and genetic analysis of the vaccinia virus d5 protein: multimerization-dependent ATPase activity is required to support viral DNA replication. J Virol 81(2):844–859
11. Mercer J, Traktman P (2003) Investigation of structural and functional motifs within the vaccinia virus A14 phosphoprotein, an essential component of the virion membrane. J Virol 77(16):8857–8871
12. Nichols RJ, Stanitsa E, Unger B, Traktman P (2008) The vaccinia virus gene I2L encodes a membrane protein with an essential role in virion entry. J Virol 82(20):10,247–10,261
13. Punjabi A, Traktman P (2005) Cell biological and functional characterization of the vaccinia virus F10 kinase: implications for the mechanism of virion morphogenesis. J Virol 79(4):2171–2190
14. Wickramasekera NT, Traktman P (2010) Structure/function analysis of the vaccinia virus F18 phosphoprotein, an abundant core component required for virion maturation and infectivity. J Virol. https://doi.org/10.1128/JVI.00399-10
15. Lackner CA, D'Costa SM, Buck C, Condit RC (2003) Complementation analysis of the Dales collection of vaccinia virus temperature-sensitive mutants. Virology 305(2):240–259
16. Condit RC, Motyczka A (1981) Isolation and preliminary characterization of temperature-sensitive mutants of vaccinia virus. Virology 113(1):224–241
17. Condit RC, Motyczka A, Spizz G (1983) Isolation, characterization, and physical mapping of temperature—sensitive mutants of vaccinia virus. Virology 128(2):429–443
18. Traktman P, Caligiuri A, Jesty SA, Liu K, Sankar U (1995) Temperature-sensitive mutants with lesions in the vaccinia virus F10 kinase undergo arrest at the earliest stage of virion morphogenesis. J Virol 69(10):6581–6587
19. Greseth MDCM, Bluma MS, Traktman P (2018) Isolation and characterization of vΔI3 confirm that Vaccinia virus SSB plays an essential role in viral replication. J Virol 92(2). https://doi.org/10.1128/JVI.01719-17
20. DeMasi J, Traktman P (2000) Clustered charge-to-alanine mutagenesis of the vaccinia virus H5 gene: isolation of a dominant, temperature-sensitive mutant with a profound defect in morphogenesis. J Virol 74(5):2393–2405
21. Punjabi A, Boyle K, DeMasi J, Grubisha O, Unger B, Khanna M, Traktman P (2001) Clustered charge-to-alanine mutagenesis of the vaccinia virus A20 gene: temperature-sensitive mutants have a DNA-minus phenotype and are defective in the production of processive DNA polymerase activity. J Virol 75(24):12,308–12,318
22. Unger B, Traktman P (2004) Vaccinia virus morphogenesis: A13 phosphoprotein is required for assembly of mature virions. J Virol 78(16):8885–8901
23. Ansarah-Sobrinho C, Moss B (2004) Role of the I7 protein in proteolytic processing of vaccinia virus membrane and core components. J Virol 78(12):6335–6343
24. Szajner P, Weisberg AS, Moss B (2001) Unique temperature-sensitive defect in vaccinia virus morphogenesis maps to a single nucleotide substitution in the A30L gene. J Virol 75(22):11,222–11,226
25. Wang S, Shuman S (1995) Vaccinia virus morphogenesis is blocked by temperature-sensitive mutations in the F10 gene, which encodes protein kinase 2. J Virol 69(10):6376–6388
26. Ishii K, Moss B (2001) Role of Vaccinia virus A20R protein in DNA replication: construction and characterization of temperature-sensitive mutants. J Virol 75(4):1656–1663
27. Li J, Pennington MJ, Broyles SS (1994) Temperature-sensitive mutations in the gene encoding the small subunit of the vaccinia virus early transcription factor impair promoter binding, transcription activation, and packaging of multiple virion components. J Virol 68(4):2605–2614
28. Mercer J, Traktman P (2005) Genetic and cell biological characterization of the vaccinia virus A30 and G7 phosphoproteins. J Virol 79(11):7146–7161
29. Wolffe EJ, Moore DM, Peters PJ, Moss B (1996) Vaccinia virus A17L open reading frame encodes an essential component of nascent viral membranes that is required to initiate morphogenesis. J Virol 70(5):2797–2808
30. Satheshkumar PS, Weisberg A, Moss B (2009) Vaccinia virus H7 protein contributes to the formation of crescent membrane precursors of immature Virions. J Virol. https://doi.org/10.1128/JVI.00877-09
31. Unger B, Mercer J, Boyle KA, Traktman P (2013) Biogenesis of the vaccinia virus membrane: genetic and ultrastructural analysis of the contributions of the A14 and A17 pro-

teins. J Virol 87(2):1083–1097. https://doi.org/10.1128/JVI.02529-12

32. Traktman P, Liu K, DeMasi J, Rollins R, Jesty S, Unger B (2000) Elucidating the essential role of the A14 phosphoprotein in vaccinia virus morphogenesis: construction and characterization of a tetracycline-inducible recombinant. J Virol 74(8):3682–3695
33. Maruri-Avidal L, Weisberg AS, Bisht H, Moss B (2013) Analysis of viral membranes formed in cells infected by a vaccinia virus L2-deletion mutant suggests their origin from the endoplasmic reticulum. J Virol 87(3):1861–1871. https://doi.org/10.1128/JVI.02779-12
34. Meng X, Wu X, Yan B, Deng J, Xiang Y (2013) Analysis of the role of vaccinia virus H7 in virion membrane biogenesis with an H7-deletion mutant. J Virol 87(14):8247–8253. https://doi.org/10.1128/JVI.00845-13
35. Boyle KA, Greseth MD, Traktman P (2015) Genetic confirmation that the H5 protein is required for Vaccinia virus DNA replication. J Virol 89(12):6312–6327. https://doi.org/10.1128/JVI.00445-15
36. Olson ATRA, Wang Z, Delhon G, Wiebe MS (2017) Deletion of the Vaccinia virus B1 kinase reveals essential functions of this enzyme complemented partly by the homologous cellular kinase VRK2. J Virol 91(15). https://doi.org/10.1128/JVI.00635-17
37. Kolli S, Meng X, Wu X, Shengjuler D, Cameron CE, Xiang Y, Deng J (2015) Structure-function analysis of vaccinia virus H7 protein reveals a novel phosphoinositide binding fold essential for poxvirus replication. J Virol 89(4):2209–2219. https://doi.org/10.1128/JVI.03073-14
38. D'Costa SM, Bainbridge TW, Kato SE, Prins C, Kelley K, Condit RC (2010) Vaccinia H5 is a multifunctional protein involved in viral DNA replication, postreplicative gene transcription, and virion morphogenesis. Virology 401(1):49–60. https://doi.org/10.1016/j.virol.2010.01.020
39. Moss B (1991) Vaccinia virus: a tool for research and vaccine development. Science 252(5013):1662–1667
40. Rempel RE, Anderson MK, Evans E, Traktman P (1990) Temperature-sensitive vaccinia virus mutants identify a gene with an essential role in viral replication. J Virol 64(2):574–583
41. Greseth MD, Boyle KA, Bluma MS, Unger B, Wiebe MS, Soares-Martins JA, Wickramasekera NT, Wahlberg J, Traktman P (2012) Molecular genetic and biochemical characterization of the vaccinia virus I3 protein, the replicative single-stranded DNA binding protein. J Virol 86(11):6197–6209. https://doi.org/10.1128/JVI.00206-12

Chapter 9

Preliminary Screening and In Vitro Confirmation of Orthopoxvirus Antivirals

Douglas W. Grosenbach and Dennis E. Hruby

Abstract

The lack of antiviral drugs for the treatment of orthopoxvirus disease represents an unmet medical need, particularly due to the threat of variola virus (the causative agent of smallpox) as an agent of biowarfare or bioterrorism (Henderson, 283:1279–1282, 1999). In addition to variola, monkeypox, cowpox, and vaccinia viruses are orthopoxviruses of concern to human health (Lewis-Jones, 17:81–89, 2004). Smallpox vaccination, using the closely related vaccinia virus, is no longer provided to the general public leading to a worldwide population increasingly susceptible not only to variola but to monkeypox, cowpox, and vaccinia viruses as well. Orthopoxviruses share similar life cycles (Fenner et al., WHO, Geneva, 1988), and significant nucleotide and protein homology, and are immunologically cross-protective against other species within the genus, which was the basis of the highly successful vaccinia virus vaccine. These similarities also serve as the basis for screening for antivirals for dangerous pathogens such as variola and monkeypox virus using generally safer viruses such as cowpox and vaccinia. Methods for preliminary screening and initial characterization of potential orthopoxvirus antivirals in vitro, using vaccinia virus as a relatively safe surrogate for more pathogenic orthopoxviruses, are described herein. They include candidate identification in a viral cytopathic effect (CPE) assay as well as evaluation of the antiviral activity in inhibition assays to determine mean effective (or inhibitory) concentrations (EC_{50} or IC_{50}). These assays were utilized in the identification and early characterization of tecovirimat (ST-246) (Yang et al., 79:13,139–13,149, 2005). These initial steps in identifying and characterizing the antiviral activity should be followed up with additional in vitro studies including specificity testing (for other orthopoxviruses and against other viruses), single-cycle growth curves, time of addition assays, cytotoxicity testing, and identification of the drug target.

Key words Antiviral, Drug development, Drug screening, Orthopoxvirus, CPE assay, EC_{50} assay, In vitro

1 Introduction

1.1 Orthopoxviruses

The orthopoxviruses are large (~200 × 400 nm) DNA viruses that replicate exclusively within the cytoplasm of host cells [1, 2]. The orthopoxvirus life cycle is complex: Upon entry, the ~200 kb genome is expressed in a temporally regulated fashion to produce early, intermediate, and late genes. Virion assembly coincides with late protein production, approximately 4 h postinfection, and

Jason Mercer (ed.), *Vaccinia Virus: Methods and Protocols*, Methods in Molecular Biology, vol. 2023,
https://doi.org/10.1007/978-1-4939-9593-6_9, © Springer Science+Business Media, LLC, part of Springer Nature 2019

continues until cell lysis 24–72 h postinfection. The first infectious virions, which are referred to as mature virus (MV), are formed as the core, which is composed of the genome, packaged enzymes and cofactors, and numerous structural proteins; condenses; and is wrapped with a lipid membrane. The MV form of the virus is fully infectious but is not released from the cell until lysis. In tissue culture systems, MVs represent the majority of virions produced, but depending on the species, strain, and host cell some MVs are further enveloped with additional membranes and released from the cell in a non-lytic fashion [2]. These multiple enveloped forms of the virus are likely of greater significance in disease, since they have been implicated in cell-to-cell spread and long-range dissemination of the virus both in vitro and in vivo [3]. In monolayer cell culture, orthopoxviruses form distinct plaques within 24–72 h if the multiplicity of infection (MOI) is sufficiently low (i.e., $<10^{-4}$ PFU/cell), but as the MOI approaches one, productive infection will lead to the destruction of the entire monolayer within 72 h. This serves as the basis for the cytopathic effect (CPE) and mean effective concentration (EC50) assays described herein to identify and preliminarily characterize compounds with antiviral activity.

Variola, monkeypox, cowpox, and vaccinia viruses are species within the orthopoxvirus genus that are of concern to human health [4, 5]. Other species, such as rabbitpox and ectromelia (mousepox) viruses, are animal host-adapted viruses that do not cause human disease although they are often used as surrogate viruses in animal model systems to study orthopoxvirus disease and develop therapeutics for smallpox [6]. In the methods described herein, vaccinia virus (VACV) is used as the prototypic orthopoxvirus as it was used as the live virus smallpox vaccine (for over 200 years) and is very well characterized both in vitro and in vivo. Antiviral activity against VACV is likely representative of activity against all orthopoxvirus species, although not all potential antiviral targets within the orthopoxviruses are 100% conserved between species. VACV is a commonly used laboratory reagent and Biosafety Level 2 practices are required for handling the virus. Personal protective equipment should include safety glasses, gloves, and protective clothing such as a lab coat. Smallpox vaccination is recommended for laboratory workers that work with VACV or other orthopoxviruses, as the vaccine is cross protective for all orthopoxviruses.

1.2 CPE Assay

The CPE assay is a very straightforward evaluation of whether or not a compound exhibits antiviral effects. The method described may be performed manually if a small number of compounds are to be tested, but can be scaled for high-throughput robotic testing [7].

VACV will destroy a cell monolayer within 72 h postinfection if inoculated at a sufficiently high MOI. In the system described below, VeroE6 cells, grown to ~90% confluency in 96-well plates, are inoculated with VACV at a MOI ranging from 0.05 to 0.1 PFU/cell. This should be optimized prior to employing the assay (*see* Subheading 4) to ensure approximately 90% destruction of the monolayer at 72 h post-infection. Compounds to be tested for antiviral activity are dissolved in dimethyl sulfoxide (DMSO) and added to cells at a single relatively high concentration (5 μM in the assay described below). The final concentration of DMSO should not exceed 0.05% when diluted in cell culture medium. Compound or virus may be added in any order, but care should be taken to avoid contamination of uninfected control wells with virus or carryover of compound to untreated wells. Uninfected and infected control wells, treated only with DMSO, serve as standards for a fully intact monolayer (0% CPE) and a destroyed monolayer (100% CPE), respectively. After a 72-h incubation, the cell monolayers are fixed, stained with crystal violet, and spectrophotometrically scanned at a wavelength of 570 nm to quantitatively determine the level of monolayer destruction or preservation in the presence of compound relative to controls. Considering the high concentration of compound used in this assay, only those compounds exhibiting >50% inhibition of virus-induced CPE should be considered for further evaluation in the EC_{50} assay described below.

1.3 EC50 Assay

Once antiviral candidates have been identified in the CPE assay, the level of virus sensitivity to the compound should be determined. Infected cells are treated with compound at various concentrations to determine the mean effective concentration (EC_{50}), which is the concentration of compound that inhibits CPE by 50% relative to untreated (DMSO treated) control wells infected with virus. The assay is set up and conducted very similarly to the CPE assay: VeroE6 cells, grown to ~90% confluency in 96-well plates, are inoculated with VACV at a MOI ranging from 0.05 to 0.1 PFU/cell. In the method described below, compound is added to infected cells at concentrations ranging from 0.0015 to 5 μM typically using eight dilutions taking advantage of the 96-well format. Uninfected and infected control wells are used as standards for 0% CPE and 100% CPE, respectively. After 72 h of incubation, the cells are fixed, stained, and scanned as above to quantitatively determine the level of CPE inhibition in the presence of various concentrations of compound. To determine the EC_{50}, the data are analyzed using curve-fitting software such as the XLFit add-in for Microsoft Excel.

1.4 Further Characterization of Antiviral Compounds In Vitro

The methods described here are by no means exhaustive, and numerous follow-up assays should be performed in the preliminary characterization of candidate antiviral compounds. The full details of these assays are not presented here. First, it will be important to determine if the compound is cytotoxic: The mean cytotoxic concentration (CC_{50}) assay is designed and conducted very similarly to the EC_{50} assay, with the exception that the range of compound concentrations evaluated is generally much higher than the range tested to determine the $EC_{50.}$ Once the CC_{50} has been determined, the therapeutic index may be calculated as CC_{50}/EC_{50}. An acceptable therapeutic index will be based on the risk/benefit ratio of using the drug to treat disease. The compound should also be tested for specificity. These tests should include evaluation of antiviral activity against other orthopoxviruses and viral species not related to the poxviruses. The mechanism of action for the compound may be investigated using single-cycle growth curve and time-of-addition assays. If cells are infected at a relatively high MOI (5–10 PFU/cell) to ensure synchronous infection, VACV will complete its full replication cycle and produce a high titer of virus both intracellularly and in the culture medium. The effect of the compound may be quantitated in terms of virus yield over 24 h by sampling virus and titering at various time points postinfection. Typically, virus is sampled hourly up to 6 or 8 h postinfection, and then at longer intervals up to 24 h. In time-of-addition assays, a synchronous 24-h infection is set up similar to the single-cycle growth curve assay with compound added prior to infection, coincident with infection, and at various time points postinfection. At 24 h postinfection, cell-associated virus and virus released into the medium is titered. This will give an indication of the step of virus replication impacted by the compound. This type of assay may also be used to investigate the impact of a compound on the level and temporal regulation of gene expression and protein production. The viral target of the compound should be identified as well. The most straightforward way to do this is by generating compound-resistant viruses. Low MOI passage of the virus in the presence of suboptimal concentrations of compound (i.e., $<EC_{50}$) that are increased with each passage may yield resistant virus. Gene mutations (potentially encoding target proteins) that mediate resistance may be identified via marker-rescue techniques, which involve transfection of wild-type virus-infected cells with genomic regions or individual genes from the resistant virus and identifying those genes that rescue the susceptible wild-type virus in the presence of inhibitory concentrations of compound.

2 Materials

1. Class II Biological Safety Cabinet (BSC).
2. Water bath 37 ± 1 °C.
3. Incubator 37 ± 1 °C with 5 ± 1% CO_2.
4. Vortex.

5. Cup horn sonicator.
6. PBS (1×).
7. Pipet aid.
8. Microplate plate reader capable of scanning 96-well plates at 570 nm, with data acquisition software.
9. Microsoft Excel with XLFit curve-fitting software package or other curve-fitting software application.
10. Micropipettes.
11. 8- or 12-well multichannel pipette.
12. Vero E6 cells (less than 50 passages).
13. Sterile 96-well cell culture plates with lids.
14. 50 mL Sterile polypropylene conical tubes.
15. Sterile serological pipettes.
16. Sterile micropipette tips.
17. Dipping container for fixation.
18. Wipes and absorbent material.
19. 70% and 100% ethanol.
20. Dimethyl sulfoxide (DMSO).
21. Dulbecco's modified Eagle medium (DMEM) + Glutamax™.
22. 200× Stocks of compound of interest (i.e., 1000 μM, 300 μM, 100 μM, 30 μM, 10 μM, 3 μM, 1 μM, and 0.3 μM).
23. 100× Stocks of penicillin/streptomycin (Pen/Strep) solution.
24. 50% Glutaraldehyde solution (w/v, in water).
25. 2.3% Crystal violet solution (w/v, commercially available as 2.3% certified crystal violet, 0.1% ammonium oxalate monohydrate, and 20% ethyl alcohol in water).
26. 0.1% Crystal violet/ethanol solution (198 mL of 1× PBS; 41 mL absolute ethanol, 11 mL of 2.3% crystal violet solution).
27. Plate seeding medium: DMEM + Glutamax™ supplemented with 10% FBS and penicillin/streptomycin solution diluted to 1× in the medium.
28. Plate seeding medium supplemented with 1.0% DMSO (v/v).
29. Fetal bovine serum (FBS; heat inactivated).
30. Sterile disposable reagent bottles.
31. Trypsin/EDTA solution suitable for cell culture.
32. VACV working stock.
33. Sterile reservoirs.

3 Methods

3.1 CPE Assay

This method outlines the assay conducted to evaluate compounds for their ability to inhibit VACV-induced cytopathic effect (CPE). The assay is described as if performed by hand, but may be scaled for robotic setup, facilitating high-throughput screening of large compound libraries. *See* Fig. 1 for a plate diagram outlining the assay setup. The assay is setup using VeroE6 cell monolayers seeded in 96-well plates. Compounds are dissolved in DMSO, diluted in medium, and added to the cell culture at a final concentration of 5 μM. The cells are then infected with VACV at a multiplicity of infection that results in the destruction of the monolayer within 72 h postinfection (if untreated). After 72 h, the cultures are fixed and stained with crystal violet and virus-induced CPE is quantified spectrophotometrically by absorbance at 570 nm. Compounds are scored as "hits" if they inhibit VACV-induced CPE by ~>/=50% relative to DMSO-treated control wells. All steps should be conducted in a Class II BSC using sterile technique.

3.1.1 96-Well Plate Seeding: Common to Both the CPE and EC_{50} Assays

1. Seed appropriate number of 96-well cell culture plates with ~1.25×10^4 VeroE6 cells per well the day prior to the assay setup. Optimize the number of cells used to seed plates, as this will determine their confluency on the day of the assay (*see* **Note 1**). The number of plates will depend on the number of compounds to be screened. The methods described below may be scaled for the number of plates needed for each assay.

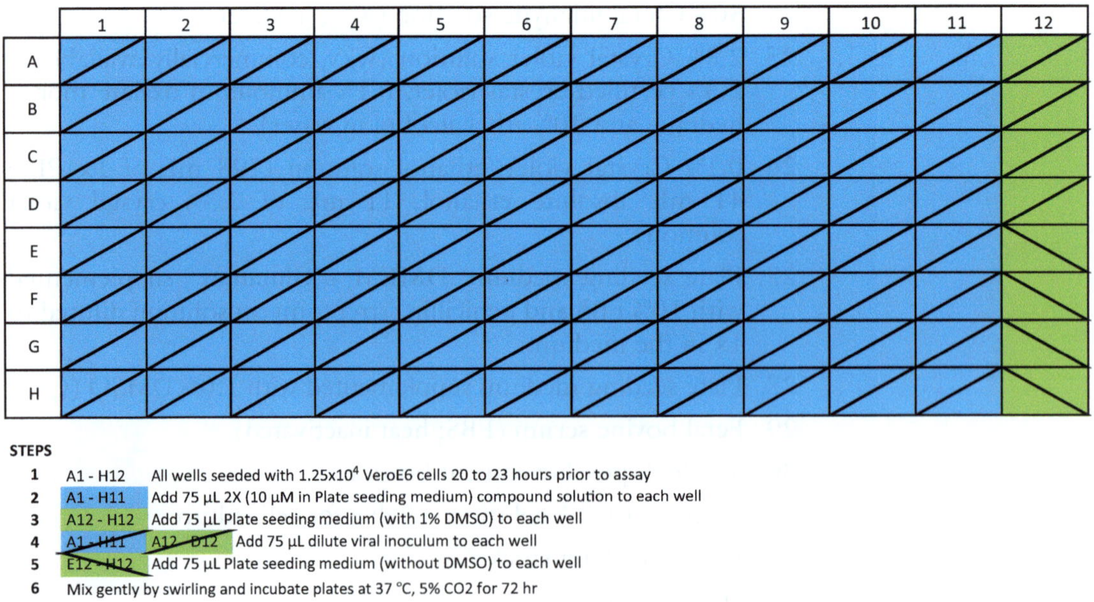

Fig. 1 CPE assay plate setup

2. Prepare plate seeding medium. The volume of medium needed will depend on the number of plates needed for each assay.
3. Trypsinize cells according to standard cell culture procedures using trypsin-EDTA solution prepared for cell culture.
4. Perform cell counts using the trypan blue dye exclusion method and a hemacytometer.
5. Dilute cells to 7.14×10^4 cells/mL using plate seeding medium (without DMSO) and add diluted cells to a sterile reservoir for plating. The dilution of cells to this concentration is dependent on the optimized cell density used to seed plates (*see* **Note 1**).
6. Using a multichannel pipette, add 175 μL of the diluted cell suspension (equal to 1.25×10^4 cells/well) to each well of a 96-well plate. Tap the sides of the plates to ensure even distribution of cells.
7. Incubate at 37 ± 1 °C and 5 ± 1% CO_2 for 20–23 h.
8. Following the incubation period and prior to the subsequent procedures, visually inspect the cell monolayer by microscopic examination to ensure that cells are greater than 90% confluent at the time of use. If less than 90% confluent, or if cells are unevenly distributed, then do not continue with the assay. If necessary, adjust cell concentration and culture conditions to ensure >90% confluence at 20–23 h post-plating.

3.1.2 Compound Preparation

1. Prepare plate seeding medium (with 1% DMSO) and mix by inversion. Volumes may be adjusted accordingly to make an appropriate volume necessary to perform the procedure.
2. Prepare 2× (i.e., 10 μM) compound working stock solutions in plate seeding medium (without DMSO) starting with 200× compound stock solutions (1000 μM in 100% DMSO). To prepare a 10 μM working stock solution that will give a final concentration of 5 μM once diluted in culture, add 100 μL of 1000 μM compound solution to 10 mL plate seeding medium (without DMSO). Prepare the compound dilutions on the day of use.

3.1.3 Compound Addition

1. Carefully remove the plate seeding medium from each well of the assay plates by decanting on a per-plate basis and proceeding to the next step.
2. For sample wells, add 75 μL of appropriate plate seeding medium supplemented with 2× concentration (10 μM) compound to wells A1 through H11. Add a different compound to each well for a preliminary screen, or add in triplicate during confirmation screens.
3. For control wells, add 75 μL of plate seeding medium (with 1% DMSO) to wells A12 through H12.

4. If not all wells are used, add 75 μL of plate seeding medium (without DMSO) to the remaining wells.
5. Incubate each plate for 1–1.5 h 37 ± 1 °C with 5 ± 1% CO_2.

3.1.4 Preparation of VACV for Infection

1. Remove the working stock of VACV from frozen storage (−70 °C or below freezer), thaw vials in a 37 ± 1 °C water bath, and then place on wet ice. Conduct the following steps in a Class II BSC.
2. Once thawed vortex each virus briefly for 3–5 s.
3. Sonicate each virus for 30 ± 2 s between 39% and 41% amplitude in a cup horn sonicator containing ice water, and then immediately place on wet ice until virus is ready to be diluted.
4. Spray the virus vials that were on ice with 70% ethanol to disinfect the surface and then dry.
5. Dilute the predetermined optimal amount of virus (*see* **Note 2**) in a total volume of 50 mL plate seeding medium.
6. Vortex briefly, label appropriately, and place on wet ice until cells are ready to be infected.

3.1.5 VACV Infection

1. After the compound incubation (1–1.5 h 37 ± 1 °C with 5 ± 1% CO_2), spray the 50 mL virus preparations (stored on ice) with 70% ethanol to disinfect the surface and dry.
2. Using a sterile reservoir and a multichannel pipet, infect wells Al through H11, and A12 through D12 using 75 μL per well of the virus dilution (Subheading 3.1.4, **step 6**).
3. Add 75 μL of plate seeding medium (without DMSO) to wells E12 through H12 as uninfected control wells.
4. Gently mix and incubate plates at 37 °C ± 1 °C with 5 ± 1% CO_2 for 72 ± 2 h.

3.1.6 Fixing and Staining Plates

1. Freshly prepare 1000 mL of a 5% glutaraldehyde solution by diluting 50% solution 1:10 in 1× PBS. Mix solution vigorously by hand for a few seconds. Volumes may be adjusted proportionally to make an appropriate volume necessary to perform the procedure. Prepare solution in a Class IIA2 or a Class IIB2 BSC. If being prepared in a ClassIIA2 BSC, wear a respirator equipped with an appropriate chemical cartridge.
2. Prepare 0.1% crystal violet/ethanol solution. Volumes may be adjusted proportionally to make an appropriate volume necessary to perform the procedure. The solution may be prepared either fresh or within 4 days prior to the actual staining of the cells.

3. Once the incubation period has ended (Subheading 3.1.5, **step 4**), in sets of up to five plates, decant medium from plates one group at a time.
4. Conduct fixation in a Class IIA2 or a Class IIB2 BSC. If being prepared in a Class IIA2 BSC, wear a respirator equipped with an appropriate chemical cartridge.
5. Using a repeater multichannel pipette, add 250 μL of 5% glutaraldehyde solution to each well.
6. Fix cell monolayer for 30–45 min at room temperature.
7. Decant the glutaraldehyde solution one set of plates at a time into an appropriate container with a wide opening. Tap plate on absorbent material until runoff is minimal.
8. One set at a time, carefully add 150 μL of 0.1% crystal violet/ethanol solution to each well using a multichannel pipette. Staining may be in a Class IIA2 or a Class IIB2 BSC.
9. Stain plates at room temperature for 30 min to 1 h.
10. Carefully decant stain and thoroughly tap dry on absorbent material. Ensure that excess stain is adequately removed and not present on the bottom of the plate. Change gloves frequently to minimize contamination of the plates with crystal violet.
11. Allow plates to air-dry upside down on the absorbent material allowing excess crystal violet to drain.
12. Once the plates have air-dried completely, read the optical density at 570 nm using the microplate reader with appropriate data acquisition software.

3.2 EC_{50} Assay

Dose-response curves are generated by measuring virus-induced cytopathic effects in the presence of a range of compound concentrations. *See* Fig. 2 for a plate diagram outlining the assay setup. In this method, eight compound concentrations are used to generate inhibition curves suitable for calculating the EC_{50} from virus-induced CPE. Compound dilutions are prepared in DMSO prior to addition to the cell culture medium. The final DMSO concentration in the culture medium placed on cells should not exceed 0.5% (*see* **Note 3**). Cell monolayers are infected with VACV at a multiplicity of infection (approximately 0.05 PFU/cell) that destroys ~90% of the monolayer of cells within 72 h postinfection. At 72 h postinfection, the assay is terminated by fixation and the level of CPE is visualized by staining the monolayers with crystal violet. Virus-induced cytopathic effects are quantified by measuring absorbance at 570 nm. EC_{50} value is calculated using curve fitting software to generate a dose-response curve. From this curve, the concentration of compound that inhibits virus-induced CPE by 50% may be calculated. All steps should be conducted in a Class II BSC using sterile technique.

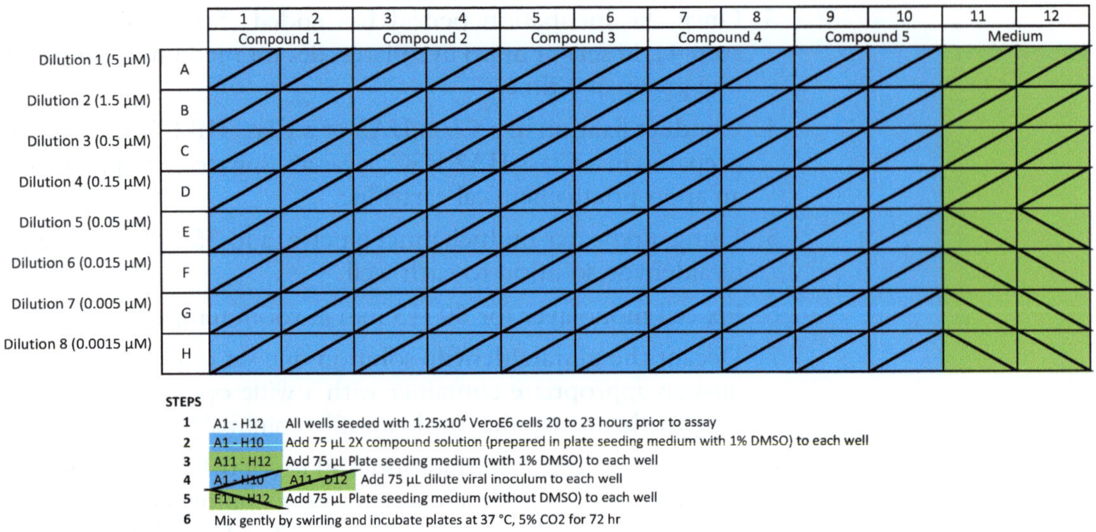

Fig. 2 EC50 assay plate setup

*3.2.1 96-Well Plate Seeding (**See** Subheading 3.1.1)*

This procedure is identical to that for the CPE assay. The number of plates used will depend on the number of compounds screened in this assay.

3.2.2 Compound Preparation

1. Prepare plate seeding medium (with 1% DMSO) and mix by inversion. Volumes may be adjusted accordingly to make an appropriate volume necessary to perform the procedure.
2. Prepare 2× compound solutions in plate seeding medium (without DMSO) using 200× compound concentrations of 1000, 300, 100, 30, 10, 3, 1, and 0.3 µM. Prepare each dose concentration according to example volumes shown in Table 1. Volumes may be adjusted proportionally to make an appropriate volume necessary to perform the procedure. Prepare the compound dilutions on the day of use.

3.2.3 Compound Addition

1. Carefully remove the plate seeding medium from each well of the plates to be used for testing by decanting on a per-plate basis and proceeding to the next step.
2. For compound sample wells, add 75 µL of each 2× compound dilution (eight total), as outlined in Table 1, in duplicate to rows A through H/columns 1 through 10. Duplicates are added in columns 1 and 2, 3 and 4, 5 and 6, 7 and 8, and 9 and 10. The eight dilutions are added highest to lowest in rows A through H. Five compounds can be evaluated per plate.
3. For the control wells, add 75 µL of plate seeding medium (with 1% DMSO) to rows A through H columns 11 and 12.

Table 1
Compound preparation for use in EC_{50} assays

Final compound concentration (1× in medium)	Compound concentration (2×)	Volume of compound (2× preparation)	Volume of medium (2× preparation)
5 μM	10 μM	100 μL of 1000 μM	10 mL
1.5 μM	3 μM	100 μL of 300 μM	10 mL
0.5 μM	1 μM	100 μL of 100 μM	10 mL
0.15 μM	0.3 μM	100 μL of 30 μM	10 mL
0.05 μM	0.1 μM	100 μL of 10 μM	10 mL
0.015 μM	0.03 μM	100 μL of 3 μM	10 mL
0.005 μM	0.01 μM	100 μL of 1 μM	10 mL
0.0015 μM	0.003 μM	100 μL of 0.3 μM	10 mL

Compounds to be evaluated in the EC_{50} assay and solubilized in DMSO at a concentration of 1000 μM as a stock dilution. The compound is further diluted in DMSO to yield concentrations of 300, 100, 30, 10, 3, 1, and 0.3 μM. These compounds are diluted into plate seeding medium (without DMSO) by addition of 100 μL into 10 mL of medium. This results in a 2× concentration of compound in medium. The compounds are added to cells at 2×

4. If some wells are unused, add 75 μL of plate seeding medium (without DMSO) to the remaining wells.
5. Incubate each plate for 1–1.5 h 37 ± 1 °C with 5 ± 1% CO_2.

*3.2.4 Preparation of VACV for Infection (**See** Subheading 3.1.4)*

VACV preparation for the EC_{50} assay is identical to that for the CPE assay.

3.2.5 VACV Infection

1. After the drug incubation, spray the 50 mL virus preparations that were on ice with 70% ethanol to disinfect the surface and then dry.
2. Using a sterile reservoir and a multichannel pipet, infect the following wells with VACV using 75 μL of the virus preparations per well.
3. To the wells containing compound (i.e., A1 through H10), add 75 μL of VACV to each well.
4. For infected control wells that do not contain compound, infect wells in rows A through D/columns 11 and 12 adding 75 μL of VACV to each well.
5. For uninfected control wells that do not contain compound, add 75 μL of plate seeding medium (without DMSO) to wells in rows E through H/columns 11 and 12.
6. Gently mix and incubate plates at 37 °C ± 1 °C with 5 ± 1% CO_2 for 72 ± 2 h.

3.2.6 Fixing and Staining Plates

Fixing and staining plates is as described for the CPE assay (*see* Subheading 3.1.6). After completing the procedure, calculate the EC_{50} using curve fitting software. Optical density at 570 nm for the uninfected control wells represents 0% CPE while infected control wells treated only with DMSO represents 100% CPE. Intermediate values from compound-treated wells should be fitted to the curve. Explore various curve-fitting parameters to identify the best fit. The EC_{50} is that concentration of compound that inhibits viral CPE by 50%.

4 Notes

1. Cells grow at different rates depending on the cell type, passage number, and specific culture conditions. In the assays described, the optimal cell density on the day of infection should be close to 90% confluency. This allows for some growth over the 72 h of the assay but also ensures that a sufficient number of cells are stained at 72 h postinfection for crystal violet staining. To ensure ~90% confluency on the day of infection, the seeding density should be optimized prior to utilizing the assay. Try seeding 96-well plates with various densities ranging from 5×10^3 cells/well to 1.5×10^4 cells/well, and culturing for 24 h. Visually inspect by microscopy and estimate the level of confluency. Once a seeding density is identified that yields ~90% confluency at 24 h post-seeding, trypsinize a few wells and perform cell counts to determine the number of cells per well. That way, it will be easier to calculate the amount of virus needed for the infection in subsequent assays as the number of cells per well will be standardized. BSC-40 or BS-C-1 cells are also appropriate for the methods described, but the seeding density and growth conditions will need to be adjusted.
2. It is important that an unimpeded infection results in ~90% destruction of the monolayer in 72 h. If the monolayer is only partially destroyed, then there will be a very narrow range of spectrophotometric values between a fully preserved healthy monolayer (0% CPE) and a monolayer in which the virus was not inhibited at all (100% CPE). Various strains of VACV replicate with differing efficiencies, which will have to be determined empirically prior to utilizing the assay. In the methods presented above, the target MOI for the infection is 0.05–0.1 PFU/cell, which typically ranges in concentration from 1×10^4 to 2×10^4 PFU/mL if inoculating individual wells with 50 μL of virus. To optimize the MOI, it is suggested that MOIs ranging from 0.01 to 0.2 be explored. Use uninfected control wells to set the baseline for intact healthy monolayers and use the higher MOI infections (0.2 PFU/cell) for complete destruction of the monolayer. Intermediate destruction of monolayers may be determined spectrophotometrically.

3. DMSO is an excellent solvent and most hydrophobic and hydrophilic compounds will dissolve in undiluted DMSO. Once a compound in DMSO is added to cell culture medium, the DMSO is diluted and compound solubility may change. Note whether precipitates form upon addition to cell culture. If precipitates form, do not increase the concentration of DMSO in the medium to try to solubilize or maintain the solubility of the compound. DMSO at a concentration of >0.5% is cytotoxic. This will confound the results of the CPE and EC_{50} assays, as they are dependent on the level of CPE caused by the virus and its inhibition by the compounds of interest. If the compound at the desired concentration is not soluble in cell culture medium with 0.5% DMSO another solvent may be required.

References

1. Fenner F, Henderson DA, Arita I, Jezek Z, Ladnyi ID (1988) Smallpox and its eradication. WHO, Geneva
2. Smith GL, Vanderplasschen A, Law M (2002) The formation and function of extracellular enveloped vaccinia virus. J Gen Virol 83(Pt 12):2915–2931
3. Payne LG (1980) Significance of extracellular enveloped virus in the in vitro and in vivo dissemination of vaccinia. J Gen Virol 50(1):89–100
4. Henderson DA (1999) The looming threat of bioterrorism. Science 283(5406):1279–1282
5. Lewis-Jones S (2004) Zoonotic poxvirus infections in humans. Curr Opin Infect Dis 17(2):81–89
6. Chapman JL, Nichols DK, Martinez MJ, Raymond JW (2010) Animal models of orthopoxvirus infection. Vet Pathol 47(5):852–870
7. Yang G, Pevear DC, Davies MH, Collett MS, Bailey T, Rippen S, Barone L, Burns C, Rhodes G, Tohan S, Huggins JW, Baker RO, Buller RL, Touchette E, Waller K, Schriewer J, Neyts J, DeClercq E, Jones K, Hruby D, Jordan R (2005) An orally bioavailable antipoxvirus compound (ST-246) inhibits extracellular virus formation and protects mice from lethal orthopoxvirus challenge. J Virol 79(20):13,139–13,149

Chapter 10

Vaccinia Virus Transcriptome Analysis by RNA Sequencing

Shuai Cao, Yongquan Lin, and Zhilong Yang

Abstract

RNA-sequencing (RNA-Seq) using next-generation sequencing (NGS) technique is a powerful tool for simultaneous analysis of global transcripts from both vaccinia virus and host cell. Here, we describe an RNA-Seq method for analyzing the vaccinia virus transcriptome from virus-infected HeLa cells. We pay particular attention to vaccinia virus-specific aspects of sample preparation, sequencing, and data analyses, but our method could be modified to analyze transcriptomes of other cells or tissues infected with different poxviruses.

Key words RNA-Seq, Transcriptome, Vaccinia virus, Poxvirus, Gene expression

1 Introduction

A transcriptome comprises all RNA species in a given cell, a population of cells, a tissue, or an organism. The transcriptome tells which genes are transcribed and their quantities—information critical to understanding how a genome's information is globally expressed. RNA-Sequencing (RNA-Seq) has become the standard method for obtaining transcriptome information. Compared to microarrays, which were widely used before the deep-sequencing era, RNA-Seq provides many advantages: low or no background noise; resolution to the single-nucleotide level; digital quantitation of transcripts; a broader range of differential expression; and greater sensitivity for detecting transcripts expressed at low levels [1]. RNA-Seq also provides a unique advantage in that it can simultaneously obtain both pathogen and host-cell transcriptomes in analyses of intracellular pathogen/host interactions (e.g., virus infection) [2].

As a group of large DNA viruses, poxviruses encode hundreds of open reading frames (ORFs). These ORFs are expressed in a three-stage cascade, with early-, intermediate-, and late-stage transcripts [3]. Transcriptome analyses were essential in classifying the genome-wide temporal expression of the genes of vaccinia virus (VACV, the

Jason Mercer (ed.), *Vaccinia Virus: Methods and Protocols*, Methods in Molecular Biology, vol. 2023,
https://doi.org/10.1007/978-1-4939-9593-6_10,

prototypic member of poxvirus) into these three stages [2, 4–8]. Since poxviruses are increasingly being used as vaccine vectors and antitumor oncolytic agents [9–14], transcriptome analysis can be a great tool for examining the global expression of viral and inserted foreign genes in both cultured cells and tissues of infected hosts. More importantly, transcriptome analysis also reflects changes in endogenous expression of host genes, providing an excellent opportunity to simultaneously analyze host responses to the infection. This is assuming that proper controls and biological replicates are included in the experimental design. Similarly, transcriptome analysis can provide both viral and host gene expression information to interrogate various aspects of poxvirus replication mechanisms.

Multiple NGS platforms can be used for RNA-Seq, and many pipelines are commercially or freely available for analyzing RNA-Seq data. Herein, we describe a workflow to analyze total mRNAs in VACV-infected HeLa cells based on the Illumina platform, and outline the workflow we commonly use to analyze sequencing reads. The whole procedure starts with cell culture and VACV infection, followed by mRNA extraction, library construction, quality assessment, sequencing, and data analysis (Fig. 1). It should be noted that many steps are common to generic RNA-Seq analysis and plenty of information can be found elsewhere. We draw particular attention to those aspects specific to VACV and describe them in greater detail. For those steps common to generic RNA-Seq, we provide fewer details or only outline the procedure.

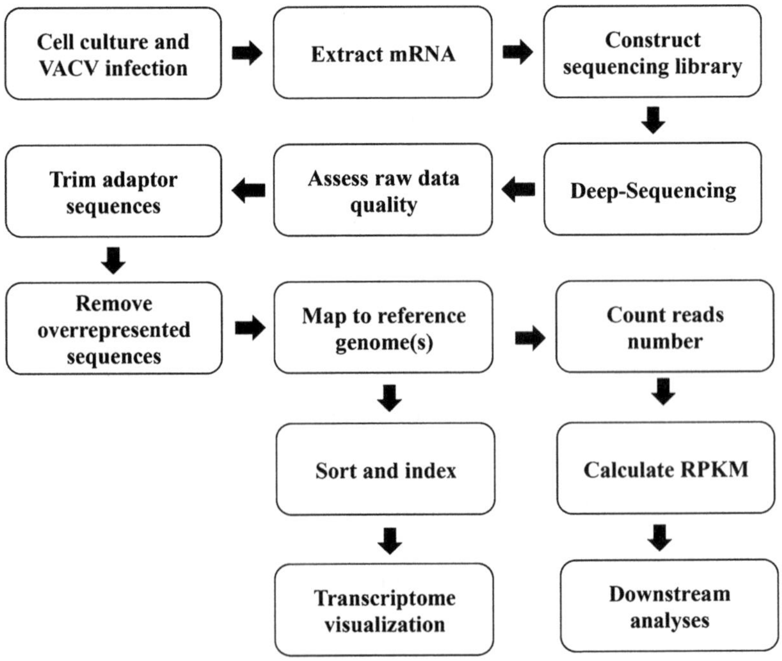

Fig. 1 Workflow of VACV transcriptome analysis by RNA-Seq

2 Materials

2.1 Cell Culture

1. HeLa cells (ATCC CCL-2).
2. Dulbecco's modified Eagle medium (DMEM) with high glucose, containing 10% fetal bovine serum (FBS), 2 mM L-glutamine, 100 U/mL penicillin, and 100 μg/mL streptomycin.
3. Phosphate-buffered saline (PBS).
4. Trypsin/EDTA.
5. T-175 cell culture flasks, T-25 cell culture flasks (or other plates/dishes with a similar surface area).

2.2 VACV Infection

1. Purified VACV Western Reserve (VACV-WR; ATCC).
2. VACV infection medium: DMEM (2.5% FBS, 2 mM L-glutamine, 100 U/mL penicillin, and 100 μg/mL streptomycin). For cell culture and VACV preparation, purification, and titration, please see methods described elsewhere [15].

2.3 Extraction of Total mRNA

1. Dynabeads mRNA DIRECTTM Kit, Thermo Fisher Scientific, Cat. # 61011.
2. DynaMag-2 Magnetic rack, Life Technologies, Cat. # 12321D.
3. Nuclease-free water.

2.4 Sequencing Library Construction and Purification (See Note 1)

1. NEBNext Ultra Directional RNA Library Prep Kit for Illumina, NEW ENGLAND BioLabs Inc. Cat. # E7420S.
2. Magnetic rack, nuclease-free water.
3. Agencourt AMPure® XP Beads, Beckman Coulter, Cat. #A63880.
4. Actinomycin D (0.1 μg/μL).

3 Methods

3.1 Prepare HeLa Cells for VACV Infection

1. HeLa cells are cultured and maintained in T-175 flasks in DMEM (containing 10% FBS, 2 mM L-glutamine, 100 U/mL penicillin, and 100 μg/mL streptomycin) at 37 °C in 5% CO_2.
2. When cells are confluent, decant culture medium and wash monolayer surface with 5 mL of 1×PBS pre-warmed at 37 °C.
3. Decant PBS and add 3–5 mL of trypsin/EDTA to flask. Gently shake flask to let trypsin/EDTA completely cover cell monolayer surface. HeLa cells will be detached from flask within 1–2 min at 37 °C.

4. After cells detach from flask, add 5 mL of DMEM (10% FBS) to neutralize trypsin activity. Separate HeLa cells to single cells by continuously and gently pipetting up and down.
5. Transfer cell-containing solution to a sterilized 15 mL tube. Collect cell pellet by centrifuging at 1000 × *g* for 5 min.
6. Remove supernatant from tube. Use 1 mL of fresh DMEM (10% FBS) to resuspend cells. Use a hemocytometer (or another device) to count cells.
7. Seed HeLa cells into T-25 flask (or another dish with a similar surface area) with 1.0×10^6 cells/flask. Continue to culture the cells at 37 °C in 5% CO_2. Normally HeLa cells will grow to 95% confluence and be ready for VACV infection within 24 h.

3.2 VACV Infection

1. HeLa cells cultured in a T-25 flask are infected with VACV at a MOI of 5, in 2 mL of VACV infection medium per well, and incubated at 37 °C in 5% CO_2 for 1 h. Gently shake flasks 1–2 times during incubation period (*see* **Note 2**).
2. After 1 h of incubation, replace VACV-containing medium with 5 mL of fresh VACV infection medium. Continue to incubate plates at 37 °C in 5% CO_2 for the desired time post-infection until mRNA extraction step (*see* **Note 3**).

3.3 Extraction and Purification of mRNA (Based on Dynabeads mRNA Direct Kit; See Note 4)

It is important to maintain an RNase-free environment in Subheadings 3.3 and 3.4 (*see* **Note 5**).

1. Pellet ~3.0×10^6 HeLa cells by trypsinization and centrifugation at 1000 × *g* for 5 min.
2. Wash cell pellets with 1 mL cold PBS and centrifuge cells again to collect cells.
3. Add 1.25 mL of lysis/binding buffer to each cell pellet and pipette up and down several times on ice to lyse cells completely.
4. Transfer 250 μL of Dynabeads Oligo $(dT)_{25}$ to an RNase-free 1.5 mL tube and place the tube on a magnetic rack.
5. Remove supernatant when clear and wash beads with 250 μL of lysis/binding buffer. Remove lysis/binding buffer when clear.
6. Incubate beads and cell lysates at room temperature for 3–5 min.
7. Put tubes on a magnetic rack and remove supernatant when clear.
8. Wash beads with 1 mL washing buffer A.
9. Wash beads with 1 mL washing buffer B.
10. Air-dry beads for 5 min (*see* **Note 6**).

11. Elute total mRNA from beads with 25 μL of nuclease-free water.
12. Measure quality and concentration of eluted mRNA (*see* **Note 7**).
13. Proceed to the next step or store mRNA at −80 °C before sequencing library construction.

3.4 Sequencing Library Construction and Purification (Protocol Based on the Commercially Available NEBNext Ultra Directional RNA Library Prep Kit for Illumina)

3.4.1 RNA Fragmentation and Priming (See Note 8)

1. Mix 10–100 ng of purified mRNA with random primers in first-strand synthesis buffer.
2. Incubate the mixture at 94 °C for 7–15 min to randomly break down mRNAs into smaller RNA fragments. Keep mixture on ice until moving to the next step. The longer the incubation time, the shorter the fragments generated will be (*see* **Note 9**).

3.4.2 First-Strand cDNA Synthesis

1. Add 1 μL of ProtoScript II Reverse Transcriptase to 10 μL of the RNA fragment mixture (from the previous step) and random primers, to synthesize cDNA (namely first-strand cDNA).
2. Add 0.5 μL of RNase inhibitors and 5 μL of actinomycin D to prevent RNA degradation and to reduce spurious antisense synthesis [16].
3. Use nuclease-free water to bring final volume to 20 μL.
4. For first-strand cDNA synthesis, run thermal cycler according to the following program: 25 °C for 10 min, 42 °C for 15 min, 70 °C for 15 min, 4 °C hold.

3.4.3 Second-Strand cDNA Synthesis

1. Add 8 μL of 10× second-strand synthesis reaction buffer, 4 μL of second-strand synthesis enzyme mix, and 48 μL of nuclease-free water to the 20 μL of first-strand cDNA synthesis product (from the previous step) to bring total volume to 80 μL.
2. Incubate mixture at 16 °C for 1 h (*see* **Note 10**).

3.4.4 Double-Stranded cDNA Purification (Based on the Agencourt® AMPure® XP Beads, from Beckman Coulter)

1. Add 144 μL (1.8×) of AMPure XP beads to 80 μL of the dscDNA product (from the previous step) and mix well by pipetting up and down at least ten times (*see* **Note 11**).
2. Incubate at room temperature for 5 min to let the dscDNAs attach to beads.
3. Place tube on a magnetic rack to separate beads from supernatant.
4. Carefully remove supernatant after it clears.

5. Use 200 μL of 80% ethanol to wash surface of beads twice, while the tube is still on the magnetic rack.
6. Air-dry beads for 5 min.
7. Add 60 μL of 10 mM Tris–HCl to resuspend beads and elute dscDNAs.
8. Place the tube back on magnetic rack.
9. Transfer 55.5 μL of dscDNA-containing supernatant into a new tube when the solution clears. Proceed to the next step or store dscDNA at −80 °C before moving on.

3.4.5 Adenylate 3′ Ends of the dscDNA

1. Add 6.5 μL of 10× NEBNext End Repair Reaction Buffer and 3 μL of NEBNext End Prep Enzyme Mix to 55.5 μL of purified dscDNA from the last step to bring total volume to 65 μL.
2. Incubate mixture in a thermal cycler as follows: 20 °C for 30 min, 65 °C for 30 min, and hold at 4 °C until next step. At this step, adenylates will be added to the three terminals of the blunt-ended dscDNA.

3.4.6 Adaptor Ligation

1. Add 15 μL of Ligase Master Mix, 1 μL of NEBNext Adaptor (1.5 μM), and 2.5 μL of nuclease-free water to 65 μL of adenylated dscDNA (from the last step). The total volume will be 83.5 μL.
2. Incubate at 20 °C for 15 min. At this step, adenylated dscDNA prepared from different samples can be ligated to adaptors with different sequences, which act as barcodes to identify the origin of the sequence reads when analyzing multiple samples in parallel.

3.4.7 Purification dscDNA with AMPure XP Beads After Adaptor Ligation

1. Add 100 μL (1.0×) of AMPure XP beads to purify dscDNA as described in Subheading 3.4.4.
2. Elute dscDNA from beads with 100 μL of 10 mM Tris–HCl.
3. Repeat Subheading 3.4.7 to purify the ligation product again; elute final purified ligation product with 19 μL of 10 mM Tris–HCl.

3.4.8 Library Enrichment by PCR

1. Transfer 17 μL of purified product from last step to a new tube.
2. Add 3 μL of NEBNext USER Enzyme, 25 μL of NEBNext Q5 Hot Start HiFi PCR Master Mix, 2.5 μL of Index (X) Primer, and 2.5 μL of Universal PCR Primer.
3. Run PCR program as follows: 37 °C for 15 min; 98 °C for 30 s; 98 °C for 10 s; 65 °C for 75 s, cycles 12–15 times; 65 °C for 5 min; hold at 4 °C until next step (*see* **Note 12**).

3.4.9 Purify PCR Product with AMPure XP Beads

1. Add 45 μL (0.9×) of AMPure XP beads to 50 μL of the PCR-enriched cDNA library for purification, as described in Subheading 3.4.4.
2. Elute dscDNA from beads in 20 μL of 10 mM Tris–HCl and store at −80 °C until the sequencing step.

3.5 Sequencing

We do not discuss this step in detail, since sequencing is usually carried out in core or commercial facilities. Work with your facility to assess the sequencing library quality and perform sequencing using an Illumina platform, for example, Hi-Seq 4000 or 2500.

3.6 Data Analysis

The raw data of RNA-Seq will be in one or several files, in a fastq format, a text-based format for high-throughput biological sequencing data. The raw data files can be downloaded from the server of a sequencing facility to a hard drive of your own computer/cluster for analysis. By mapping sequencing reads to a reference genome of interest, the relative abundance of mRNAs can be shown in the form of reads per kilobase of transcript per million mapped reads (RPKM) for single-end sequencing or fragments per kilobase of transcript per million mapped reads (FPKM) for paired-end sequencing. In addition to information generated from mRNAs, the raw data also contains some non-mRNA-related sequences that result from the library construction process, such as sequences generated from rRNAs, tRNAs, and cDNA library adaptors; these can complicate the data analysis. In addition to adaptors, rRNA and tRNA sequences are sometimes overrepresented and should be removed from raw data before mapping to a reference genome. After the mapping step, data interpretation may need further processing, including counting read numbers, calculating RPKM, finding differentially expressed genes, and visualization of transcriptome on a browser. Here, we provide a pipeline for analyzing RNA-Seq data focusing on viral mRNAs of VACV infection (Fig.1). We describe the analysis tools but do not describe scripts, as they may be updated with new versions of software and may be slightly different when using different computer systems. Plus, different parameters may be used in different experiments. In general, these are described in corresponding software manuals. For those unfamiliar with computer programming languages and environments, please consult a bioinformatician. Alternatively, the Galaxy, an online data analysis website, incorporates many useful tools for each step of RNA-Seq data analysis on a friendly interface (*see* Subheading 3.7).

3.6.1 Software, Tools, and Reference Sequences (See Note 13)

FastQC, Trimmomatic, Bowtie, Tophat, FeatureCounts, Excel, Mochiview, SAMtools, VACV genome annotation file (NC_006998.1), Bowtie index files containing rRNA and tRNA sequences.

3.6.2 Download RNA-Seq Data from Server of Sequencing Agency

Raw sequencing data are in a fastq format. Save data on the hard drive of a local computer.

3.6.3 Quality Assessment of Raw Data Using FastQC

The general quality of raw data (in a fastq format) can be assessed in FastQC, which has both command-line and GUI versions. For example, the lengths and quality scores of reads, presence of adaptor sequences, and overrepresented rRNA and tRNA sequences can be assessed in this step. FastQC takes .fastq files as input. The analysis results are presented in an .html file. The presence of adaptor sequences and overrepresented rRNA and tRNA sequences (if any) can be observed in the results. Usually, no further action is needed at this step unless read lengths are unexpectedly short or read quality is low. FastQC is more often used as a tool to show overrepresented sequences or residual adaptor sequences (see later steps).

3.6.4 Trim Adaptor Sequences Using Trimmomatic

Raw data will contain many adaptor sequences that were generated when the sequencing library was constructed. Trimmomatic removes them from the RNA-Seq data because they contain no real information. The software uses a JAVA environment to read .fastq or.fastq.gz files and trim the adaptor sequence, save clean reads to a new .fastq file, and give a summary. Most of the common adaptor sequences are packaged in the software.

3.6.5 Remove Overrepresented rRNA and tRNA Sequences Using Bowtie

The amounts of overrepresented rRNA and tRNA sequences in the RNA-Seq data can vary. The most abundant overrepresented sequences are usually from rRNAs. To remove those sequences, Bowtie can be used to map the total sequencing reads to a Bowtie index containing rRNA and tRNA sequences. The sequences that do not match to rRNA and tRNA sequences will be saved as a new file. The input files include a .fastq file and reference index files. The output files include a .fastq file that stores the sequences not mapped to rRNA or tRNA for further analysis and an .aln file that stores rRNA and tRNA matching sequences.

Bowtie uses index files as a reference but does not use a sequence file directly. A Bowtie index file can be generated using Bowtie-build. A reference sequence file is usually used as input. The output files will be stored in a folder that contains four files with the name of *.ebwt, and two files with the name of *.rev.*.ebwt.

3.6.6 Mapping to the VACV Genome Using Tophat

Tophat can work with a large genome and complex annotations (as with the human genome data). It can process junctions better than Bowtie. The input file is a .fastq file generated from Subheading 3.6.5, mapping the reads to VACV genome reference files (.gtf and index files for VACV). The output data will be stored in a .bam file.

3.6.7 Read Counts and Calculation of RPKM Using Featurecounts

This step counts the number of reads mapped to individual VACV ORFs. Featurecounts can be used either as software or as a package in R. It uses the .bam file from the last step as input and a .gtf file (with VACV genome annotations) as a reference. Since the VACV genome has inverted terminal repeats (ITRs), one read must be allowed to be mapped to at least two genomic locations so the reads of ORFs in the ITRs can be mapped. The output is a .txt file containing several columns of information including gene name, gene length, number of reads mapped to it, etc. The .txt file can be opened in Excel and converted to an Excel file. By definition, RPKM can be calculated with the following equation: RPKM of gene A = number of reads mapped to gene A × (length of gene A/1000) × (number of total reads/1,000,000).

3.6.8 Transcriptome Visualization Using Mochiview

The transcriptome profiling result can be visualized with Mochiview or other genome browsers. As both the positive and negative strands of the VACV genome encode viral genes, the reads must be separated according to the positive or negative strand after mapping. SAMtools can achieve this. SAMtools can also sort and index the mapping results. The .bam files will then be converted into .wig files. The .wig files can be opened in Mochiview and the distribution and abundance of reads on the VACV genome visualized. Mochiview also takes multiple other formats and can display several features. Figure 2 is an example of our results analyzing the genome-wide VACV transcriptome at the late infection stage.

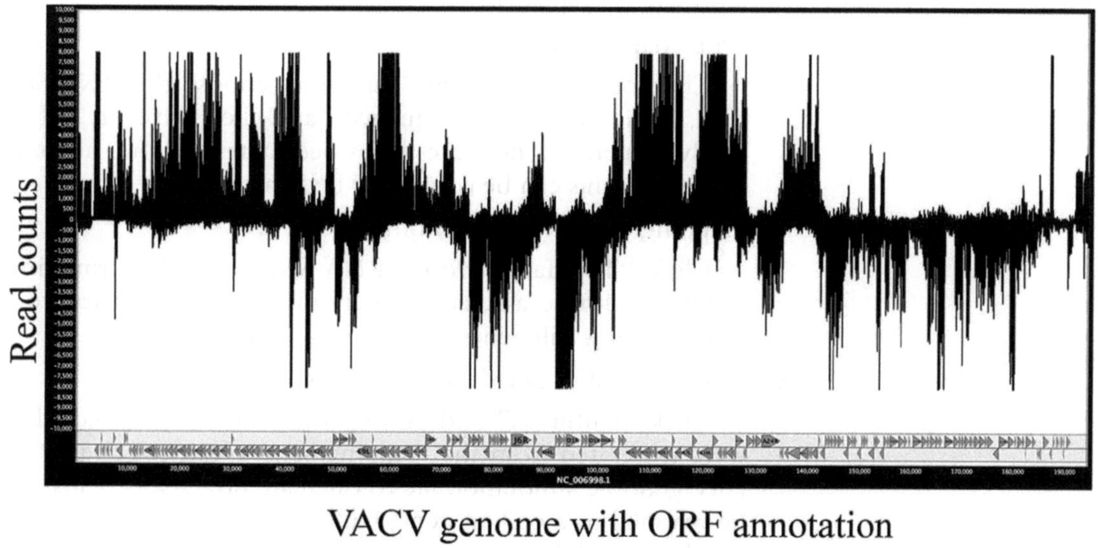

Fig. 2 VACV transcriptome visualization in Mochiview. A genome-wide transcriptome at late stage of infection is shown. The number of reads per nucleotide was displayed over the VACV genome with ORF annotation. The reads above the line map to the upper (rightward) strand and counts below the line map to the lower (leftward) strand of the DNA genome. The highest read counts are off-scale due to high read counts

3.6.9 Host-Cell Transcriptome Analysis

The host-cell transcriptome can be analyzed simultaneously. If experiments are designed with proper controls and biological replicates, host-cell response to VACV infection, under different conditions, can be studied by differential gene expression analysis (*see* **Note 14**). The overall procedure is similar to the analysis of VACV reads, but uses the host genome annotation files, accordingly.

3.7 Analyze RNA-Seq Data with Galaxy

RNA-Seq data can also be analyzed using Galaxy: www.usegalaxy.org. We briefly describe the procedure below.

1. Download RNA-Seq data from a sequencing server (in a fastq format).
2. Upload RNA-Seq data to the Galaxy website.

   Click on "Get Data" and upload the RNA-Seq data in a fastq format to your Galaxy user account. Your uploaded files will be displayed on the right side bar of the web page, under the history column. Many tools are available on the Galaxy website for RNA-Seq data analysis, e.g., FastQC, Cutadapt, Tophat, and Cufflinks. The Galaxy website has very detailed explanations of each of those functions.
3. Quality control of raw data.

   The general quality of RNA-Seq data can be visualized in FastQC. Click "C" on "FastQC," choose the data files for analysis, and click "Execute." The analysis results will show the length and quality of reads, adaptor sequences, and over-represented sequences.
4. Trim adaptor sequences.

   Click on the "Cutadapt" function, choose the RNA-Seq data, input the adaptor sequences, and click "Execute." The newly generated files are RNA-Seq data without adaptor sequences; this can be confirmed by "FastQC."
5. Mapping to annotated genomes.

   Click "TopHat," choose RNA-Seq data after trimming adaptors, choose a reference genome, and click "Execute." The mapping will generate a file in a .bam format.
6. Read count and calculation of RPKM.

   Click "Cufflinks" and choose the .bam format files that were generated from the mapping step. Next, select a reference genome annotation file to calculate how many reads can be mapped to a certain gene. Cufflinks can also calculate the Reads per kilobase of transcript per million mapped reads (RPKMs), to estimate the relative abundance of mRNAs and test for significantly differential gene expression in RNA-Seq samples.

4 Notes

1. Other protocols and commercially available kits are available for sequencing library construction. It is important to choose one that maintains strand-specific information as the VACV ORFs are very closely spaced and transcription read-throughs are extensive, such that almost every nucleotide is transcribed [2, 7].
2. A MOI of 5 ensures that almost all cells are infected synchronously. If using suspension HeLa S3 cells, a MOI of 20 is desired.
3. Each sample will need ~100 ng mRNA. Sufficient mRNAs can be extracted from 3.0×10^6 HeLa cells (cells from one T-25 flask or from three wells of a 6-well plate). If using other cell types, determine the number of cells needed based on how much mRNA can be extracted.
4. A Dynabeads mRNA Direct kit can extract mRNAs from samples, using magnetic beads coated with oligo(dT). If other types of RNA are desired, alternative methods of RNA extraction should be used, for example, Trizol-based methods to extract total RNA.
5. The quality of extracted RNA is crucial for RNA-Seq. The working area should be clean enough to handle RNA. Dust and aerosols should be removed as much as possible from the working surface during RNA extraction and sequencing library construction steps. All pipettes, tubes, and water used for RNA extraction and sequencing library construction should be nuclease free. Keep samples on ice if possible. RNA quality can be determined by measuring RNA concentration with Nanodrop 2000 or Bioanalyzer. A260/280 and A260/230 should be around 2.0 and 2.2. A low reading at A260/280 usually indicates contamination with proteins or phenols, while low A260/230 indicates contamination with guanidine salts, which may affect construction of cDNA.
6. Do not let magnetic beads dry completely when air-drying the beads. Find the point where the color of beads changes from "wet-brown" to "semi-wet-brown." At this point, the beads are about to dry and are ready to move to the next step. Completely dried beads may impair elution and production of mRNA or cDNA.
7. Good-quality RNA can be indicated by absorption ratio at 260 nm/280 nm (A260/280) around 2.0 and 260 nm/230 nm (A260/230) around 2.2.
8. If an additional RNA spike-in is desired (used for calibrating measurements of RNA-Seq reads in applications such as com-

paring reads among different samples), it should be added before this step.

9. RNA can be fragmented in multiple ways, for example physical shearing with sonication, or RNase III digestion.
10. The RNA template strand used for first-strand cDNA synthesis in last step will be degraded by RNase H, to produce RNA fragments for use as primers for DNA polymerase I to synthesize Okazaki fragments of the second cDNA strand. The Okazaki fragments will be ligated into the complete second-strand cDNA by DNA ligase and the first and second cDNA strands will form double-stranded cDNA (dscDNA). Uracil (U) is used instead of thymine (T) for synthesis of the second-strand cDNA, to ensure that for each adaptor barcode only the first-strand cDNA will be amplified and sequenced.
11. The AMPure XP bead ratio is defined as the volume of resuspended beads (in the commercial product) versus the volume of cDNA samples. For example, "1×" means equal volumes of the resuspended beads and cDNA sample.
12. In the cDNA library enrichment step, the recommended number of PCR cycles ranges from 12 to 15. Minimize the PCR cycles if yield is enough to produce an adequate amount for DNA sequencing. Fewer PCR cycles reduce artifacts generated by PCR amplification.
13. Sources of software and other index or annotation files:

    FastQC can be downloaded from

    http://www.bioinformatics.babraham.ac.uk/projects/download.html#fastqc

    Trimmomatic can be downloaded from [17]

    http://www.usadellab.org/cms/uploads/supplementary/Trimmomatic/Trimmomatic-0.36.zip

    Bowtie can be downloaded from [18]

    https://sourceforge.net/projects/bowtie-bio/files/bowtie/1.2.1.1

    Tophat can be downloaded from [19]

    https://ccb.jhu.edu/software/tophat/tutorial.shtml

    Tophat depends on Bowtie; please install it per the website instruction.

    Featurecounts can be downloaded from [20]

    http://sourceforge.net/projects/subread/files/subread-1.5.3/

    Mochiview can be downloaded from [21]

    http://www.johnsonlab.ucsf.edu/mochiview-downloads

SAMtools can be downloaded from [22]

https://sourceforge.net/projects/samtools/files/

The computer environment may ask JAVA, Perl, or Python.

JAVA: https://www.java.com/en/download/win10.jsp

Perl: https://www.perl.org/get.html

Python: https://www.python.org/

VACV genome sequence file can be downloaded from

https://www.ncbi.nlm.nih.gov/nuccore/NC_006998.1?report fasta

VACV genome annotation files can be downloaded from

ftp://130.14.250.10/genomes/Viruses/Vaccinia_virus_uid15241/NC_006998.gff

The gff file was converted to gtf file using the script in

https://github.com/vipints/GFFtools-GX/blob/master/gff_to_gtf.py

Human genome sequence files can be downloaded from [23]

ftp://ftp.ensembl.org/pub/release-96/fasta/homo_sapiens/dna/

Human genome annotation files can be downloaded from [23]

ftp://ftp.ensembl.org/pub/release-96/gtf/homo_sapiens/

Predefined rRNA and tRNA sequence files are generated by Bowtie

tRNA sequence: http://gtrnadb.ucsc.edu/genomes/eukaryota/Hsapi19/hg19-tRNAs.fa

Human ribosomal DNA sequence: https://www.ncbi.nlm.nih.gov/nuccore/555853/

Human 5S DNA sequence: https://www.ncbi.nlm.nih.gov/nuccore/23898/

14. Differential expression can be analyzed with multiple software programs, for example, Cuffdiff in Cufflinks (as mentioned in data analysis with Galaxy) [24]. http://cole-trapnell-lab.github.io/cufflinks/releases/v2.2.1/

References

1. Wang Z, Gerstein M, Snyder M (2009) RNA-Seq: a revolutionary tool for transcriptomics. Nat Rev Genet 10:57–63
2. Yang Z, Bruno DP, Martens CA, Porcella SF, Moss B (2010) Simultaneous high-resolution analysis of vaccinia virus and host cell transcriptomes by deep RNA sequencing. Proc Natl Acad Sci U S A 107:11,513–11,518
3. Moss B (2013) Poxviridae: the viruses and their replication. Knipe DM, Howley PM (eds) Fields Virology 2:2129–2159
4. Yang Z, Moss B (2015) Decoding poxvirus genome. Oncotarget 6:28,513–28,514
5. Yang Z, Martens CA, Bruno DP, Porcella SF, Moss B (2012) Pervasive initiation and 3′-end formation of poxvirus postreplicative RNAs. J Biol Chem 287:31,050–31,060

6. Yang Z, Maruri-Avidal L, Sisler J, Stuart CA, Moss B (2013) Cascade regulation of vaccinia virus gene expression is modulated by multistage promoters. Virology 447:213–220
7. Yang Z, Reynolds SE, Martens CA, Bruno DP, Porcella SF, Moss B (2011) Expression profiling of the intermediate and late stages of poxvirus replication. J Virol 85:9899–9908
8. Bengali Z, Satheshkumar PS, Yang Z, Weisberg AS, Paran N, Moss B (2011) Drosophila S2 cells are non-permissive for vaccinia virus DNA replication following entry via low pH-dependent endocytosis and early transcription. PLoS One 6:e17248
9. Albelda SM, Thorne SH (2014) Giving oncolytic vaccinia virus more BiTE. Mol Ther 22:6–8
10. Chan WM, McFadden G (2014) Oncolytic poxviruses. Annu Rev Virol 1(1):191–214
11. Draper SJ, Heeney JL (2010) Viruses as vaccine vectors for infectious diseases and cancer. Nat Rev Microbiol 8:62–73
12. Altenburg AF, Kreijtz JH, de Vries RD, Song F, Fux R, Rimmelzwaan GF, Sutter G, Volz A (2014) Modified vaccinia virus Ankara (MVA) as production platform for vaccines against influenza and other viral respiratory diseases. Viruses 6:2735–2761
13. Izzi V, Buler M, Masuelli L, Giganti MG, Modesti A, Bei R (2014) Poxvirus-based vaccines for cancer immunotherapy: new insights from combined cytokines/co-stimulatory molecules delivery and "uncommon" strains. Anti Cancer Agents Med Chem 14:183–189
14. Moss B (2013) Reflections on the early development of poxvirus vectors. Vaccine 31:4220–4222
15. Earl PL, Cooper N, Wyatt LS, Moss B, Carroll MW 2001 Preparation of cell cultures and vaccinia virus stocks. Curr Protoc Mol Biol Chapter 16:Unit16 16
16. Head SR, Komori HK, LaMere SA, Whisenant T, Van Nieuwerburgh F, Salomon DR et al (2014) Library construction for next-generation sequencing: overviews and challenges. BioTechniques 56:61–64. 66, 68, passim
17. Bolger AM, Lohse M, Usadel B (2014) Trimmomatic: a flexible trimmer for Illumina sequence data. Bioinformatics 30:2114–2120
18. Langmead B, Trapnell C, Pop M, Salzberg SL (2009) Ultrafast and memory-efficient alignment of short DNA sequences to the human genome. Genome Biol 10:R25
19. Trapnell C, Pachter L, Salzberg SL (2009) TopHat: discovering splice junctions with RNA-Seq. Bioinformatics 25:1105–1111
20. Liao Y, Smyth GK, Shi W (2014) featureCounts: an efficient general purpose program for assigning sequence reads to genomic features. Bioinformatics 30:923–930
21. Homann OR, Johnson AD (2010) MochiView: versatile software for genome browsing and DNA motif analysis. BMC Biol 8:49
22. Li H, Handsaker B, Wysoker A, Fennell T, Ruan J, Homer N, Marth G, Abecasis G, Durbin R, 1000 Genome Project Data Processing Subgroup (2009) The sequence alignment/map format and SAMtools. Bioinformatics 25:2078–2079
23. Yates A, Akanni W, Amode MR, Barrell D, Billis K, Carvalho-Silva D et al (2016) Ensembl 2016. Nucleic Acids Res 44:D710–D716
24. Trapnell C, Roberts A, Goff L, Pertea G, Kim D, Kelley DR et al (2012) Differential gene and transcript expression analysis of RNA-seq experiments with TopHat and Cufflinks. Nat Protoc 7:562–578

Chapter 11

Ribosome Profiling of Vaccinia Virus-Infected Cells

Yongquan Lin, Wentao Qiao, and Zhilong Yang

Abstract

Ribosome profiling is a method that determines genome-wide mRNA translation through measuring ribosome-protected mRNA fragments by deep sequencing. This method can be used to quantify gene expression at the translational level and precisely pinpoint ribosome loading onto mRNA with codon-level resolution. Genome-wide regulation of mRNA translation can also be determined if RNA-Sequencing (RNA-Seq) is carried out in parallel. Here, we describe a protocol for simultaneously performing ribosome profiling and RNA-Seq in cells infected with vaccinia virus.

Key words Poxvirus, Vaccinia virus, Translation, Gene expression, Ribosome profiling, RNA-Seq

1 Introduction

Ribosomes are a type of cellular machinery that decodes the genetic information of mRNA into protein. Once bound to mRNA, one ribosome molecule can protect a ~28–30 nucleotide (nt) mRNA sequence from nuclease digestion [1]. To measure ribosome-protected mRNA fragments, ribosome profiling, a deep sequencing method, was developed by Weissman and colleagues [1, 2]. In comparison to RNA-Sequencing (RNA-Seq), which quantifies total mRNA levels, ribosome profiling measures gene expression undergoing active translation, which correlates more closely with protein levels [1, 2]. Ribosome profiling also precisely determines ribosome-binding sites on mRNA with codon-level resolution, identifying actively translated mRNA and implicating sequences as translational regulatory elements [1–3]. A comparative analysis combining RNA-Seq with ribosome profiling allows us to calculate the ratio of ribosome-protected mRNA to total mRNA, and for any individual mRNA its relative translational efficiency [4–7]. The vast amount of information generated by ribosome profiling provides a unique advantage when studying the gene expression of poxviruses, including the vaccinia virus (VACV, the prototype poxvirus). For VACV, the open reading frames (ORFs) are closely

Jason Mercer (ed.), *Vaccinia Virus: Methods and Protocols*, Methods in Molecular Biology, vol. 2023,
https://doi.org/10.1007/978-1-4939-9593-6_11,

spaced and post-replicative mRNAs often contain extensive read-throughs, so by determining which regions in the VACV are translated we dramatically improved our ability to decode the entire VACV genome [5, 8, 9]. Additionally, by comparing cellular ribosome profiling and RNA-Seq data in VACV-infected and mock-infected cells, the effect of VACV infection on host cell mRNA translation can be revealed [5].

This chapter presents a protocol for performing ribosome profiling and RNA-Seq in parallel, in VACV-infected cells (Fig. 1). We based the protocol on the original method developed by the Weissman Laboratory [2] and, for many of the steps, suggest kits when they are available. The overall procedure includes cell culture and infection, library construction, deep sequencing, and data analysis. Similar to other high-throughput approaches, the results should be validated by corresponding conventional methods before going forward with further investigation.

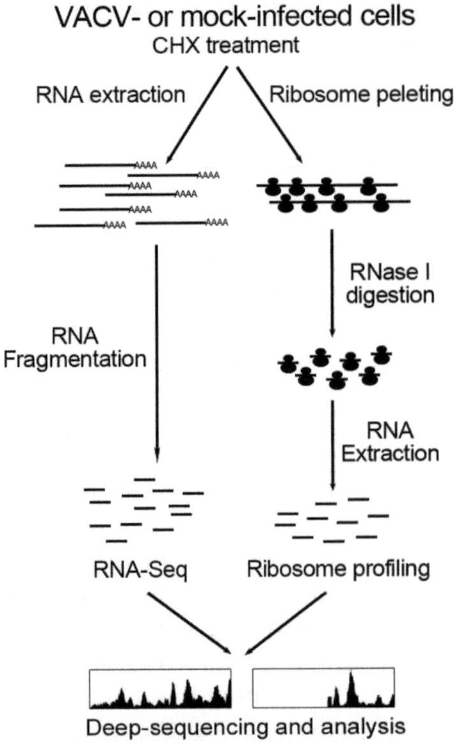

Fig. 1 Schematic of simultaneous ribosome profiling and RNA-Seq. HeLa S3 cells were infected with VACV or mock infected. The cells were treated with cycloheximide and harvested at indicated time postinfection. For RNA-Seq, mRNAs were purified and fragmented. For ribosome profiling, mRNAs were treated with RNase, and ribosome-protected fragments (RPFs) were purified. Both mRNA and RPFs were used for library construction, sequencing, and analysis.

2 Materials

2.1 Cell Culture and Infection (See Note 1)

1. HeLa S3 cells (American Type Culture Collection (ATCC, no. CCL2.2)).
2. VACV Western Reserve (WR) strain (ATCC, VR-1354).
3. Minimum essential medium spinner modification (S-MEM).
4. Equine serum.

2.2 Library Construction

1. * Indicates that the reagent is included in the TruSeq Ribo Profile (Mammalian) Kit (Illumina).
2. # Denotes that alternative reagents can be used (*see* **Note 2**).

2.2.1 Cell Lysis and RNA Extraction

1. Phosphate-buffered saline (PBS).
2. 5× Mammalian polysome buffer*#.
3. 100 mM Dithiothreitol (DTT)*.
4. 10% Sodium dodecyl sulfate (SDS)*.
5. 10% Triton X-100*.
6. DNase I (1 U/μL)* #.
7. 10% NP-40.
8. 50 mg/mL Cycloheximide (CHX) in ethanol.
9. Nuclease-free water.
10. Mammalian lysis buffer (1× mammalian polysome buffer, 1% Triton X-100, 1 mM DTT, 10 U/mL DNase I, 0.1 mg/mL CHX, and 0.1% NP-40).
11. S-MEM supplemented with 0.1 mg/mL CHX.

2.2.2 Purification of Ribosome-Protected RNA Fragments and Purification of mRNA

1. 10 U/μL TruSeq Ribo Profile Nuclease*#.
2. SUPERase In RNase Inhibitor (Thermo Fisher).
3. Illustra MicroSpin S-400 HR Columns (GE Healthcare).
4. TruSeq Ribo Profile RNA Control*#.
5. 10% SDS*.
6. Glycogen*#.
7. 5 M Ammonium acetate*#.
8. 3 M Sodium acetate (pH 5.2).
9. 15% Polyacrylamide/7–8 M urea/TBE gel.
10. Denaturing gel loading dye.
11. 1 ng/μL, 20/100 Oligo Ladder.
12. SYBR Gold (Thermo Fisher).
13. Tris-phenol:chloroform solution (1:1 ratio).
14. Chloroform.

15. Isopropanol.
16. 80% (v/v) Ethanol (freshly prepared).
17. Nuclease-free water.
18. Sucrose (optional, *see* **Note 5**).
19. $MgCl_2$ (optional, *see* **Note 5**).
20. Ribo-Zero Magnetic Gold Kit (Human/Mouse/Rat; Illumina) (optional, *see* **Note 6**).
21. Ultracentrifuge buffer (1× mammalian polysome buffer, 50% (w/v) sucrose, 0.5 mM DTT, 3.5 mM $MgCl_2$, 100 U/mL SUPERase In RNase Inhibitor, and 0.1 mg/mL CHX)#.
22. Dynabeads mRNA DIRECT Kit (Thermo Fisher).

2.2.3 RNA Modification and Reverse Transcription

1. TruSeq Ribo Profile PNK Buffer*#.
2. TruSeq Ribo Profile PNK*#.
3. TruSeq Ribo Profile 3′ Adapter*#.
4. TruSeq Ribo Profile Ligation Buffer*#.
5. TruSeq Ribo Profile Ligase*#.
6. TruSeq Ribo Profile AR Enzyme*#.
7. 100 mM DTT*.
8. TruSeq Ribo Profile RT Reaction Mix*#.
9. EpiScript RT*#.
10. TruSeq Ribo Profile Exonuclease*#.
11. TruSeq Ribo Profile RNase Mix*#.
12. TruSeq Ribo Profile CL Reaction Mix*#.
13. CircLigase*#.
14. ATP*#.
15. $MnCl_2$*#.
16. 10% SDS*.
17. Glycogen*#.
18. 5 M Ammonium acetate*#.
19. 3 M Sodium acetate (pH 5.2).
20. 10% Polyacrylamide/7–8 M urea/TBE gel (Bio-Rad).
21. Denaturing Gel Loading Dye (Thermo Fisher).
22. 1 ng/μL 20/100 Oligo Ladder (IDT).
23. SYBR Gold (Thermo Fisher).
24. Tris-phenol:chloroform solution (1:1 ratio).
25. Chloroform.
26. Isopropanol.
27. Freshly prepared 80% (v/v) ethanol.
28. Nuclease-free water.

2.2.4 Ribosomal RNA (rRNA) cDNA Depletion

1. 2× SSC buffer: 0.3 M NaCl, 0.03 M sodium citrate (pH 7.0).
2. Dynabeads MyOne Streptavidin C1 (Thermo Fisher).
3. MinElute PCR Purification Kits (QIAGEN).
4. 5′-Biotin modification by a C6 spacer Oligos (check the oligo sequences in Table 1 and *see* **Note 3**).

2.2.5 PCR Amplification

1. 2× Phusion MasterMix (New England BioLabs (NEB)).
2. TruSeq Ribo Profile Forward PCR Primer*#.
3. TruSeq Ribo Profile Index PCR Primers*#.
4. 10% SDS*.
5. Glycogen*#.
6. 5 M Ammonium acetate*#.
7. 3 M Sodium acetate (pH 5.2).
8. 8% Native polyacrylamide gel.
9. 6× Native loading buffer.
10. 10 bp DNA Ladder.
11. SYBR Gold (Thermo Fisher).
12. Isopropanol.
13. Freshly prepared 80% (v/v) ethanol.
14. Nuclease-free water.

2.2.6 Quantification and Quality Control

1. 2100 Bioanalyzer Instrument (Agilent Technologies).
2. KAPA Library Quantification Kits for Illumina sequencing platforms (Kapa Biosystems).

2.3 Data Analysis

2.3.1 Software

1. FastQC (http://www.bioinformatics.babraham.ac.uk/projects/download.html#fastqc).
2. FASTX Toolkit (http://hannonlab.cshl.edu/fastx_toolkit/index.html).
3. Bowtie (https://sourceforge.net/projects/bowtie-bio/files/bowtie/1.2.1.1) [10].
4. Tophat (https://ccb.jhu.edu/software/tophat/tutorial.shtml) [11].
5. FeatureCounts (http://sourceforge.net/projects/subread/files/subread-1.5.3/) [12].
6. Mochiview (http://www.johnsonlab.ucsf.edu/mochiview-downloads) [13].
7. SAMtools (https://sourceforge.net/projects/samtools/files/) [14].
8. EdgeR (https://bioconductor.org/packages/release/bioc/html/edgeR.html) [15].

Table 1
Sequence and volume of biotin-modified oligos

Sequence	Volume of 100 μM stock (μL)
tcgtgggggcccaagtccttctgatcgaggccc	224.6
tcctcccggggctacgcctgtctgagcgtcgct	80.0
tccagtgcgccccgggcgggtcgcgccgtcgggcccgggg	60.0
ctgcataatttgtggtagtgggg	52.0
actcgccgaatcccggggccgagggag	44.0
ggggggatgcgtgcatttatcagatca	28.2
actgacccggtgaggcgggg	23.5
aggggctctcgcttctggcgccaagcgt	20.0
cgggaccggggtccggtgcggagtgcccttcgtcc	20.0
cggcggatctttcccgcccccgttcctcccgacccct	20.0
cgtgggggggccgggccacccctcccacggcgcgacc	20.0
agggggtctcccccgcgggggcgcgccggcg	20.0
tcaccgcccgtccccgccccttgcctctcggc	20.0
cgcgcgcgcgggagggcgcgtgccccgccgcgcg	20.0
gaacttgactatctagaggaagtaaaagtcgt	17.6
agagcgaaagcatttgccaagaatgttttc	16.0
tccgccgagggcgcaccaccggcccgtctcgcc	12.0
gtgcgccgcgaccggctccgggacggctggg	11.5
tcccggggctacgcctgtct	10.5
cccagtgcgccccgggcgtcgtcgcgccgtcgggtcccggg	10.5
taaaccattcgtagacgacctgctt	10.0
ggctctcgcttctggcgcca	10.0
ttggtgactctagataacctcgggccgatcgcacg	10.0
gagccgcctggataccgcagctaggaataatggaat	10.0
ggggccgggccgcccctcccacggcgcg	10.0
gagcctcggttggccccggatagccgggtccccgt	10.0
tcgctgcgatctattgaaagtcagccctcgacaca	10.0
aactttcgatggtagtcgccgtgcctaccatggtgacc	7.0
ggatggtttagtgaggccctcggatcggc	3.0
cggccgaggtgggatcccgaggc	2.0
ccgccacgcagttttatccggtaaagc	1.5
acgattaaagtcctacgtgatctgagt	1.3
ccccccgagtgttacagcccccc	1.1

The computer environment may require JAVA, Perl, R, or Python:

9. JAVA (https://www.java.com/en/download/).
10. Perl (https://www.perl.org/get.html).
11. Python (https://www.python.org/).
12. R (https://www.r-project.org/).

2.3.2 Reference Files

1. VACV genome sequence file (https://www.ncbi.nlm.nih.gov/nuccore/NC_006998.1?report=fasta).
2. VACV genome annotation file (ftp://130.14.250.10/genomes/Viruses/Vaccinia_virus_uid15241/NC_006998.gff).
   This .gff file was converted to a .gtf file using the script in https://github.com/vipints/GFFtools-GX/blob/master/gff_to_gtf.py
3. Human genome sequence file (ftp://ftp.ensembl.org/pub/release-96/fasta/homo_sapiens/dna/).
4. Human genome annotation file (ftp://ftp.ensembl.org/pub/release-96/gtf/homo_sapiens).

2.3.3 Pre-defined Overrepresented Sequences

1. Transfer RNA (tRNA) sequence (http://gtrnadb.ucsc.edu/genomes/eukaryota/Hsapi19/hg19-tRNAs.fa).
2. Human rRNA sequence (https://www.ncbi.nlm.nih.gov/nuccore/555853/).
3. Human 5S RNA sequence (https://www.ncbi.nlm.nih.gov/nuccore/23898/).

3 Methods

3.1 Cell Culture and VACV Infection

1. HeLa S3 cells were cultured in S-MEM supplemented with 5% equine serum at 37 °C and 5% CO_2.
2. HeLa cells (1×10^7 cells/mL) were infected with VACV at a multiplicity of infection (MOI) of 20 for 30 min and then diluted 100-fold. For more details on basic cell culture and preparation, purification, and titration of the virus stock, please *see* ref. 16.

3.2 Library Construction

3.2.1 Cell Lysis and RNA Extraction

1. Prepare 1 mL of mammalian lysis buffer (for approximately 2×10^7 cells) for each sample. Prepare fresh buffer before every use and chill the prepared buffer to 4 °C.
2. Treat cells using culture medium supplemented with 0.1 mg/mL CHX for 9 min before the desired collection time.
3. Wash cells with 10 mL of ice-cold PBS supplemented with 0.1 mg/mL CHX.

4. Lyse cells with 800 μL of mammalian lysis buffer and pass through a sterile 25-gauge needle 13 times to lyse the cells completely. Transfer the cell lysate to a fresh, ice-chilled tube.
5. Incubate for 10 min on ice with gentle shaking every 2 min.
6. Centrifuge for 10 min at 20,000 × *g* at 4 °C. Transfer the supernatant to a fresh, ice-cold tube. Approximately 1 mL of clarified lysate should be recovered (*see* **Note 4**).

3.2.2 RNA Purification: Purification of Ribosome-Protected RNA Fragments (RPFs)

1. Dilute the collected lysate and 1× mammalian lysis buffer ten-fold using nuclease-free water. Assess nucleic concentration by measuring A260 on a Nanodrop, using water as a blank. Calculate the value of A260/mL using the following equation: (A260 cell lysate − A260 1× mammalian lysis buffer) × 10 = A260/mL. This number will be used to calculate the amount of TruSeq Ribo Profile Nuclease required in the next step.
2. Add TruSeq Ribo Profile Nuclease to 200 μL clarified lysate (Subheading 3.2.1, **step 6**) at a final concentration of 5 U/A260.
3. Incubate at room temperature for 45 min with gentle shaking.
4. Add 15 μL of SUPERase In RNase inhibitor and chill on ice to stop the reaction.
5. For each sample, prepare 3 mL 1× mammalian polysome buffer.
6. Resuspend MicroSpin S-400 columns by gentle inverting. Make sure that there are no bubbles in the resin.
7. Open caps at both ends of the column to remove the liquid.
8. Wash the column with 500 μL 1× mammalian polysome buffer six times. DO NOT centrifuge the column during the first five washes. For the final wash, centrifuge for 4 min at 600 × *g* at room temperature.
9. Immediately load 100 μL of the nuclease-digested RPF sample (**step 4** above) onto the column, centrifuge for 2 min at 600 × *g*, and collect the flow-through.

   (For **steps 5–9**, an alternative method based on ultracentrifugation can be used; *see* **Note 5**).
10. Add 10 μL of 10% SDS and incubate at 65 °C for 5 min.
11. Add ~200 μL of water to adjust the volume to 300 μL. Add 300 μL of Tris-phenol/chloroform and centrifuge at 17,900 × *g* for 15 min at 4 °C.
12. Perform **step 11** once with phenol/chloroform and once with chloroform.

13. Add 2 μL of glycogen, 670 μL of isopropanol, and 70 μL of 3 M sodium acetate (pH 5.2), and place at −80 °C for at least 2 h.
14. Centrifuge at 20,000 × *g* for 30 min at 4 °C. Discard the supernatant.
15. Wash the pellet with fresh, ice-cold 80% alcohol once. Air-dry.
16. Dissolve the pellet in 10 μL of nuclease-free water.

    (An alternative method can be performed to deplete rRNA after this step; *see* **Note 6**).
17. Add an equal volume of denaturing gel-loading dye to the TruSeq Ribo Profile RNA Control, samples, and ladder.
18. Heat the samples and ladder at 95 °C for 5 min, and then chill on ice. Keep the TruSeq Ribo Profile RNA Control on ice during denaturing.
19. Load 15 μL of each sample along with 8 μL of ladder and 10 μL of TruSeq Ribo Profile RNA Control onto a 15% urea-polyacrylamide gel.
20. Run the gel at 180 V until the bromophenol blue reaches the bottom (~70 min).
21. Dip the gel into SYBR Gold solution (1:10,000 diluted in gel-running buffer) and gently shake for 15 min at 4 °C.
22. Visualize the RNA using a dark-field blue light transilluminator.
23. Cut the gel (using a blade) between the 28 nt and 30 nt using the TruSeq Ribo Profile RNA Control as a reference (*see* **Notes 7** and **8**).
24. Transfer each gel to a 0.5 mL tube that has been punctured at the bottom using a sterile 20-gauge needle (*see* **Note 8**). Centrifuge for 20 min at 13,000 x *g* at room temperature to shred the gel slices.
25. Add 400 μL of water, 40 μL of 5 M ammonium acetate, and 2 μL of 10% SDS to each tube.
26. Gently rock the samples at room temperature for 2 h or at 4 °C overnight.
27. Transfer the slurry to 1.5 mL filter tubes (provided in the kit). Centrifuge for 3 min at 2000 × *g*.
28. Transfer the solution to a new 1.5 mL tube. Add 2 μL of glycogen and 700 μL of 100% isopropanol. Store at −80 °C for at least 2 h.
29. Centrifuge for 30 min at 20,000 × *g* at 4 °C. Discard the supernatant.

30. Wash the pellet once with fresh, ice-cold 80% alcohol. Air-dry.
31. Dissolve the pellet in 64.5 μL of water and dissolve the control RNA in 8 μL of water.

3.2.3 Total mRNA Purification and Fragmentation Based on the Dynabeads mRNA DIRECT Kit

1. Resuspend Dynabeads® Oligo(dT)$_{25}$ before use.
2. Transfer 250 μL of beads to a RNase-free 1.5 mL tube. Place the tube on a magnet, wait until the suspension is clear, and then discard the supernatant.
3. Wash the beads once with 250 μL of fresh lysis/binding buffer.
4. Add 950 μL of fresh lysis/binding buffer and 300 μL of lysate (Subheading 3.2.1, **step 6**) to the beads and mix well.
5. Incubate for 3–5 min at room temperature.
6. Place the tubes on a magnet until the suspension is clear, and then discard the supernatant.
7. Wash the beads twice with 1 mL of washing buffer A at room temperature.
8. Wash the beads twice with 1 mL of washing buffer B at room temperature.
9. Add 20 μL of water, mix well, and incubate at 65 °C for 2 min.
10. Immediately place the tube on a magnet and collect the clear supernatant.

    An optional rRNA depletion step can be performed after this step (*see* **Note 6**).
11. Add 7.5 μL of TruSeq Ribo Profile PNK Buffer to each mRNA sample (from (**step 10** above). Heat at 94 °C for 25 min to fragment the RNA, and then chill on ice.

3.2.4 RNA Modification and Reverse Transcription

In the following steps, the RPF (Subheading 3.2.2) and total mRNA (Subheading 3.2.3) samples should be treated the same unless otherwise stated.

PNK treatment of RNA samples

1. To each mRNA sample (Subheading 3.2.3, **step 11**), add 44.5 μL of water and 3 μL of TruSeq Ribo Profile PNK.

    To each RPF sample (Subheading 3.2.2, **step 31**), add 7.5 μL of TruSeq Ribo Profile PNK Buffer and 3 μL of TruSeq Ribo Profile PNK.
2. Mix samples well and incubate at 37 °C for 1 h.
3. Purify the samples as described (Subheading 3.2.2, **steps 11–15**). Dissolve all samples in 8 μL of nuclease-free water.

Ligate RNA with a 3′ adapter

4. Add 1 μL of the TruSeq Ribo Profile 3′ Adapter to all samples and the control. Heat for 2 min at 65 °C, and then hold at 4 °C in a thermal cycler. The positive control should be the control RNA (Subheading 3.2.2, **step 31**) and the negative control should be water.
5. Prepare a ligation mixture by adding 3.5 μL of TruSeq Ribo Profile Ligation Buffer, 1 μL of 100 mM DTT, and 1.5 μL of TruSeq Ribo Profile Ligase to each sample. Incubate at 23 °C for 2 h.
6. Add 2 μL of TruSeq Ribo Profile AR Enzyme to remove excessive adapters. Mix well and incubate at 30 °C for 2 h.

Reverse transcription and purification of the cDNAs

7. Prepare a reverse transcription mixture by adding 4.5 μL of TruSeq Ribo Profile RT Reaction Mix, 1.5 μL of 100 mM DTT, 6 μL of nuclease-free water, and 1 μL of EpiScript RT to each sample. Incubate at 50 °C for 30 min.
8. Add 1 μL of TruSeq Ribo Profile Exonuclease to each sample and mix well. Incubate at 37 °C for 30 min, 80 °C for 15 min, and then hold at 4 °C.
9. Add 1 μL of TruSeq Ribo Profile RNase Mix to each sample. Mix well, incubate at 55 °C for 5 min, and then chill on ice.
10. Purify the samples as described (Subheading 3.2.2, **steps 11–15**). Dissolve all the samples in 7.5 μL of nuclease-free water.
11. Prepare samples for separation on a urea-polyacrylamide gel (Subheading 3.2.2, **steps 17** and **18**).
12. Load 15 μL of each sample along with 8 μL of ladder onto a 10% urea-polyacrylamide gel.
13. Run the gel at 180 V until the bromophenol blue reaches the bottom (~45 min).
14. Dip the gel into SYBR Gold solution (1:10,000 diluted in gel running buffer) and gently shake for 15 min at 4 °C.
15. Visualize the DNA using a dark-field blue light transilluminator.
16. Using a blade cut the gel corresponding to ~80–100 nt for mRNA samples and at ~70–80 nt for RPF samples and control (*see* **Notes 7** and **8**).
17. Extract the DNA as in Subheading 3.2.2, **steps 24–30**, and dissolve the cDNA pellet in 10 μL of nuclease-free water.
18. To perform single-stranded DNA circulation, prepare a CircLigase mixture by adding 4 μL of TruSeq Ribo Profile CL Reaction Mix, 2 μL of ATP, 2 μL of $MnCl_2$, and 2 μL of CircLigase to each sample. Incubate at 60 °C for 2 h, and then chill on ice or store at −20 °C.

3.2.5 rRNA cDNA Depletion

1. Prepare 100 μM stock of each biotin-labeled oligo using 2× SSC buffer.
2. Prepare oligo mix (*see* Table 1 and **Note 3**).
3. Adjust each sample volume to 30 μL with 10 μL of 2× SSC buffer.
4. Add 3.35 μL of mixed oligos to each sample.
5. Heat at 94 °C for 5 min, and then ramp down to room temperature at a speed of 2.5 °C/min.
6. Resuspend MyOne Streptavidin beads before use.
7. Transfer 135 μL of beads to a 1.5 mL tube, and wash the beads with 200 μL of 2× SSC buffer three times.
8. Resuspend beads in 20 μL of 2× SSC buffer. Mix the beads with the hybridized samples. Incubate at room temperature for 15 min with gentle rotation.
9. Put each tube on a magnet for 2 min and recover the supernatant.
10. Add 250 μL of Buffer PB (from the MinElute PCR purification kit) and 10 μL of 3 M sodium acetate (pH 5.2) to each reaction, and mix well.
11. Place a MinElute column in a 2 mL collection tube. Add the sample to the column, and centrifuge for 1 min at 17,900 × *g*. Discard the flow-through.
12. Wash the MinElute column with 750 μL of Buffer PE.
13. Centrifuge the empty column for 1 min at maximum speed to dry.
14. Elute the DNA by adding 50 μL of water to the center of the membrane. Let the column stand for 1 min at room temperature and then centrifuge for 1 min at maximum speed. The elution collected will be used as the PCR template.

3.2.6 PCR Amplification

1. Prepare a PCR mixture by adding 9.5 μL of nuclease-free water, 1 μL of TruSeq Ribo Profile Forward PCR Primer, 1 μL of the TruSeq Ribo Profile Index PCR Primer of choice, 1 μL of PCR template (Subheading 3.2.5, **step 14**), and 12.5 μL of 2× Phusion Master Mix (*see* **Note 9**).
2. Place samples in a thermal cycler and run the following program: 98 °C for 30 s, 9 cycles (*see* **Note 9**) of 94 °C for 15 s, 55 °C for 5 s, and 65 °C for 10 s and then hold at 4 °C.
3. Mix the PCR product with 5 μL of 6× native loading buffer.
4. Load each entire sample along with 10 μL of 10 bp DNA ladder onto an 8% native polyacrylamide gel.
5. Run the gel at 200 V until the bromophenol blue reaches the bottom (~30 min).

6. Dip the gel into SYBR Gold solution (1:10,000 diluted in gel-running buffer) and gently shake for 15 min at 4 °C.
7. Visualize the DNA using a dark-field blue light transilluminator.
8. Using a blade cut the gel corresponding to ~140–160 nt for all the samples (*see* **Notes** 7 and **8**).
9. Extract the DNA as in Subheading 3.2.2, **steps 24–30**. Dissolve the DNA pellet in 25 μL of nuclease-free water.

3.3 Quantification, Quality Control, and Deep Sequencing

Please work with your sequencing service provider to precisely quantify and assess the quality of your sequencing libraries. You may use Kapa Library Quantification Kits and/or an Agilent 2100 Bioanalyzer system following the manufacturer's instructions. For sequencing, Illumina platforms such as the MiSeq or HiSeq systems can be used. The sequencing depths and biological replicates are dependent on individual projects.

3.4 Data Analysis (See Note 10)

A typical pipeline usually includes quality control, removal of overrepresented sequences, alignment, and quantification. This is typically followed by additional downstream analyses, for example, translation efficiency analysis, data visualization, new translation unit identification, and others. Generally, the RPFs can be analyzed in the same way as the RNA-Seq samples unless otherwise stated. Multiple pipelines and tools are available. Here, we present the workflow we have implemented. We do not provide scripts as they are provided in the manuals of the corresponding software and can vary based on the running environment. Plus, the parameters likely vary for different projects.

1. Quality assessment: FastQC has both a command line and a graphical user interface (GUI) version available. FastQC requires .fastq files or .fastq.gz files as input and outputs the quality control results as a .html file. Adaptor sequences and overrepresented sequences are displayed in the results. This quality-control step should also be carried out after adaptor trimming and removal of overrepresented sequences.
2. Adaptor trimming: The FASTX Toolkit is a series of command-line tools with multiple functions for next-generation sequencing. The *fastx_clipper* function allows trimming of RNA-seq and ribosome profiling read adaptors. Both input and output files must be .fastq. The adaptor sequence can be found in the TrueSeq library construction kit manuals.
3. Removal of overrepresented sequences: Usually, some tRNA and rRNA sequences are overrepresented within the data and should be removed before mapping. This can be achieved using Bowtie and input files including a .fastq file and reference index files. The total sequencing reads should be mapped

to an index file containing all the overrepresented sequences (generated by *Bowtie-build*). The sequences that are not mapped will be saved in the output .fastq file that will then be used in the next step. The unwanted, overrepresented sequences will be stored in an .aln file.

4. Mapping: The reads can then be mapped to the VACV genome (GenBank, NC_006998.1) or the human genome (Ensembl, GRCh38 [17]) using Tophat. Tophat uses the .fastq file from last step as input, and the .gtf file containing the VACV gene annotation and the VACV genome index file as references. The output data will be stored in a .bam file.
5. Quantification of reads: Reads of protein-coding genes are quantified by FeatureCounts. FeatureCounts can be used either as independent software or as a package in R, and uses the .bam file from the last step as input and the .gtf file with VACV gene annotation as the reference. Because of the inverted terminal repeats (ITRs) in the VACV genome, please allow reads to be mapped to at least two sites on the VACV genome. The counting result is formatted as a .txt file, which will contain a file head with tab-separated columns. The file can be opened by R or Excel and RPKMs (reads per kilobase of transcript per million reads) can be calculated by edgeR.
6. Downstream analyses: Many downstream analyses can be carried out from here. For example, this data can be used for analyzing differential gene expression (at both the RNA and translation levels) or translation efficiency, or identifying translation initiation sites. Nevertheless, it is important to note that although both harringtonine and lactimidomycin are known to cause ribosomes to accumulate at the translation initiation regions of cellular mRNAs, only harringtonine causes ribosomes to accumulate on VACV post-replicative mRNAs [2, 5, 18].

3.5 Data Visualization (See Note 10)

Visual browsers are available that can be used for visualizing the data (e.g., Integrative Genomics Viewer (IGV) [19] or Mochiview). Follow the user manuals to prepare files for visualization of the sequencing data.

4 Notes

1. If using other cell types for VACV infection, please use the corresponding cell culture medium. Alternative recombinant viruses may also be used for infection. During the infection, various drug treatments can be added, for example, cytosine arabinoside (AraC) which allows for only VACV early gene expression [20].

Table 2
Alternative reagents if not using TruSeq Ribo Profile Kit

Content in the kit	Alternatives
5× Mammalian polysome buffer	100 mM Tris (pH = 7.4), 1.25 M NaCl, 75 mM $MgCl_2$
DNase I (1 U/μL)	TURBO DNase (2 U/μL) (Thermo Fisher)
TruSeq Ribo Profile Nuclease (10 U/μL)	Ambion™ RNase I, cloned, 100 U/μL (Thermo Fisher)
Glycogen	Glycogen, RNA grade (Thermo Fisher)
5 M Ammonium acetate	Ammonium acetate (5 M) (Thermo Fisher)
TruSeq Ribo Profile PNK	T4 Polynucleotide Kinase (NEB)
TruSeq Ribo Profile PNK Buffer	Supplied with T4 Polynucleotide Kinase
TruSeq Ribo Profile Ligase	T4 RNA Ligase 2, truncated (NEB)
TruSeq Ribo Profile Ligation Buffer	Supplied with T4 RNA Ligase
TruSeq Ribo Profile AR Enzyme	Not required (this step can be skipped)
EpiScript RT	SuperScript™ III Reverse Transcriptase (Thermo Fisher)
TruSeq Ribo Profile RT Reaction Mix	Supplied with SuperScript™ III Reverse Transcriptase
TruSeq Ribo Profile Exonuclease	RNase H (NEB)
TruSeq Ribo Profile RNase Mix	RNase H (NEB)
CircLigase	CircLigase™ ssDNA Ligase (Epicentre)
TruSeq Ribo Profile CL Reaction Mix	Supplied with CircLigase™ ssDNA Ligase
ATP	Supplied with CircLigase™ ssDNA Ligase
$MnCl_2$	Supplied with CircLigase™ ssDNA Ligase

2. Alternative reagents are listed in Table 2 if the TruSeq Ribo Profile (mammalian) Kit (Illumina) is not used. The oligonucleotides needed can be synthesized based on the sequences described in the TruSeq Ribo Profile (mammalian) Kit.
3. Different oligos and their ratios may be used based on the sequences and amounts of overrepresented non-mRNA sequences found in different cell types and conditions.
4. A typical procedure will require only 500 μL. Flash freeze the remaining lysate with liquid nitrogen and store at −80 °C as a backup.
5. An alternative method based on ultracentrifugation can be used to separate the monosome. If using the following method, skip Subheading 3.2.2, **steps 5–9**, and perform the following steps:

(a) For each sample, prepare 1 mL of ultracentrifuge buffer chilled on ice.

(b) Add 200 μL of lysate (Subheading 3.2.2, **step 4)** to 1 mL of ultracentrifuge buffer. Centrifuge at 4 °C for 4 h at ~143,000 × *g*.

(c) Resuspend the pellet in 100 μL nuclease-free water and continue to Subheading 3.2.2, (**step 10**).

6. An alternative rRNA depletion step can be performed using the Ribo-Zero Magnetic Gold Kit (human/mouse/rat). Please skip Subheading 3.2.5 if using this method for rRNA depletion. This will result in a more unbiased rRNA depletion compared to the method described in Subheading 3.2.5. Nevertheless, this method will also result in a lower concentration of RNA for the subsequent steps and, therefore, the bands may not be visible in the gel for size-selection step(s). The alternative rRNA depletion step is as follows:

(a) Prepare 225 μL of beads for each sample. Place tubes on a magnet until the liquid is clear. Discard the supernatant.

(b) Wash the beads twice with 225 μL of nuclease-free water. Tubes may need to be vortexed to mix the beads and water completely.

(c) Resuspend the beads in 65 μL of magnetic bead resuspension solution.

(d) Set up a hybridized mixture containing 1–5 μg RNA (Subheading 3.2.2, **step 15** and Subheading 3.2.3, **step 10**), 4 μL of Ribo-Zero Reaction Buffer, 10 μL of Ribo-Zero Removal Solution, and an appropriate volume of nuclease-free water to bring the total volume to 40 μL.

(e) Incubate at 68 °C for 10 min, and then room temperature for 5 min.

(f) Combine the beads from **step 3** and the mixture from **step 5** above and immediately mix well.

(g) Incubate at room temperature for 5 min.

(h) Place the samples on a magnet and collect the supernatant when it becomes clear.

(i) Purify the RNA as described (Subheading 3.2.2, **steps 11–15**). Dissolve RPF samples in 10 μL of nuclease-free water. Proceed to Subheading 3.2.2, **step 17**. For total mRNA samples dissolve in 20 μL of nuclease-free water. Proceed to Subheading 3.2.3, **step 11**.

7. Cut the gel even if you cannot see the bands.

8. Be careful!

9. The amount of PCR template and PCR cycles used should be optimized. Too much template or too many PCR cycles may result in overamplification, the appearance of higher than expected molecular weight bands, smeared PCR products, and adapter-dimer-derived products. Not enough PCR template or cycles may result in not enough quantification, quality control, and deep sequencing.
10. None of the software used in Subheadings 3.4 and 3.5 is exclusive. Other software and tools may be used, for example, Trimmomatic for trimming adaptor [21], HTSeq for read quantification [22], RSeQC for converting file types [23], UCSC genome browser for visualization [19, 24], and others.

Acknowledgments

This work is supported, in part, by NIH grants to ZY, including a project of P20GM113117. This work is also supported, in part, by National Science Foundation of China grant NSFC81571988 to WQ.

References

1. Ingolia NT, Ghaemmaghami S, Newman JR, Weissman JS (2009) Genome-wide analysis in vivo of translation with nucleotide resolution using ribosome profiling. Science 324:218–223
2. Ingolia NT, Lareau LF, Weissman JS (2011) Ribosome profiling of mouse embryonic stem cells reveals the complexity and dynamics of mammalian proteomes. Cell 147:789–802
3. Ingolia NT (2014) Ribosome profiling: new views of translation, from single codons to genome scale. Nat Rev Genet 15:205–213
4. Stern-Ginossar N (2015) Decoding viral infection by ribosome profiling. J Virol 89:6164–6166
5. Yang Z, Cao S, Martens CA, Porcella SF, Xie Z, Ma M, Shen B, Moss B (2015) Deciphering poxvirus gene expression by RNA sequencing and ribosome profiling. J Virol 89:6874–6886
6. Dhungel P, Cao S, Yang Z (2017) The 5′-poly(A) leader of poxvirus mRNA confers a translational advantage that can be achieved in cells with impaired cap-dependent translation. PLoS Pathog 13:e1006602
7. Cao S, Dhungel P, Yang Z (2017) Going against the tide: selective cellular protein synthesis during virally induced host shutoff. J Virol 91. https://doi.org/10.1128/JVI.00071-17
8. Yang Z, Bruno DP, Martens CA, Porcella SF, Moss B (2010) Simultaneous high-resolution analysis of vaccinia virus and host cell transcriptomes by deep RNA sequencing. Proc Natl Acad Sci U S A 107:11513–11518
9. Yang Z, Reynolds SE, Martens CA, Bruno DP, Porcella SF, Moss B (2011) Expression profiling of the intermediate and late stages of poxvirus replication. J Virol 85:9899–9908
10. Langmead B, Trapnell C, Pop M, Salzberg SL (2009) Ultrafast and memory-efficient alignment of short DNA sequences to the human genome. Genome Biol 10:R25
11. Trapnell C, Pachter L, Salzberg SL (2009) TopHat: discovering splice junctions with RNA-Seq. Bioinformatics 25:1105–1111
12. Liao Y, Smyth GK, Shi W (2014) featureCounts: an efficient general purpose program for assigning sequence reads to genomic features. Bioinformatics 30:923–930
13. Homann OR, Johnson AD (2010) MochiView: versatile software for genome browsing and DNA motif analysis. BMC Biol 8:49
14. Li H, Handsaker B, Wysoker A, Fennell T, Ruan J, Homer N, Marth G, Abecasis G,

Durbin R, Genome Project Data Processing Subgroup (2009) The sequence alignment/map format and SAMtools. Bioinformatics 25:2078–2079

15. Robinson MD, McCarthy DJ, Smyth GK (2010) edgeR: a Bioconductor package for differential expression analysis of digital gene expression data. Bioinformatics 26:139–140
16. Earl PL, Cooper N, Wyatt LS, Moss B, Carroll MW (2001) Preparation of cell cultures and vaccinia virus stocks. Curr Protoc Mol Biol Chapter 16:Unit16 16
17. Yates A, Akanni W, Amode MR, Barrell D, Billis K, Carvalho-Silva D, Cummins C, Clapham P, Fitzgerald S, Gil L, Giron CG, Gordon L, Hourlier T, Hunt SE, Janacek SH, Johnson N, Juettemann T, Keenan S, Lavidas I, Martin FJ, Maurel T, McLaren W, Murphy DN, Nag R, Nuhn M, Parker A, Patricio M, Pignatelli M, Rahtz M, Riat HS, Sheppard D, Taylor K, Thormann A, Vullo A, Wilder SP, Zadissa A, Birney E, Harrow J, Muffato M, Perry E, Ruffier M, Spudich G, Trevanion SJ, Cunningham F, Aken BL, Zerbino DR, Flicek P (2016) Ensembl 2016. Nucleic Acids Res 44:D710–D716
18. Lee S, Liu B, Lee S, Huang SX, Shen B, Qian SB (2012) Global mapping of translation initiation sites in mammalian cells at single-nucleotide resolution. Proc Natl Acad Sci U S A 109:E2424–E2432
19. Robinson JT, Thorvaldsdottir H, Winckler W, Guttman M, Lander ES, Getz G et al (2011) Integrative genomics viewer. Nat Biotechnol 29:24–26
20. Oda KI, Joklik WK (1967) Hybridization and sedimentation studies on "early" and "late" vaccinia messenger RNA. J Mol Biol 27:395–419
21. Bolger AM, Lohse M, Usadel B (2014) Trimmomatic: a flexible trimmer for Illumina sequence data. Bioinformatics 30:2114–2120
22. Anders S, Pyl PT, Huber W (2015) HTSeq – a Python framework to work with high-throughput sequencing data. Bioinformatics 31:166–169
23. Wang L, Wang S, Li W (2012) RSeQC: quality control of RNA-seq experiments. Bioinformatics 28:2184–2185
24. Kent WJ, Sugnet CW, Furey TS, Roskin KM, Pringle TH, Zahler AM, Haussler D (2002) The human genome browser at UCSC. Genome Res 12:996–1006

Chapter 12

Quantitative PCR-Based Assessment of Vaccinia Virus RNA and DNA in Infected Cells

Moona Huttunen and Jason Mercer

Abstract

Quantitative PCR-based methods have proven to be easy-to-use, cost-effective procedures for the quantification of viral gene expression and viral genome numbers. Quantitative PCR (qPCR) and quantitative reverse transcriptase-PCR (qRT-PCR) are rapid and sensitive approaches that can be used to pinpoint defects in viral DNA replication and transcriptional activity, respectively. Due to the significant nucleotide overlap between Poxviridae these methods can be employed across a wide range of viruses from this family. Here we provide methods for the quantification of vaccinia DNA replication by qPCR and quantification of the three classes of vaccinia gene transcription by qRT-PCR.

Key words Vaccinia gene expression, Vaccinia DNA replication, Quantitative PCR (qPCR), Quantitative reverse transcriptase PCR (qRT-PCR)

1 Introduction

Vaccinia virus (VACV) gene expression occurs in three temporal stages: early, intermediate, and late. These gene classes are characterized by divergent promoters and different expression kinetics with early genes expressed between 20 min and 2 h postinfection (hpi), intermediates from 1.5 to 2.5 hpi, and late genes from 2.5 hpi onwards [1]. Furthermore, early genes are transcribed from within viral cores prior to genome release, while intermediate and late genes are only transcribed after DNA replication, which begins between 1 and 2 hpi [1].

Given this complexity, when assessing the impact of targeted genomic mutations or potential inhibitors on VACV infection, it may be desirable to quantify viral DNA replication and transcriptional activity. Several strategies have been developed to analyze VACV virus DNA replication including (1) assessing the rate of DNA synthesis by monitoring 3H-thymide incorporation into newly formed DNA [2], (2) quantifying the steady-state levels of viral DNA accumulation by Southern dot-blot hybridization [2],

Jason Mercer (ed.), *Vaccinia Virus: Methods and Protocols*, Methods in Molecular Biology, vol. 2023,
https://doi.org/10.1007/978-1-4939-9593-6_12,

(3) visualizing the subcellular sites of DNA replication by incorporation of bromodeoxyuridine (BrdU) with DNA [2], and (4) using quantitative PCR (qPCR) [3, 4]. For analysis of viral transcription, classical northern blot analysis has been used to investigate the expression of individual VACV genes [5], while RNA sequencing has been recently applied for the global analysis of VACV gene expression [6].

Over the past several years, quantitative PCR (qPCR) has become the leading tool for the detection and quantification of DNA and RNA because of its advantages over traditional endpoint PCR methods. These include single-tube amplification and detection, and no need for post-PCR manipulations (i.e., gel runs/detection). By measuring the amount of DNA after each PCR cycle using fluorescent dyes that yield increasing signal in direct proportion to the number of PCR product molecules generated, qPCR provides extremely accurate quantification of the starting material (*see* Fig. 1).

The protocols described here provide methods for the quantification of VACV DNA replication by qPCR as well as quantification of early, intermediate, and late gene transcription activity using quantitative reverse transcriptase-PCR (qRT-PCR) from infected cell culture samples.

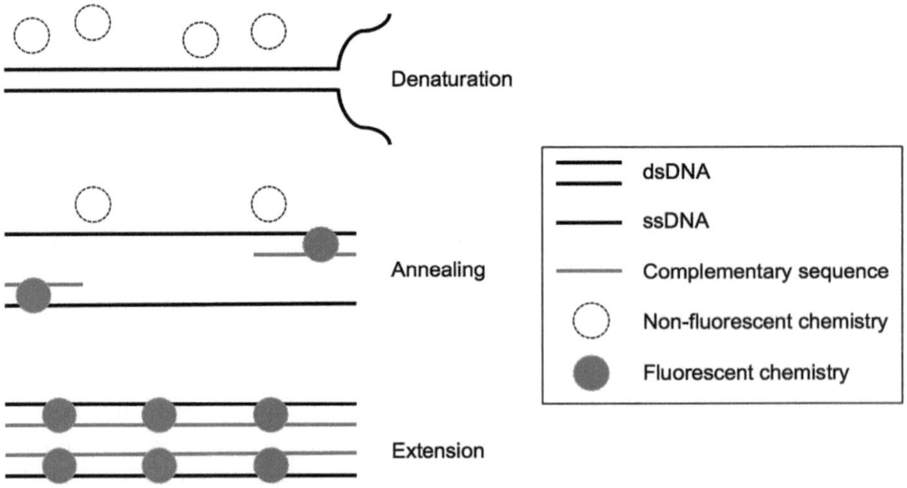

Fig. 1 qPCR reaction. Just like in endpoint PCR, there are three major steps that make up each cycle in qPCR reaction: denaturation, annealing, and extension. During denaturation high temperature "melts" double-stranded DNA into single strands and loosens secondary structure of single-stranded DNA. Complementary sequences (primers) will hybridize during annealing, and at the same time fluorescent dye binds to newly formed hybrids. At optimal temperature DNA polymerase becomes active, and the primer extension occurs

2 Materials

1. Cell culture medium: Dulbecco's modified Eagle's medium (DMEM) supplemented with 10% heat-inactivated fetal bovine serum (FBS), 1% GlutaMAX, 1% penicillin/streptomycin (Pen/Strep).
2. Infection medium: DMEM without supplements.
3. VACV virus, Western Reserve (WR) strain (wild type [WT]) (*see* **Note 1**).
4. HeLa cells (American Type Culture Collection) (*see* **Note 2**).
5. 35 mm Cell culture dishes.
6. Phosphate-buffered saline (PBS).
7. Cell scrapers.
8. 1.5 mL Tubes.
9. Nuclease-free water.
10. White 96-well plates for PCR and cap strips (*see* **Note 3**).
11. qPCR detection chemistry (*see* **Note 4**).
12. qPCR machine (*see* **Note 4**).
13. Cell culture incubator (37 °C, 5% CO_2).
14. Tabletop centrifuges for 1.5 mL tubes, PCR tubes, and PCR 96-well plates.
15. Pipettes and tips.
16. Ice.

2.1 Specific Materials for qPCR Reaction

1. Samples: Uninfected (UI), WT infected (WT), and WT-infected cytosine arabinoside (AraC)-treated HeLa cells, 8 hpi.
2. 40 mM AraC diluted in water (4000 × stock).
3. DNA extraction kit (*see* **Notes 4** and **5**).
4. VACV genomic DNA (gDNA) with known DNA concentration (*see* **Note 6**).
5. Template DNA (DNA extracted from samples).
6. VACV early gene C11-specific primers [4]:
   (a) Forward:5′-AAACACACACTGAGAAAC AGCATAAA-3′
   (b) Reverse:5′-ACTATCGGCGAATGATCT GATTA-3′
7. qPCR reaction buffer (for one reaction): 7.6 μL Nuclease-free water, 0.2 μL forward primer (10 μM), 0.2 μL reverse primer

(10 μM), 10 μL MESA Blue qPCR MasterMix Plus for SYBR® Assay MasterMix.

8. Thermomixer (56 °C).

2.2 Specific Materials for Two-Step qRT-PCR

1. Samples: Uninfected (UI), WT infected (WT), and WT-infected Hoechst-treated HeLa cells, 2, 4, and 8 hpi.
2. 16 mM Hoechst diluted in water (80,000 × stock) (*see* **Note 7**).
3. RNA extraction kit (*see* **Notes 4** and **5**).
4. 96–100% Ethanol.
5. 14.3 M Beta-mercaptoethanol (BME).
6. Blunt 20-gauge needles.
7. RNase-free syringes.
8. Reverse transcriptase enzyme (*see* **Note 4**).
9. Oligo(dT)$_{12-18}$ (500 μg/mL).
10. dNTP mix (10 mM each).
11. Extracted sample mRNA.
12. PCR tubes.
13. Endpoint PCR machine (*see* **Note 8**) or sterile water bath/heat block capable of heating 42 °C, 65 °C, and 70 °C.
14. Mixture of Oligo(dT)$_{12-18}$ and dNTP mix (for one reaction): 1 μL Oligo(dT)$_{12-18}$ (500 μg/mL), 1 μL dNTP mix (10 mM each), 9 μL nuclease-free water.
15. Buffer mixture (for one reaction): 4 μL 5× First-strand buffer (provided by the kit), 2 μL 0.1 M DTT (provided by the kit), and 1 μL nuclease-free water.
16. VACV early gene J2-specific primers [4]:
    (a) Forward: 5′-TACGGAACGGGACTATGGAC-3′
    (b) Reverse: 5′-GTTTGCCATACGCTCACAGA-3′
17. VACV intermediate gene G8-specific primers [4]:
    (a) Forward: 5′-AATGTAGACTCGACGGATGAGTTA-3′
    (b) Reverse:5′-TCGTCATTATCCATTACGATTCTAGTT-3′
18. VACV late gene F17 specific primers [4]:
    (a) Forward: 5′-ATTCTCATTTTGCATCTGCTC-3′
    (b) Reverse: 5′-AGCTACATTATCGCGATTAGC-3′
19. Glyceraldehyde 3-phosphate dehydrogenase (GAPDH)-specific primers (housekeeping gene) [4]:
    (a) Forward: 5′-AAGGTCGGAGTCAACGGATTTGGT-3′
    (b) Reverse: 5′-ACAAAGTGGTCGTTGAGGGCAATG-3′

3 Methods

Make sure that all reagents are sterile, pure, and of PCR grade. Use sterile technique.

3.1 Quantifying VACV DNA Replication Using qPCR

3.1.1 VACV Virus Infection

1. Infect confluent 35 mm dishes of HeLa cells (*see* **Notes 2** and **9**) at multiplicity of infection (MOI) of 10 with WT VACV in infection medium.
2. Incubate in cell culture incubator for 1 h. UI sample is treated similarly but without virus.
3. Remove infection medium and replace with 1 mL cell culture medium.
4. For AraC sample replace infection medium with cell culture medium containing 1× AraC.
5. Place samples in cell culture incubator for 7 h.
6. Scrape cells and place in 1.5 mL tubes.
7. Centrifuge for 5 min at 300 × *g*.
8. Discard the supernatant and freeze the cell pellets (−20 °C or −70 °C) or proceed with total DNA extraction (Subheading 3.1.2) (*see* **Note 10**).

3.1.2 Total DNA Extraction

Extract total DNA from samples with Qiagen DNeasy® DNA extraction kit (*see* **Notes 4** and **5**). **Steps 3–7** should be performed in virus hood after which it is safe to move to a PCR-quality lab bench.

1. Prepare buffers AL, AW1, and AW2 according to the manufacturer's instructions.
2. Preheat thermomixer to 56 °C.
3. Resuspend cell pellets from Subheading 3.1.1, **step 8**, in 200 μL PBS (*see* **Note 10**).
4. Lyse samples by adding 20 μL proteinase K (*see* **Note 11**).
5. Add 200 μL buffer AL, and mix by vortexing (*see* **Note 12**).
6. Incubate at 56 °C for 10 min.
7. Add 200 μL ethanol (96–100%), and mix by vortexing (*see* **Note 13**).
8. Pipette cell lysates into provided spin columns.
9. Centrifuge at 6000 × *g* for 1 min.
10. Discard the flow-through and collection tube.
11. Place the spin column into a new collection tube and add 500 μL buffer AW1.
12. Centrifuge for 1 min at 6000 × *g*.
13. Discard the flow-through and collection tube.
14. Place the spin column into a new collection tube and add 500 μL buffer AW2.

15. Centrifuge for 3 min at 20,000 × *g* (*see* **Note 14**).
16. Place the spin column in a clean 1.5 mL tube and add 200 μL nuclease-free water directly onto the spin column membrane.
17. Incubate at room temperature for 1 min.
18. Centrifuge for 1 min at 6000 × *g* to elute the DNA (*see* **Note 15**).
19. Continue straight to qPCR (Subheading 3.1.3) or store the samples at −80 °C or −20 °C.

3.1.3 qPCR

Quantify the amount of viral DNA in samples using commercial detection chemistry (for example MESA Blue qPCR MasterMix Plus for SYBR® Assay, Eurogentec, *see* **Note 4**).

1. Thaw all reagents and keep them on ice. Mix all reagents well and spin them down prior to pipetting.
2. Plan your pipetting/plate layout (*see* **Note 16** and Table 1).
3. Prepare VACV gDNA dilution series (*see* **Notes 6** and **17**).
4. Prepare the template DNA (extracted in Subheading 3.1.2) by diluting them 1:200 in nuclease-free water (*see* **Note 18**).
5. Prepare qPCR reaction buffer being sure to mix thoroughly by inversion and spin down (*see* **Note 19**).
6. Pipette 2 μL of diluted template DNA into the bottom of the wells as planned (*see* Table 1).
7. Add 18 μL of qPCR reaction buffer into each well (final volume 20 μL, *see* **Note 20**).
8. Spin samples down (3 min, 300 × *g*).
9. Run qPCR according to qPCR reagent's instructions (*see* **Note 21**).

Table 1
Layout plan for qPCR plate. Vaccinia gDNA dilution series (DNA dil. 1–8), uninfected (UI), vaccinia wild-type infected (WT), vaccinia WT infected and AraC treated (AraC), negative control (NT, "no template"), technical triplicates (roman numerals)

	1	2	3	4	5	6	7	8	9	10	11	12
A	DNA dil. 1	II	III	UI	II	III						
B	DNA dil. 2	II	III	WT	II	III						
C	DNA dil. 3	II	III	AraC	II	III						
D	DNA dil. 4	II	III	NT	II	III						
E	DNA dil. 5	II	III									
F	DNA dil. 6	II	III									
G	DNA dil. 7	II	III									
H	DNA dil. 8	II	III									

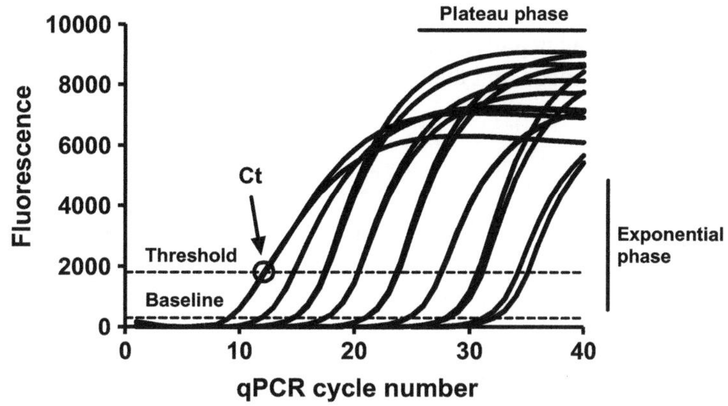

Fig. 2 Amplification curves. The amplification curve shows the increase of detected fluorescence on the *y*-axis and the PCR run cycle number on the *x*-axis. The threshold is the level of fluorescence above the baseline at which the signal can be considered above background. The threshold cycle (Ct) is the cycle number at which the fluorescent signal of the reaction crosses the threshold. In the plateau phase one of the reagents in the reaction becomes limited

3.1.4 Quantification of qPCR Results

All qPCR machines report amplification results in a similar way. Usually the first graph to look at is the amplification curve (*see* Fig. 2). The threshold cycle (Ct) is the cycle number at which the fluorescent signal of the reaction crosses the threshold (*see* **Note 22**).

Since several factors can affect qPCR result analysis and comparison (*see* **Note 23**), in order to make high-quality data comparison, it is important to use correct control samples, data normalization, and quantification methods (*see* **Note 24** and Table 2). This section describes how to do absolute quantification from the raw qPCR data (*see* **Note 25**).

1. To form a standard curve, plot the VACV gDNA concentration dilutions against their corresponding Ct values (*see* Fig. 3).
2. Determine the function for the curve. The function for this example standard curve is $y = -1.492 \ln(x) + 25.415$.
3. Calculate the absolute amount of starting material in your samples by plugging the average sample Ct values from your raw data into standard curve equation (*see* **Note 26** and Table 3).
4. Check your positive and negative controls and evaluate the quality of the results (*see* Table 3 and Subheading 3.1.5).

3.1.5 Evaluation of qPCR Results

This section describes how to evaluate the qPCR results using the standard and melting curves. The overall PCR efficiency can be checked from the slope value of the standard curve (*see* Fig. 3b). The slope of the standard curve should be between −3.1 and −3.58. The linearity of the standard curve (R^2) should be higher than 0.985 (*see* Fig. 3b). When using detection method based on

Table 2
(A) Commonly used negative and positive controls and their aims. (B) Commonly used qPCR normalization methods and their advantages and disadvantages

(A)	
Negative controls	
No template	Detection of primer dimers and contamination
No reverse transcription	Detection of genomic DNA contamination
Positive controls	
Same sample, different target	Quality of reagents and normalization
Same target, different sample	Quality of reagents
(B)	
Normalization to the original cell number/sample quantity	
+ Can minimize variability	- Only approximation - Does not account biases in DNA/RNA extraction - Only for cell culture/blood samples
Normalization to the total DNA/RNA mass	
	- DNA/RNA mass determined by photospectrometer is not accurate - Does not control for differences in efficiency in reverse transcription or qPCR reactions - Very sensitive to operator variation
Normalization to one or more housekeeping genes	
+ Controls for DNA/RNA quality and quantity, and differences in both reverse transcriptase and qPCR efficiencies	- The gene(s) must be present at consistent level in all samples

SYBR Green I, the system specificity should be tested by running a melting curve analysis at the end of the qPCR run (*see* Fig. 4). A specific melting curve should show one, unique melting peak per primer pair. If additional peaks appear, the primer pair is annealing with itself and you should avoid using it.

3.2 Quantifying VACV Transcriptional Activity Using qRT-PCR

3.2.1 VACV Virus Infection

1. Infect confluent 35 mm dishes of HeLa cells (*see* **Notes 2** and **9**) at MOI 10 with WT VACV in infection medium.
2. Incubate in cell culture incubator for 1 h. UI sample is treated similarly but without virus.
3. Remove infection medium and replace with 1 mL cell culture medium.
4. For Hoechst sample replace infection medium with cell culture medium containing 1× Hoechst.
5. Place samples in cell culture incubator for 1, 3, or 7 h.
6. At indicated time points, scrape cells and transfer to 1.5 mL tubes.

A

Vaccinia gDNA (pg)	Average Ct
0.004	34.88
0.04	29.49
0.4	26.37
4	23.09
40	19.74
400	16.12
4000	13.22
40000	10.13

B

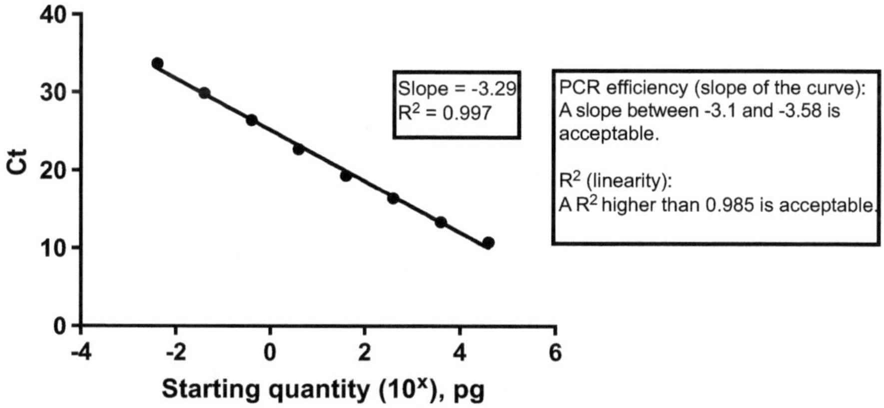

Fig. 3 A standard curve. (**a**) A standard curve is established by a dilution series of known template concentrations. (**b**) To plot the standard curve each known concentration in the dilution series (*x*-axis) is plotted against the Ct value for that concentration (*y*-axis). From the standard curve, the initial starting amount of the target template in experimental samples can be determined. Additional information about the performance of the qPCR reaction and various reaction parameters (including efficiency and R^2) can be determined from this plot

Table 3
Calculating the absolute amounts of starting material. The amount of starting material was calculated by placing the average sample Ct values into equation gained from standard curve. In this example experiment the equation was $x = e^{((25.415 - y)/1.492)}$, where x is the calculated DNA mass and y is the average Ct of the sample. Uninfected sample (UI), vaccinia wild-type infection (WT), vaccinia DNA replication inhibitor-treated infected samples (AraC), negative control (NT, "no template"), non-determined (N.D.)

Sample	Average Ct	Calculated DNA mass (pg)
UI	N.D.	–
WT	19.00	73.83
AraC	28.17	0.16
NT	N.D.	–

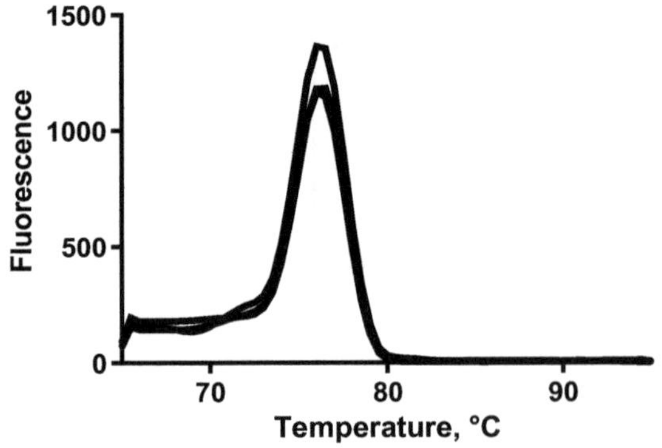

Fig. 4 Melting curve. One peak in melting curve means there is no primer dimerization

7. Centrifuge for 5 min at 300 × *g*.
8. Determine the number of cells before continuing to the next step (*see* **Note 27**).
9. Discard the supernatant and proceed with total RNA extraction (Subheading 3.2.2) or freeze cell pellets at −80 °C (*see* **Notes 10** and **28**).

3.2.2 Total RNA Extraction

Harvest total RNA from infected cells using a commercial kit according to the manufacturer's instructions (for example RNeasy®, Qiagen) (*see* **Notes 4** and **5**). Steps 3–6 should be performed in a virus hood after which it is safe to move to a PCR-quality lab bench.

1. Adfd BME to buffer RLT according to the manufacturer's protocol (10 μL BME per 1 mL buffer RLT) (*see* **Note 29**).
2. Prepare buffer RPE according to the manufacturer's instructions.
3. Disrupt cell pellets from Subheading 3.2.1, **step 8**, by adding 350 μL buffer RLT (*see* **Notes 30–32**).
4. Mix by vortexing or pipetting (*see* **Note 35**).
5. Homogenize the lysate by passing it through a 20-gauge needle at least five times (*see* **Notes 33** and **34**).
6. Add 350 μL 70% ethanol to the homogenized lysate (*see* **Note 35**). After this step it is safe to move from virus hood to PCR-quality lab bench.
7. Mix well by pipetting (*see* **Note 36**).
8. Transfer up to 700 μL of the sample (including any precipitate) to a supplied spin column.
9. Centrifuge for 15 s at 8000 × *g*.
10. Discard the flow-through (*see* **Note 37**).
11. Add 700 μL buffer RW1 to the spin column.

12. Centrifuge for 15 s at 8000 × *g*.
13. Discard the flow-through (*see* **Note 38**).
14. Add 500 μL buffer RPE to the spin column (*see* **Note 39**).
15. Centrifuge for 15 s at 8000 × *g*.
16. Discard the flow-through.
17. Add 500 μL buffer RPE to the spin column.
18. Centrifuge for 2 min at 8000 × *g* (*see* **Note 40**).
19. Place the spin column in a new 1.5 mL tube.
20. Add 30 μL nuclease-free water directly to the spin column membrane (*see* **Note 41**).
21. Centrifuge for 1 min at 8000 × *g* to elute the RNA (*see* **Note 42**).
22. Continue straight to qRT-PCR step (Subheading 3.2.3) or store samples at −80 °C or −20 °C.

3.2.3 Two-Step qRT-PCR (Reverse Transcription Step)

After the extraction, a portion of total RNA is reverse transcribed into cDNA with reverse transcriptase enzyme (for example SuperScript II, Invitrogen, *see* **Notes 4** and **43**) and primers according to the manufacturer's protocol.

1. Plan your final qPCR plate layout (*see* **Notes 16** and **44**, and Table 4).
2. Thaw 5× first-strand buffer and 0.1 M DTT (both provided with the kit) at room temperature just prior to use and refreeze immediately.
3. Measure total RNA concentrations of your samples from Subheading 3.2.2, **step 22**, with Nanodrop or other spectrometric method.
4. Calculate the total volume of RNA needed for the reverse transcription reaction (*see* **Note 45**).
5. Prepare mixture of Oligo(dT)$_{12–18}$ and dNTP Mix in nuclease-free PCR tube. Make enough for all reactions (*see* **Notes 46** and **47**).
6. Pipette 11 μL of Oligo(dT)$_{12–18}$, dNTP Mix in pre-labeled PCR tubes.
7. Add 1 μL of extracted RNA sample in each tube (*see* **Note 45**).
8. Spin samples down briefly.
9. Heat samples to 65 °C for 5 min and quickly chill on ice (*see* **Note 8**).
10. Collect the contents by brief spin.
11. Prepare enough buffer mixture for all reactions.
12. Add 7 μL buffer mixture into each reaction tube.
13. Mix tube gently and spin briefly.

Table 4
Layout plan for qRT-PCR. Uninfected (UI), vaccinia wild-type infected (WT), vaccinia WT infected and Hoechst treated (H), negative control (NT, "no template"), housekeeping gene (GAPDH), vaccinia early gene (J2), vaccinia intermediate gene (G8), vaccinia late gene (F17), technical triplicates (Roman numerals)

	1	2	3	4	5	6	7	8	9	10	11	12
A	Ctrl UI 2 h	II	III	Ctrl UI 4 h	II	III	Ctrl UI 8 h	II	III			
B	Ctrl WT 2 h	II	III	Ctrl WT 4 h	II	III	Ctrl WT 8 h	II	III			
C	Ctrl H 2 h	II	III	Ctrl H 4 h	II	III	Ctrl H 8 h	II	III			
D	Ctrl NT	II	III	Ctrl NT	II	III	Ctrl NT	II	III			
E	J2 UI 2 h	II	III	G8 UI 4 h	II	III	J2 UI 8 h	II	III			
F	J2 WT 2 h	II	III	G8 WT 4 h	II	III	J2 WT 8 h	II	III			
G	J2 H 2 h	II	III	G8 H 4 h	II	III	J2 H 8 h	II	III			
H	J2 NT	II	III	G8 NT 4 h	II	III	J2 NT	II	III			

14. Incubate at 42 °C for 2 min (*see* **Notes 8** and **48**).
15. Add 1 μL (200 units) SuperScript II reverse transcriptase into each tube (*see* **Note 49**).
16. Mix gently and spin briefly.
17. Incubate at 42 °C for 50 min (*see* **Notes 8** and **50**).
18. Inactivate the reaction by heating at 70 °C for 15 min (*see* **Note 8**).
19. With newly synthesized cDNA (final volume 20 μL), continue to qPCR step (Subheading 3.2.3, *see* **Note 51**).

3.2.4 Two-Step qRT-PCR (qPCR Step)

Quantify the amount of viral cDNA in samples using commercial detection chemistry (for example MESA Blue qPCR MasterMix Plus for SYBR® Assay, Eurogentec, *see* **Note 4**).

1. Thaw all reagents and keep them on ice. Mix all reagents well and spin them down prior to pipetting.
2. Plan your pipetting/plate layout (*see* **Note 16** and Table 4).
3. Prepare qPCR reaction buffer for all samples (*see* **Notes 19** and **52**).
4. Mix thoroughly by inversion and spin down.
5. Pipette 2 μL of template cDNA into the bottom of the wells as planned (*see* **Note 16** and Table 4).
6. Add 18 μL of qPCR reaction buffer to each well (final volume 20 μL, *see* **Note 20**).
7. Spin samples down (3 min, 300 × *g*).
8. Run qPCR according to qPCR reagent's instructions (*see* **Note 21**).

3.2.5 Quantification of qRT-PCR Results

The last step of qRT-PCR is essentially a normal qPCR reaction as already outlined in Subheading 3.1.4. This chapter describes how to calculate the relative amount of gene of interest normalized against the housekeeping gene (*see* **Note 53**). Start by calculating ΔCt values for qRT-PCR results. This is done by subtracting the Ct value of the housekeeping gene from the Ct value of the same "gene of interest" sample (*see* Table 5A, B). For example $Ct_{UI\ J2} - Ct_{UI\ GAPDH}$: 31.86–18.04 = 13.82. Next, ΔΔCt values are calculated by subtracting the ΔCt value of the chosen control sample from the ΔCt value of the gene of interest (*see* Table 5C). For example: $\Delta Ct_{UI\ J2} - \Delta Ct_{WT\ J2}$: 13.82 − (−5.65) = 19.47. Finally calculate the final relative quantification from the ΔΔCt values with equation $0.5^{(\Delta\Delta Ct)}$ (*see* Table 5D). For example for UI 2h J2 sample: $0.5^{19.47} = 1.38 \times 10^{-6}$.

4 Notes

1. Besides WR, many other strains of VACV (e.g., Copenhagen, IHD-J) can be used in qPCR assays.
2. Besides HeLa, other cell lines can also be used.
3. Depending on the number of samples, individual or strips of PCR tubes can also be used. Check the compatibility between PCR tubes and qPCR machine.
4. Several companies provide qPCR machines, qPCR detection chemistries, nucleic acid extraction kits, and reverse transcriptase enzymes. It is important to make sure that they are compatible. Examples of such reagents and machines:
   (a) qPCR machine: CFX Connect Real-Time System (BioRad).

Table 5

(A) Average Ct results from qRT-PCR run. (B) Calculated ΔCt values ($Ct_{Gene\ of\ interest} - Ct_{Housekeeping\ gene}$) for qRT-PCR results. Values were calculated by subtracting the Ct value of the housekeeping gene from Ct value of the same "gene of interest" sample. (C) Calculated ΔΔCt values ($\Delta Ct_{Gene\ of\ interest} - \Delta Ct_{Control}$) for qRT-PCR results. Values were calculated by subtracting the ΔCt value of the chosen control sample (WT) from ΔCt of the gene of interest. (D) Final relative quantification. Values were calculated by equation $0.5^{(\Delta\Delta Ct)}$. Uninfected (UI), vaccinia wild-type infected (WT), vaccinia WT infected and Hoechst treated (H), negative control (NT, "no template"), housekeeping gene (GAPDH), vaccinia early gene (J2), vaccinia intermediate gene (G8), vaccinia late gene (F17)

(A)			
Ct	2 h, GAPDH	4 h, GAPDH	8 h, GAPDH
UI	18.04	17.36	17.39
WT	18.59	18.86	21.57
H	18.38	18.40	19.65
NT	N.D.	N.D.	N.D.
	2 h, J2	4 h, G8	8 h, F17
UI	31.86	30.00	29.50
WT	12.93	21.33	15.28
H	12.49	28.30	16.22
NT	N.D.	N.D.	N.D.
(B)			
ΔCt	2 h, J2	4 h, G8	8 h, F17
UI	13.82	12.64	12.11
WT	−5.65	2.47	−6.29
H	−5.88	9.90	−3.42
(C)			
ΔΔCt	2 h, J2	4 h, G8	8 h, F17
UI	19.47	10.16	18.39
WT	0	0	0
H	−0.23	7.42	2.87
(D)			
Final	2 h, J2	4 h, G8	8 h, F17
UI	1.38×10^{-6}	8.72×10^{-4}	2.91×10^{-6}
WT	1	1	1
H	1.17	5.82×10^{-3}	0.14

(b) qPCR detection chemistry: MESA Blue qPCR MasterMix Plus for SYBR® Assay (Eurogentec).

(c) DNA extraction kit: DNeasy® Blood and Tissue extraction kit (Qiagen).

(d) RNA extraction kit: RNeasy® RNA extraction kit (Qiagen).

(e) Reverse transcriptase enzyme: SuperScript™ II reverse transcriptase (Invitrogen).

5. It is highly recommended to use commercially available kits for extraction of nucleic acids and qPCR detection. The nucleic acid extraction step is probably the most critical in any qPCR assay. The quality of the extraction will influence the quality of the DNA detection and quantification. With biological samples it is also very important to ensure the reproducibility of the extraction and any following steps. For qPCR detection, companies usually provide so-called Core Kits that contain all essential components (except the template nucleic acid and primers) in separate tubes or MasterMixes in which all essential components are already mixed in an optimized way. The advantage of Core Kits is that each component can be separately optimized. However, MasterMixes can save time and provide high reproducibility and ease of use.

6. VACV gDNA [7] concentration is measured by spectrometric method (Nanodrop).

7. Hoechst is an inhibitor of VACV intermediate and late gene transcription [4].

8. Endpoint PCR machine can be used in heating and cooling steps of two-step qRT-PCR.

9. In order to obtain optimum nucleic acid yield and quality, it is important to ensure that an appropriate number of cells are used in the extraction step. See the manufacturer's instructions.

10. Fresh or frozen cell pellets can be used in nucleic acid extraction. Best results are obtained with fresh material or material that has been immediately frozen and stored at −20 °C or −70 °C. With DNA samples, avoid freeze/thaw cycles, since this reduces DNA size. If using frozen cell pellets, allow cells to thaw before adding PBS (Subheading 3.1.2, **step 3**).

11. The activity of proteinase K provided with this kit is 600 mAU/mL solution.

12. Mix sample and buffer AL immediately and thoroughly by vortexing or pipetting to yield a homogenous solution.

13. Sample and ethanol should be mixed thoroughly to yield a homogenous solution.

14. It is important to dry the membrane of the spin column by centrifuging, since residual ethanol may interfere with subsequent reactions. Remove the spin column carefully so that the column does not come into contact with the flow-through.
15. DNA can also be eluted into provided buffer AE. Elution with 100 μL instead of 200 μL increases the final DNA concentration in the eluate, but also decreases the overall DNA yield. For maximum DNA yield, repeat elution once as described in Subheading 3.1.2, **step 18**. If the eluate volume exceeds 200 μL, use 2 mL tube.
16. Planning the pipetting and plate layout helps you to determine what samples and the quantity of reagents are needed. It is important to make a technical duplicate, preferably triplicate, for each sample and include all necessary control samples (positive and negative). The reproducibility of an experiment is indicated by replicates (technical and biological). An example plate layout for qPCR can be found in Table 1 and for qRT-PCR in Table 4.
17. Dilution series for absolute quantification needs to cover the whole range of concentrations within your samples. In this example experiment we initially diluted stock VACV gDNA (100 ng/μL) to 20 ng/μL solution. This dilution was used as our first gDNA dilution in the series. From there we made the following dilutions (all 1:10): 2 ng/μL, 0.2 ng/μL, 20 pg/μL, 2 pg/μL, 0.2 pg/μL, 0.02 pg/μL, and 0.002 pg/μL (dilutions 2–8). It is recommended to make enough dilutions for all your biological replicates at once.
18. In this example experiment the samples were normalized by volume (all samples were diluted 1:200 and equal volumes were used in the qPCR assay).
19. Calculate the required amount of qPCR reaction buffer and prepare at least 10% extra for avoid running short. In this example experiment you will need 18 μL qPCR reaction buffer per sample.
20. Pipette onto walls of the well, and do not touch the bottom/template DNA; this way you can use the same tip throughout the plate.
21. Use cycle temperatures and times suggested by qPCR detection chemistry provider. The qPCR machine might do the melting curve automatically or you have to activate it yourself. For MESA Blue qPCR MasterMix Plus:
    (a) 5 min 95 °C (nuclease activation)
    (b) 15 s 95 °C followed by 1 min 60 °C (40 cycles)
    (c) Perform a melting curve (or hold at 50 °C forever).

22. Low Ct values means that the fluorescence crosses the threshold early, and that the amount of target in the sample is high. It is important to quantify the qPCR reaction in the early part of the exponential phase as opposed to when the reaction reaches the plateau phase when one of the reagents in the reaction becomes limited (*see* Fig. 2). As a rule of thumb technical replicates that have more than 0.5 Ct difference should be avoided.
23. Several factors such as the amount of starting material, enzymatic efficiencies, and differences in experimental conditions may have an effect on analysis and qPCR result comparison.
24. The two major methods of quantification of qPCR results are the absolute quantification and the relative quantification. In absolute quantification, a dilution series of known template concentrations can be used to establish a standard curve (*see* Fig. 3). This can then be used for determining the initial starting amount of the target template in experimental samples (*see* Table 3). In relative quantification the expression of a gene of interest in one sample (i.e., treated) is compared to the expression of the same gene in another sample (i.e., untreated). The results are presented as fold change. In relative quantification a housekeeping gene is used as a normalizer for experimental variability (Subheading 3.2.4).
25. Controls in this example experiment are "no template" (negative) and "same target, different sample" (positive).
26. For absolute quantification of the starting amount of VACV genome in samples, change the equation to $x = e^{((25.415 - y)/1.492)}$.
27. It is essential to use the correct number of cells to obtain optimal RNA yield and purity. See the manufacturer's instructions.
28. Incomplete removal of infection medium will inhibit lysis in the following steps and dilute the lysate, reducing the RNA yield.
29. Dispense BME in a fume hood wearing appropriate protective clothing. Buffer RLT containing BME can be stored at room temperature for up to 1 month.
30. Frozen pellets should be thawed before moving onto next step.
31. The volume of buffer RLT depends on the initial number of cells. See the manufacturer's instructions.
32. For direct lysis of cell monolayer, add the appropriate volume of buffer RLT to the cell culture dish. Collect the lysate by scraping the cells and pipetting suspension into 1.5 mL tube. For pelleted cells, loosen the cell pellet thoroughly by flicking the tube. Incomplete loosening of the cell pellet or insufficient

mixing leads to inefficient lysis and reduced RNA yields. Add appropriate volume of buffer RLT.

33. Other homogenization methods include commercial homogenizing spin columns and mechanical homogenizers. See more info in the manufacturer's protocol.
34. Incomplete homogenization leads to significantly reduced RNA yields and can also cause clogging of spin columns in subsequent steps.
35. You should add 1 volume of 70% ethanol. The volume of lysate may be less than the original 350 μL due to the loss during homogenization. Precipitates may be visible after addition of ethanol. This does not affect the procedure.
36. Do not centrifuge.
37. If the sample volume exceeds 700 μL, repeat Subheading 3.2.2, **steps 9–11**. Discard the flow-through after each centrifugation.
38. Remove the spin column from the collection tube carefully so that the column does not contact the flow-through. Empty the collection tube completely.
39. Ensure that ethanol is added to buffer RPE before use. See the manufacturer's instructions.
40. The long centrifugation step dries the spin column membrane (residual ethanol may interfere with downstream reactions). Remove the spin column from the collection tube carefully so that the column does not contact the flow-through.
41. Add 30–50 μL nuclease-free water.
42. If the expected RNA yield exceeds 30 μg, repeat Subheading 3.2.2, **steps 21** and **22**, using another 30–50 μL nuclease-free water or using the elute form **step 22**. If using the elute from **step 22**, the RNA yield will be less when obtained using a second volume of nuclease-free water, but the final RNA concentration will be higher.
43. The qRT-PCR can be done in one or two steps, both of which have their own pros and cons. One-step qRT-PCR combines the cDNA synthesis reaction and qPCR reaction in the same tube, simplifying reaction setup and reducing the possibility of contamination between the reverse transcription and qPCR steps. Gene-specific primers are required for one-step qRT-PCR, and it is usually less sensitive than two-step qRT-PCR. In this example experiments were using the two-step qRT-PCR method. Two-step qRT-PCR starts with the reverse transcription of total mRNA into cDNA using a reverse transcriptase. Usually the cDNA synthesis reaction is primed using a mixture of commercially available random primers or a mixture of Oligo(dT) and random primers. These should give an equal

representation of all targets in later qPCR applications. The two-step qRT-PCR also gives the possibility to stock cDNA and is the more flexible of the two options as the two reactions (reverse transcription and qPRC) can be optimized separately. One disadvantage with the two-step qRT-PCR is that as it requires more pipetting, there is a higher risk of RNase inhibitors to influence the qPCR reaction.

44. Since this example experiment is going to use relative quantification, quantifications of genes of interest (J2, G8, or F17) and the housekeeping gene (GAPDH) have to be made from each sample (*see* Table 4).

45. In this example we are pipetting identical mRNA volumes of different samples and we decided to reverse transcribe ~500 ng of RNA per sample. In all samples RNA concentrations (measured with Nanodrop) were 450–590 ng/μL, so we decided to continue with 1 μL of each sample to reverse transcription reaction. Manufacturer's protocol gives a working range for the enzyme. With SuperScript™ II enzyme the 20 μL reaction volume can be used for 1 ng to 5 μg of total RNA or 1–500 ng of mRNA.

46. Calculate your need reaction buffer and prepare at least 10% extra for avoid running out. In this example experiment you will need 11 μL reaction buffer per sample.

47. Instead of Oligo$(dT)_{12-18}$ (500 μg/mL), random primers (50–250 ng) or gene-specific primer (2 pmol) can be used.

48. If random primers are used, incubate at 25 °C for 2 min.

49. If less than 1 ng of RNA is used, reduce the amount of reverse transcriptase enzyme to 0.25 μL and add nuclease-free water to obtain the final volume of 20 μL.

50. If random primers are used, incubate tube at 25 °C for 10 min prior to 50-min incubation.

51. Amplification of some large PCR targets (>1 kb) may require removal of RNA complementary to the cDNA. This can be done by adding 1 μL (2 units) of *E. coli* RNase H to the tube and incubating at 37 °C for 20 min.

52. Forward and reverse primers for genes of interest (VACV early, intermediate, and late genes {J2, G8, F17} and housekeeping gene {GAPDH}) are diluted in nuclease-free water (10 μM). In this example experiment J2 expression was checked from 2-h sample, G8 from 4-h sample, and F17 from 8-h sample. Since GAPDH serves as a housekeeping gene (and normalizing factor) in this experiment, its expression should be checked from all samples.

53. In the example experiment controls are "no template" (negative) and "same sample, different target" (positive).

Acknowledgments

HM and JM are supported by core funding to the MRC Laboratory for Molecular Cell Biology at University College London (J.M.), the European Research Council (649101-UbiProPox), and the UK Medical Research Council (MC_UU12018/7).

References

1. Moss B (2001) Poxviridae: the viruses and their replication. In: Knipe DM, Howley PM (eds) Fields virology, vol 2. Lippincott William & Wilkins, Philadelphia, PA, pp 2849–2883
2. Traktman P, Boyle K (2004) Methods for analyzing of poxvirus DNA replication. In: Isaacs SN (ed) Methods in molecular biology, vol 1. Humana Press, Totowa, NJ, pp 169–185
3. Senkevich TG, Koonin EV, Moss B (2009) Predict poxvirus FEN1-like nuclease required for homologous recombination, double-strand break repair and full-size genome formation. PNAS 106:17921–17926
4. Yakimovich A, Huttunen M, Zehnder B, Coulter LJ, Gould V, Schneider C et al (2017) Inhibition of poxvirus gene expression and genome replication by bisbenzimide derivatives. J Virol 91:e00838–e00817
5. Jones EV, Moss B (1985) Transcriptional mapping of the vaccinia virus DNA polymerase gene. J Virol 53:312–315
6. Yang Z, Bruno DP, Martens GA, Porcella SF, Moss B (2010) Simultaneous high-resolution analysis of vaccinia virus and host cell transcriptomes by deep RNA sequencing. PNAS 107:11513–11518
7. Roper R (2004) Rapid preparation of vaccinia virus DNA template for analysis and cloning by PCR. In: Isaacs SN (ed) Methods in molecular biology, vol 1. Humana Press, Totowa, NJ, pp 113–118

Chapter 13

Click Chemistry-Based Labeling of Poxvirus Genomes

Harriet Mok and Artur Yakimovich

Abstract

Vaccinia virus packages its dsDNA genome inside its core for protection during the extracellular phases of its life cycle. In the cytoplasm of a newly infected cell the viral genome is released from the core so the viral DNA replication machinery can access it and initiate DNA replication. Vaccinia virus replication sites in the cell cytosol can be detected with conventional DNA staining methods; these, however, do not provide enough specificity to be used for quantitative image analysis or further probing of the replication step. Likewise, the ability to generate recombinant vaccinia viruses with fluorescently tagged proteins has provided insight into many stages of the viral life cycle, but many of the early steps involving the viral genome remain to be elucidated. Nucleotide and nucleoside analogs are traditionally used for probing the cell cycle and investigating other changes in cellular DNA, with the more novel nucleoside analogs providing a better way to label with click chemistry. Here we demonstrate how nucleoside analogs and click chemistry can be used for tracking poxvirus replication in the viral factories, and tracking single viral genomes in infected cells.

Key words Poxvirus, Vaccinia virus, Nucleoside analogs, Click chemistry, Genome, DNA replication

1 Introduction

Nucleotide analogs have been recognized as invaluable research compounds for cell proliferation assays [1]. For example, the thymidine analog, bromodeoxyuridine (5-bromo-2′-deoxyuridine, BrdU), is known to incorporate into newly synthesized DNA during the S phase of the cell cycle [2] with its degree of incorporation being assessed by immunolabeling [3, 4]. Although the most widely used nucleoside analog, it has various unwanted effects, including invoking conformational changes in DNA, alteration of cell cycle progression, and cell toxicity [5–7]. Furthermore, BrdU labeling requires harsh treatments with heat and acid to unwind the DNA and allow for immunostaining, and therefore risks sample destruction and low sign-to-noise ratio in microscopy.

A number of novel alkyne-containing nucleoside analogs, such as 5-ethynyl-2′-deoxyuridine (EdU), 7-deaza-7-ethynyl-2′-deoxyadenosine (EdA), and deoxy-5-ethynylcytidine (EdC) [8,

Jason Mercer (ed.), *Vaccinia Virus: Methods and Protocols*, Methods in Molecular Biology, vol. 2023, https://doi.org/10.1007/978-1-4939-9593-6_13,

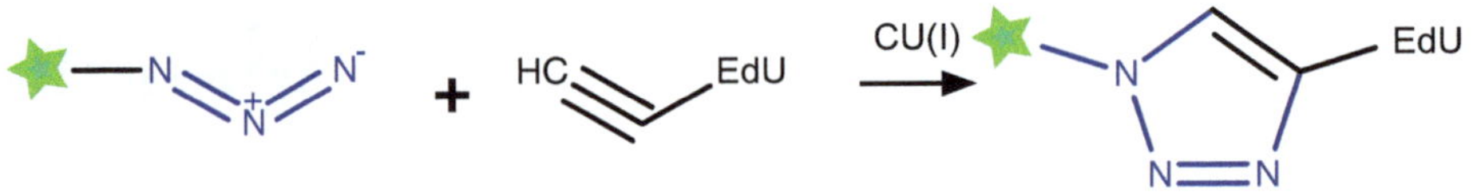

Fig. 1 Principle reaction scheme of copper(I)-catalyzed azide-alkyne cycloaddition used for fluorescent labeling of clickable nucleoside analogs

9], can be directly detected using click chemistry with an azide (N_3^-)-containing fluorescent dye [10–13]. This type of copper(I)-catalyzed azide-alkyne cycloaddition (CuAAC) [14, 15] uses a fluorescent azide reagent with a terminal alkyne to covalently form a five-membered heteroatom ring, without quenching the label (Fig. 1).

The low molecular mass of reagents involved in the reaction of nucleoside analogs and fluorescent azides enables click chemistry labeling to be much less sensitive to spatial restrictions of the crowded environment of a cell. Furthermore, the rarity of terminal alkyne residues within a living cell also allows for high specificity of labeling.

In addition to their use in cell cycle assays, clickable nucleoside analogs can be used to directly visualize viral replication sites [16]. During vaccinia virus (VACV) replication, formation of viral replication sites, or factories, concludes the early stage of viral infection. As a large double-stranded DNA virus which packages its own replication machinery, VACV replicates its genome exclusively in the cytoplasm [17]. This process occurs in the replication factories which, due to their rich dsDNA content, can be visualized by staining VACV-infected cells with DNA intercalators or minor groove binders, such as DAPI or Hoechst, respectively [16, 18]. DNA dyes may often be unspecific, however, and due to the large size and dense structure of the viral replication sites, they may often be mistaken for cellular nuclei. This poses problems when carrying out quantitative image analysis of viral replication.

Another use for these nucleoside analogs in poxvirology is tracking single viral genomes in early stages of the virus life cycle. Poxviruses are the largest, most complex mammalian viruses [17]. They package an ~180 Kbp which is condensed and packaged into the virus core for protection during the extracellular phase of the VACV life cycle. While advances in the generation of fluorescent recombinant viruses [19–21] have allowed for further investigation of many steps in the life cycle, those involving visualization of viral dsDNA remain largely undefined.

Incorporation of nucleoside analogs into the viral genome allows for closer investigation of the steps surrounding viral DNA uncoating and replication, while facilitating tracking of single VACV genomes during the early stages of the virus life cycle, even when still packaged within the viral core (*see* Fig. 2). In this chapter we outline how to use the clickable nucleoside analog EdU to

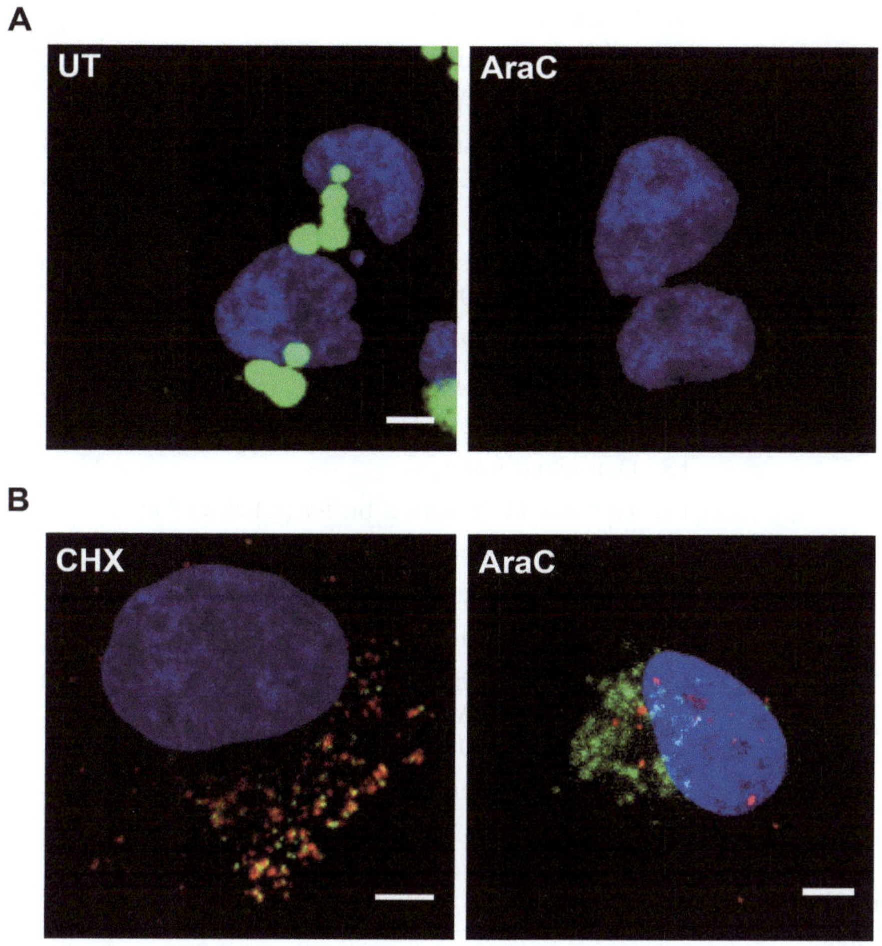

Fig. 2 Examples of click chemical EdU staining. (**a**) HeLa cells infected with WR VACV stained for VACV replication centers 8 h postinfection. The click chemistry reaction to couple an Alexa Fluor™ 488 azide to EdU was carried out on fixed infected cells. AraC was used as a negative control compared to untreated (UT). Scale bar, 5 μm. (**b**) Click chemistry of HeLa cells infected with WR VACV core-mCherry EdU-DNA. The click chemistry reaction to couple an Alexa Fluor™ 488 azide to EdU was carried out on fixed infected cells. CHX and AraC were used as controls to demonstrate the visualization of the viral genome without viral cores, respectively. Scale bar 5 μm

visualize virus DNA replication sites and generate EdU-DNA VACV viruses that can be used to visualize and track single viral genomes.

2 Materials

2.1 Click Chemistry of VACV Replication Sites

1. Infection medium: Dulbecco's modified Eagle medium (DMEM).
2. Phosphate-buffered saline (PBS, pH 7.4).
3. Cell culture medium: DMEM, 10% fetal bovine serum, nonessential amino acids, penicillin/streptomycin, sodium pyruvate.

4. HeLa cells (ATCC).
5. Tissue culture incubator (37 °C, 5% CO_2).
6. Band-purified Western Reserve VACV (VACV WR) [22].
7. Cytarabine (AraC, cytosine b-D-arabinofuranoside) 1 mM stock in ddH_2O.
8. Microscopy-compatible dishes or #1.5 coverslips.
9. 2× Working solution of EdU (2-ethynyl-2′-deoxyuridine, 2 μM, from 10 mM stock in DMSO) (*see* **Note 1**).
10. 4% w/v Paraformaldehyde (PFA) solution in PBS.
11. 50 mM NH_4Cl in PBS.
12. 0.5% v/v Triton® X-100 solution.
13. Deionized water.
14. 1× Click-iT® reaction buffer (Click-iT® kit).
15. 1× CuSO4 (component E of the Click-iT® kit).
16. 1× Alexa Fluor® Azide (Click-iT® kit).
17. 1× Reaction buffer additive (Click-iT® kit).
18. Click-iT® reaction cocktail: Click-iT® reaction buffer/$CuSO_4$/Alexa Fluor® Azide/Reaction buffer additive mixed in a ratio of 21.4/1/0.062/2.5 (*see* Click-iT® kit manual).
19. Blocking buffer: 5% w/v BSA in PBS.
20. Hoechst 33342 nuclear dye (*see* **Note 2**).

2.2 Recombinant EdU-DNA Virus Generation

1. Infection medium: Dulbecco's modified Eagle medium (DMEM).
2. Cell culture medium: DMEM, 10% fetal bovine serum, nonessential amino acids, penicillin/streptomycin, sodium pyruvate.
3. BSC-40 (African green monkey kidney) cells.
4. Roller bottles or 15 cm tissue culture-treated dishes (*see* **Note 3**).
5. Tissue culture incubator (37 °C, 5% CO_2), with roller if bottles are used.
6. 10 mM Stock solution EdU in DMSO stored at −20 °C
7. Western Reserve VACV core-mCherry BSC-40 crude extract [21].
8. Cell scrapers.
9. 50 mL Conical centrifuge tubes.
10. 15 mL Conical centrifuge tubes.
11. PBS (pH 7.4).
12. 20 mM Tris, pH 9.0.
13. 10 mM Tris, pH 9.0.

14. 1 mM Tris, pH 9.0.
15. 36% Sucrose solution (36% sucrose w/v in 20 mM Tris pH 9.0).
16. 25% Sucrose solution (25% sucrose w/v in 20 mM Tris pH 9.0).
17. 40% Sucrose solution (36% sucrose w/v in 20 mM Tris pH 9.0).
18. 15 mL Tight-fitting Dounce homogenizer.
19. DNase I.
20. Centrifuge (pre-cooled to 4 °C).
21. Ultracentrifuge (pre-cooled to 4 °C).
22. SW32 and SW40 ultracentrifuge tubes and rotors.
23. Mineral oil.
24. 12 cm Cannula (*see* **Note 4**).
25. BioComp Gradient Master and gradient pouring accessories.
26. Water bath or horn sonicator.
27. 21 G Needle.
28. Scotch tape.
29. BD Plastipak 2.5 mL syringes.

2.3 Click Chemistry of EdU-DNA Viral Genomes

1. Tissue culture-treated dish.
2. HeLa cells.
3. Clean coverslips high performance (Zeiss).
4. Band-purified WR VACV core-mCherry EdU-DNA.
5. Infection medium: Dulbecco's modified Eagle medium (DMEM).
6. Cell culture medium: DMEM, 10% fetal bovine serum, nonessential amino acids, penicillin/streptomycin, sodium pyruvate.
7. Tissue culture incubator (37 °C, 5% CO_2).
8. AraC (cytarabine).
9. CHX (cycloheximide).
10. Dark box, for click chemistry and staining (must fit tissue culture dish).
11. PBS (pH 7.4).
12. 4% PFA diluted in PBS.
13. Permeabilization solution: 0.1% v/v Triton® X-100 diluted in PBS.
14. Blocking solution: 5% BSA diluted in PBS.

15. Deionized water.
16. 1× Click-iT® EdU reaction buffer in deionized water stored at 2–6 °C.
17. 10× Click-iT® EdU buffer additive in deionized water stored in aliquots at −20 °C.
18. Click-iT reaction cocktail (per coverslip): 322.5 μL 1× Click-iT™ reaction buffer, 15 μL $CuSO_4$, 0.9 μL Alexa Fluor™ 488 azide, and 37.5 μL 1× EdU buffer additive (*see* **Note 5**).
19. Working solution of Alexa Fluor™ 488 in DMSO stored at −20 °C.
20. Hoechst (5 μg/mL) diluted in PBS stored at 4 °C.

3 Methods

3.1 Visualizing VACV Genome Replication by Nucleoside Analog Labeling

The following procedures have been optimized using Click-IT® EdU labeling kit, but its core principles can be extrapolated to similar kits.

1. Seed HeLa cells in microscopy-compatible dishes or on coverslips.
2. Place dishes in tissue culture incubator overnight.
3. Prepare 2× working solution of EdU (*see* **Note 6**).
4. Perform the infection assay by adding VACV WR inoculum at a desired multiplicity of infection (e.g., MOI 10).
5. Add EdU working solution at 1× final concentration.
6. Remove the VACV inoculum and incubate cells according to the assay desired.
7. Fix cells with 4% PFA. It is crucial to use azide-free reagent throughout the entire experiment (*see* **Note** 7).
8. Upon fixation wash cells with PBS three times.
9. Quench the fixed cells with 50 mM NH_4Cl in PBS for 5 min.
10. Permeabilize fixed cells by incubating them with 0.5% Triton® X-100 (*see* **Note 8**) for 15 min.
11. Prepare Click-iT® reaction cocktail being sure that the final volume is sufficient to cover the cells in all sample wells. Use Click-iT® reaction cocktail within 15 min of mixing.
12. Wash cells twice with blocking buffer.
13. Add Click-iT® reaction cocktail to the cells.
14. Incubate the cells for 30 min at room temperature, protecting the sample from the direct sunlight during the incubation.
15. Remove the reaction cocktail.
16. Wash samples twice with blocking buffer.

17. Proceed to nuclear staining if desired (*see* **Note 2**).
18. Image using preferred imaging technique compatible with wavelengths of fluorophores used.

3.2 Tracking Single Viral Genomes by Nucleoside Analog-Labeled Virus

Depending on assay requirements any recombinant vaccinia virus can be used, but as an example we used a virus with an mCherry-tagged core protein allowing for labeling of the genome with AF488 after EdU incorporation (*see* Fig. 3). The protocol for band-purifying VACV has been optimized for BSC-40 cells but other cell lines may be used [21].

3.2.1 Recombinant EdU-DNA Virus Generation

1. Infect confluent cells with VACV crude extract, 100 μL extract in 10 mL infection medium per roller bottle, and incubate for 30 min in tissue culture incubator.
2. Add 40 mL of cell culture medium with a final concentration of 1 μM EdU.
3. Incubate for 48 h in tissue culture incubator.
4. After incubation, scrape cells into conical centrifuge tube (*see* **Note 9**).
5. Pellet cells by centrifugation at 300 × *g* for 5 min.
6. Aspirate the supernatant and discard.
7. Resuspend the cell pellets from three roller bottles in 50 mL PBS and repeat **steps 4** and **5** to rinse cells of media.

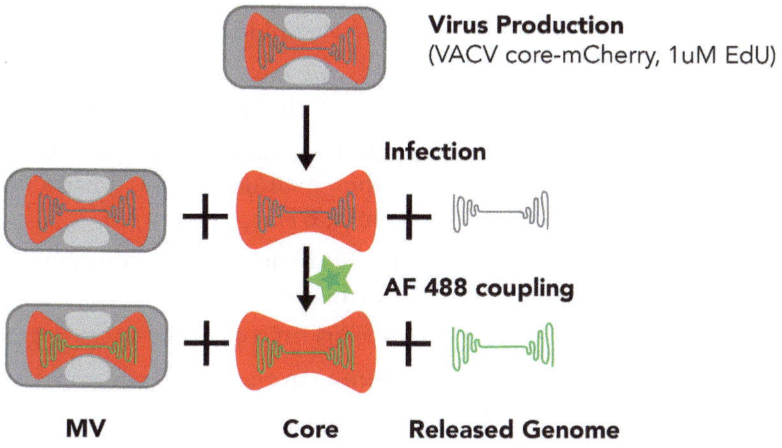

Fig. 3 Click chemistry of EdU-DNA virus. EdU incorporated into a VACV genome can be detected using click chemistry and coupling to an azide. When used in a recombinant virus with a fluorescent core, the clickable genome can be visualized when released, but also when still in a mature virion or viral core

8. Continue with purification protocol, or alternatively store the cell pellet at −80 °C.
9. Resuspend the pellet in 12 mL 10 mM Tris, pH 9.0.
10. Leave on ice for 10 min.
11. Disrupt cells using 25 strokes in a tight-fitting Dounce.
12. Transfer the extract to a 15 mL conical centrifuge tube.
13. Spin at 2000 × *g* for 10 min at 4 °C.
14. Transfer the supernatant to a 15 mL conical being careful not to disturb the pellet.
15. Spin at 2000 × *g*, 10 min, 4 °C.
16. Transfer the supernatant to a 50 mL conical being careful not to disturb the pellet (*see* **Note 10**).
17. Treat with 2 mg DNase I for 30 min at 37 °C (*see* **Note 11**).
18. Adjust supernatant to 20 mL using 10 mM Tris, pH 9.0.
19. Prepare a SW32 tube with 16 mL 36% sucrose solution.
20. Carefully layer the supernatant from **step 18** on top of the sucrose.
21. Carefully layer 2 mL mineral oil on top of the supernatant.
22. Ultracentrifuge tubes in an SW32 rotor at 38,000 × *g*, 4 °C, for 1 h 20 min.
23. Carefully aspirate the supernatant.
24. Resuspend the sedimented virus in 300 μL 1 mM Tris, pH 9.0. Freeze virus solution at −80 °C or proceed with band purification.
25. For band purification mark the middle of an SW40 tube and fill up to the middle with 25% sucrose solution (approx. 6 mL).
26. Using a 12 cm cannula carefully underlay the 40% sucrose solution to the marked midpoint.
27. Prepare the gradient using the BioComp Gradient Master as per the manufacturer's instructions (BioComp settings: 3:00 min, 81.5°, 18 rpm).
28. Vortex and sonicate the resuspended virus from **step 23** for 10 s each.
29. Load virus on top of the sucrose gradient.
30. Centrifuge in the SW40 rotor at 12,000 × *g*, 4 °C, 52 min.
31. To pull the band, cover the side of the tube where the virus band resides with scotch tape.
32. Carefully puncture the side of the tube below the virus band using a 21 G needle.

33. Aspirate the purified virus. This will be approximately 2 mL (*see* **Note 12**).
34. Transfer the virus band to a new SW40 tube.
35. Add 1 mM Tris, pH 9.0, to a final volume of 12 mL.
36. Spin virus at 38,000 × *g*, 4 °C, for 40 min in ultracentrifuge.
37. Carefully aspirate supernatant.
38. Resuspend virus pellet in 300 μL 1 mM Tris, pH 9.0.
39. The virus titer can then be determined as normal [23].

3.2.2 Tracking Single Viral Genomes with Click Chemistry

1. Seed cells for experiments on coverslips the day before (*see* **Note 13**).
2. Dilute band-purified WR core-mCherry EdU-DNA virus (*see* Subheading 3.2.1) in infection medium. The MOI used will depend on the assay requirements (*see* **Note 14**).
3. Leave cells at RT for 30 min to allow for virus binding.
4. Feed cells with cell culture medium (*see* **Note 15**).
5. Incubate cells at 37 °C for 4.5 h.
6. Aspirate the medium and replace with 4% PFA.
7. Incubate at RT for 20 min.
8. Wash cells three times with blocking solution.
9. Incubate in permeabilization solution for 15 min.
10. Carry out three washes of the coverslips using blocking buffer.
11. Incubate coverslips in blocking buffer at RT for a minimum of 15 min.
12. Meanwhile prepare 1× solution of Click-iT™ EdU buffer additive by diluting 10× in deionized water (*see* **Note 16**).
13. Then prepare the Click-iT reaction cocktail being sure to prepare enough for all coverslips.
14. Remove the blocking solution from the coverslips and add 375 μL of the Click-iT reaction cocktail to each coverslip, rocking to ensure that the whole coverslip is coated evenly.
15. Incubate the plate at RT for 30 min, protected from light.
16. Remove the reaction cocktail and wash immediately with blocking solution three times.
17. Replace blocking buffer with Hoechst and incubate for 30 min at RT protected from light.
18. Wash with PBS three times for 5 min each wash.
19. Mount coverslips for imaging and analysis as required.

4 Notes

1. Higher EdU concentrations may lead to the interference with VACV replication [16].
2. Nuclear staining may be performed using Hoechst 33342, 33480, 33258, or other dyes like DAPI and DRAQ5. The procedure may also be compatible with a subsequent immunofluorescent staining. It is, however, subject to specific epitope stability and hence may or may not work for certain antibody combinations.
3. One cell culture roller bottle is equivalent to 5 × 15 cm dishes.
4. As an alternative to a 12 cm cannula a needle can be used with the tip slightly bent so as not to puncture the ultracentrifuge tube.
5. Azides with other emission spectra can be used depending on other fluorophores used in the specific assays.
6. The protocol presented has been designed to visualize VACV by addition of EdU nucleoside analog. The final concentration of EdU in the culturing medium or infectious inoculum should be 1 μM. As a negative control for VACV replication site formation, a sample containing 10 μM AraC can be included. If total VACV replication site visualization is desired, EdU should be added in the very beginning of the infection and included for the duration of infection until fixation. If visualization of virus life cycle kinetics is desired, pulse labeling of cells upon VACV infection by brief exposures to EdU can be used.
7. Make sure to use azide-free solutions; azide-containing solutions (e.g., PBS containing NaN_3) will interfere with the click chemical staining procedure by competing for fluorophore cycloaddition (*see* click reaction scheme in Fig. 1).
8. Methanol, saponin, or other permeabilization agents can be used and do not interfere with the Click-iT® kit reaction.
9. If the entire virus purification step is to be carried out in one go, pre-cool all centrifuges to 4 °C.
10. After this spin, the pellet of cell debris should be hard to see and distinguish from the color of the supernatant.
11. A DNase I treatment is required on the post-nucleic extract to remove any contaminating cellular DNA which may have also incorporated EdU.
12. The band of purified virus will sit close to the center of the tube and appear white opalescent.
13. To confirm incorporation of EdU into the viral genome, virions can be analyzed using click chemistry methods. For this,

1% v/v Triton® X-100 in 5% BSA should be used to permeabilize the virions for 30 min at RT on an orbital shaker before performing the Click-iT reaction.

14. For example, in our core stabilization assays (Fig. 2) we used an MOI of 10.
15. Use AraC (cytarabine) and CHX (cycloheximide) as controls, with final concentrations of 10 μM and 1 mM, respectively (Fig. 2).
16. It is important to prepare the solution fresh and use it on the same day. The 10× aliquots of buffer additive should not be reused once thawed as the buffer is very sensitive.

Acknowledgments

AY, HM, and JM were supported by core funding to the MRC Laboratory for Molecular Cell Biology at University College London (J.M.), the European Research Council (649101-UbiProPox), and the UK Medical Research Council (MC_UU12018/7).

References

1. Lehner B, Sandner B, Marschallinger J, Lehner C, Furtner T, Couillard-Despres S, Rivera FJ, Brockhoff G, Bauer H-C, Weidner N (2011) The dark side of BrdU in neural stem cell biology: detrimental effects on cell cycle, differentiation and survival. Cell Tissue Res 345(3):313
2. Nowakowski R, Lewin S, Miller M (1989) Bromodeoxyuridine immunohistochemical determination of the lengths of the cell cycle and the DNA-synthetic phase for an anatomically defined population. J Neurocytol 18(3):311–318
3. Cooper-Kuhn CM, Kuhn HG (2002) Is it all DNA repair?: methodological considerations for detecting neurogenesis in the adult brain. Dev Brain Res 134(1):13–21
4. Kuhn HG, Cooper-Kuhn CM (2007) Bromodeoxyuridine and the detection of neurogenesis. Curr Pharm Biotechnol 8(3):127–131
5. Goz B (1977) The effects of incorporation of 5-halogenated deoxyuridines into the DNA of eukaryotic cells. Pharmacol Rev 29(4):249–272
6. Poot M, Hoehn H, Kubbies M, Grossmann A, Chen Y, Rabinovitch PS (1994) Cell-cycle Analysis using continuous bromodeoxyuridine labeling and Hoechst 33358—ethidium bromide bivariate flow cytometry. Methods Cell Biol 41:327–340
7. Caldwell MA, He X, Svendsen CN (2005) 5-Bromo-2′-deoxyuridine is selectively toxic to neuronal precursors in vitro. Eur J Neurosci 22(11):2965–2970
8. Guan L, van der Heijden GW, Bortvin A, Greenberg MM (2011) Intracellular detection of cytosine incorporation in genomic DNA by using 5-ethynyl-2′-deoxycytidine. Chembiochem 12(14):2184–2190
9. Qu D, Wang G, Wang Z, Zhou L, Chi W, Cong S, Ren X, Liang P, Zhang B (2011) 5-Ethynyl-2′-deoxycytidine as a new agent for DNA labeling: detection of proliferating cells. Anal Biochem 417(1):112–121
10. Neef AB, Samain F, Luedtke NW (2012) Metabolic labeling of DNA by purine analogues in vivo. Chembiochem 13(12): 1750–1753
11. Salic A, Mitchison TJ (2008) A chemical method for fast and sensitive detection of DNA synthesis in vivo. Proc Natl Acad Sci 105(7):2415–2420
12. Cavanagh BL, Walker T, Norazit A, Meedeniya AC (2011) Thymidine analogues for tracking DNA synthesis. Molecules 16(9):7980–7993
13. Kolb HC, Finn M, Sharpless KB (2001) Click chemistry: diverse chemical function from a few good reactions. Angew Chem Int Ed 40(11):2004–2021

14. Rostovtsev VV, Green LG, Fokin VV, Sharpless KB (2002) A stepwise huisgen cycloaddition process: copper(I)-catalyzed regioselective "ligation" of azides and terminal alkynes. Angew Chem 114(14):2708–2711
15. Tornøe CW, Christensen C, Meldal M (2002) Peptidotriazoles on solid phase:(1, 2, 3)-triazoles by regiospecific copper(I)-catalyzed 1, 3-dipolar cycloadditions of terminal alkynes to azides. J Org Chem 67(9):3057–3064
16. Wang I, Suomalainen M, Andriasyan V, Kilcher S, Mercer J, Neef A, Luedtke NW, Greber UF (2013) Tracking viral genomes in host cells at single-molecule resolution. Cell Host Microbe 14(4):468–480
17. Moss B (2013) Poxviridae. In: Fields BN, Knipe DM, Howley PM et al (eds) Fields Virology, vol 1. 6 edn. Lippincott Williams & Wilkins, a Wolters Kluwer business, Philadelphia, PA, p 2664
18. Yakimovich A, Huttunen M, Zehnder B, Coulter LJ, Gould V, Schneider C, Kopf M, McInnes CJ, Greber UF, Mercer J (2017) Inhibition of poxvirus gene expression and genome replication by bisbenzimide derivatives. J Virol 91(18):e00838-00817
19. Domínguez J, del Lorenzo MM, Blasco R (1998) Green fluorescent protein expressed by a recombinant vaccinia virus permits early detection of infected cells by flow cytometry. J Immunol Methods 220(1–2):115–121
20. Smith GL, Moss B (1983) Infectious poxvirus vectors have capacity for at least 25 000 base pairs of foreign DNA. Gene 25(1):21–28
21. Mercer J, Helenius A (2008) Vaccinia virus uses macropinocytosis and apoptotic mimicry to enter host cells. Science 320(5875): 531–535
22. Kilcher S, Schmidt FI, Schneider C, Kopf M, Helenius A, Mercer J (2014) siRNA screen of early poxvirus genes identifies the AAA+ ATPase D5 as the virus genome-uncoating factor. Cell Host Microbe 15(1):103–112
23. Condit RC, Motyczka A (1981) Isolation and preliminary characterization of temperature-sensitive mutants of vaccinia virus. Virology 113(1):224–241

Chapter 14

Visualizing Poxvirus Replication and Recombination Using Live-Cell Imaging

Quinten Kieser, Patrick Paszkowski, James Lin, David Evans, and Ryan Noyce

Abstract

A modernized version of an old saying goes that "If a picture is worth a thousand words, then a video is worth a million." Although made with reference to "YouTube", the quotation also has relevance for microbiologists when one considers how modern microscopes can be used to track biological fluorophores for hours without bleaching or phototoxicity. Confocal fluorescence microscopy provides a powerful tool for capturing dynamic processes within a cellular context that are better understood when viewed using time-lapse videos. In our laboratory we have long been interested in the links between poxvirus DNA replication and recombination and, since these are cytoplasmic viruses, such DNA-dependent processes are easily imaged throughout the virus life cycle without interference from signals coming from nuclear DNA. In this chapter we outline methods that can be used to follow the movement and replication of vaccinia virus DNA, and to also detect the products of poxvirus-catalyzed recombination reactions. We describe how to use the bacteriophage lambda DNA-binding protein, cro, as a way of labeling DNA within a cell when it is conjugated to fluorescent proteins. When used in conjunction with other fluorescent reagents, new labeling technologies, and tagged reporter constructs, these approaches can generate visually appealing and highly informative insights into diverse aspects of vaccinia virus biology.

Key words Recombination, Poxvirus, Live-cell imaging, Replication, DNA synthesis

1 Introduction

Genetic recombination is a molecular process that produces new combinations of traits through the physical exchange of two (or more) strands of nucleic acid. Recombination between coinfecting viruses has the potential to generate new virus variants with different biological properties including altered virulence. It also provides an essential route by which virus populations can be purged of the defective mutants that would otherwise accumulate over time due to damage and replication errors ("Muller's

Quinten Kieser and Patrick Paszkowski contributed equally to this work.

Jason Mercer (ed.), *Vaccinia Virus: Methods and Protocols*, Methods in Molecular Biology, vol. 2023, https://doi.org/10.1007/978-1-4939-9593-6_14,

ratchet"). From a technical perspective, homologous recombination has been used extensively in poxvirus research, ranging from studies concerning the function of individual virus proteins to the development of poxviruses as vectors capable of expressing multiple foreign antigens [1–3].

We have been studying how poxvirus recombinants are formed at a molecular level as well as trying to integrate these processes into the larger picture of the virus life cycle. Interestingly, recombination between two coinfecting poxviruses seems to be constrained by the fact that each infecting virus particle initiates the formation of a separate replication site called a virus "factory" [4–7]. We have shown that the virus genomes making up these membrane-enclosed structures mix inefficiently and that the products of virus recombination are not detected until well after late gene expression is established [6, 8]. Our working hypothesis is that, although the recombination machinery is assembled early in infection, it is only after these structures start to break down late in the infection cycle that different DNAs can mix well enough to permit recombinant formation.

Here we describe how live time-lapse fluorescence microscopy can be used to track intracellular recombination between two coinfecting viruses. These methods employ the bacteriophage lambda DNA-binding protein, cro, fused to enhanced green fluorescent protein (EGFP). Although cro is best known for being a sequence-specific repressor protein [9], it also binds to DNA in a nonspecific manner and can thus be used to localize any DNA in an infected cell. The EGFP-cro fusion protein thus permits tracking the developing virus factories during vaccinia virus (VACV) infection. We will outline how to construct a BSC-40 cell line that stably expresses EGFP-cro protein as well as two recombinant VACV viruses that harbor overlapping fragments of the mCherry-cro gene. These fluorescent cell lines and recombinant viruses enable us to track both virus factory development and recombinant virus production using live-cell microscopy. More specifically, we can track virus-by-virus recombination between two coinfecting viruses by looking for the appearance of a mCherry-cro protein. The production of this protein is of course dependent upon otherwise optically invisible recombination reactions to reconstruct a full-length mCherry-cro gene.

We and others have noted that there is strong connection between poxvirus replication and genetic recombination [10, 11] in part because poxvirus DNA polymerases can catalyze the single-strand annealing reactions that underpin poxvirus recombination reactions [12, 13]. In order to show that virus replication is also observed at these sites of recombination, we can use recently developed "click chemistry" DNA labeling methods to tag newly synthesized virus DNA [14, 15]. We show here how 5-ethynyl-2′-deoxyuridine (EdU) can be taken up by cells at different times during the course of virus infection, and then used to detect DNA synthesis occurring at sites where one also detects recombinant protein (mCherry) production. Collectively, the methods outlined in this chapter show how correlative fluorescence microscopy, applied to both live and fixed cells, can be used to characterize the links between poxvirus replication and recombination and where and when these processes are active.

2 Materials

2.1 Cell Culture

1. BSC-40 cells (ATCC CRL-2761).
2. Complete minimum essential medium (MEM): MEM supplemented with 5% fetal bovine serum (FBS), 1% nonessential amino acids, 1% sodium pyruvate.
3. OPTI-MEM.
4. Lipofectamine 2000™ transfection reagent.
5. G418 disulfate salt.
6. Sterile cloning cylinders.
7. Sterile silicone vacuum grease.

2.2 Generation of Recombinant Plasmids and Viruses

1. Bacteriophage λ DNA.
2. pEGFP-C1 and pmCherry-C1 plasmids.
3. TOPO pCR 2.1 plasmid.
4. pTM3 plasmid [16].
5. Taq DNA polymerase.
6. 1.0% Agarose gel in 1× Tris-acetate buffer (TAE).
7. QIAEX II gel extraction kit.
8. VACV (strain WR) (ATCC VR-1354).
9. Recombinant VACV liquid selection medium: MEM supplemented with 5% FBS, 25 μg/mL mycophenolic acid (MPA), 15 μg/mL hypoxanthine, and 250 μg/mL xanthine.
10. 2× Plaque selection medium: 2× MEM supplemented with 10% FBS, 50 μg/mL mycophenolic acid (MPA), 30 μg/mL hypoxanthine, and 500 μg/mL xanthine.
11. 1.7% Noble agar (Becton Dickinson) in distilled water.

2.3 Live-Cell and Fixed-Cell Imaging

1. Phenol-red-free FluoroBrite Dulbecco's modified Eagle's medium (DMEM) supplemented with 10 mM HEPES, 1% nonessential amino acids, and 5% FBS (*see* **Note 1**).
2. Optically clear 35 mm high-grid-500 glass-bottom μ-dish.
3. Phosphate-buffered saline (PBS) pH 7.2.
4. 4% Paraformaldehyde in PBS (pH 7.2)

5. 10 mM 5-Ethynyl-2′-deoxyuridine (EdU).
6. Click chemistry reaction (100 mM Tris pH 9.0, 100 mM copper(II) sulfate, Alexa Fluor® azide, 100 mM (+)-sodium L-ascorbate) (*see* **Note 2**).
7. 100 mM Glycine and 0.1% Triton-X-100 in PBS.
8. 1 mg/mL 4′,6-Diamidino-2-phenylinodole (DAPI).
9. Glycerol-based mounting medium with anti-fade reagent.
10. Inverted spinning disk confocal microscope: Olympus IX-81 motorized microscope base with Yokogawa CSU X1 spinning disk confocal scan head, an oil immersion plan-Neofluar 40×, 1.3 NA objective, and a single Hamamatsu EMCCD for image acquisition. The microscope is equipped with a 44 mW 405 nm pumped diode laser (for blue dyes, i.e., DAPI), a 50 mW 491 nm pumped diode laser (for green dyes, i.e., GFP), a 50 mW 561 nm pumped diode laser (for red dyes, i.e., mCherry), and a 45 mW 642 nm pumped diode laser (for far-red dyes, i.e., Alexa 647).
11. Filter sets: Our system is fitted with two interchangeable filter wheels, one for imaging standard "fixed" samples (DAPI, 460/50; Cy5, 690/50, to detect DNA and AF647-conjugated EdU in these experiments), and one for imaging fluorescent proteins in "live" cell experiments (GFP, 520/40; RFP, 595/50, to detect EGFP-cro and mCherry expression in these experiments).
12. Top-stage incubator system: Heated cell chamber, temperature controller including lens warmer, CO_2 air/gas mixer, and humidity controller.
13. Volocity 6.3 imaging acquisition software (Perkin Elmer).
14. FIJI imaging analysis software.
15. Protein-blocking solution: 3% Bovine serum albumin in PBS containing 0.1% Tween-20.

3 Methods

3.1 Construction of Recombinant Cro Plasmids

3.1.1 Construction of a Reporter Cell Line Expressing EGFP Fused to λ Cro

1. The λ *cro* gene is amplified by high-fidelity PCR using bacteriophage λ DNA as template. The two PCR primers (cro-*Hind*III-fwd and cro-*Bam*HI-rev; *see* Table 1) spanning the start and stop codons of *cro* were used to amplify the gene using typical thermocycler conditions of 94 °C for 30 s, 56 °C for 30 s, and 72 °C for 30 s for a total of 30 cycles.
2. The DNA is cloned into pCR2.1, and then digested with *Hind*III and *Bam*HI and sub-cloned into pEGFP-C1 (Clontech). This creates a gene encoding the λ cro protein (66 amino acids) fused to the C-terminus of EGFP.

Table 1
Primers used to construct recombinant cro plasmids[a]

Plasmid	Primer	Sequence (5′ to 3′)
cro-pEGFPC1	cro-*Hind*III-fwd	AAGCTTGTATGGAACAACGCATAACCCTGAAAG
	cro-*Bam*HI-rev	GGATCCTATTATGCTGTTGTTTTTTTGTTACTCGGGA
pTM3-pE/L-mCherry(t)	*Xho*I-pE/L-mCherry-fwd	CGATCACTCTCGAGAAAAATTGAAATTTTATTTTTTTTTTTTGGAATATAAATGGTGAGCAAGGGCGAGG
	*Eco*RI-mCherry(t)-rev	CTAGCTGAGAATTCCTACTGCTTGATCTCGCCCTTC
pTM3-mCherry-cro	*Xho*I-mCherry-fwd	CGATCACTCTCGAGATGGTGAGCAAGGGCGAGG
	*Eco*RI-cro-rev	CTAGCTGAGAATTCTTATGCTGTTGTTTTTTTGTTAC

[a]Restriction endonuclease sites encoded in each primer are underlined

3. Linearize 1 μg cro-pEGFPC1 plasmid with *Mlu*I. Remove salts and restriction enzyme by purification with the QIAEX II purification kit.
4. Transfect the linearized cro-pEGFPC1 plasmid into a subconfluent (~70%) 100 mm dish of BSC-40 cells using Lipofectamine 2000™.
5. Twenty-four hours post-transfection, medium is replaced with fresh medium containing G418 (2 mg/mL) until visible colonies are observed (*see* **Note 3**).
6. Isolate individual cell colonies that have a uniform pattern of predominantly nuclear expression of green fluorescence with cloning cylinders (*see* **Note 4**).

3.1.2 Construction of pTM3-pE/L-mCherry(t) Plasmid

1. Digest the cro fragment from cro-pEGFPC1 with *Hind*III and *Bam*HI. Also digest the pmCherry-C1 plasmid (Clontech) with these restriction enzymes.
2. Separate the digested products by gel electrophoresis and extract DNA fragments using a gel extraction kit.
3. Sub-clone the isolated cro fragment into pmCherry-C1 to create an N-terminal mCherry fused to cro (pmCherry-croC1).
4. Amplify a truncated fragment of mCherry with the *Xho*I-pE/L--mCherry-fwd and *Eco*RI-mCherry(t)-rev primers (*see* Table 1) and using the pmCherry-C1 (Clontech) plasmid as a DNA template using typical thermocycler conditions of 94 °C for 30 s, 63 °C for 30 s, and 72 °C for 30 s for a total of 30 cycles.
5. Digest the truncated mCherry PCR product with *Xho*I and *Eco*RI.
6. Digest the pTM3 plasmid with *Xho*I and *Eco*RI.
7. Gel-purify both the PCR product and the digested plasmid on an agarose gel.
8. Ligate the PCR product and plasmid to create pTM3-pE/L-mCherry(t).

3.1.3 Construction of pTM3-mCherry-cro Plasmid

1. The cro fragment is digested from cro-pEGFPC1 with *Hind*III and *Bam*HI. Digest the pmCherry-C1 plasmid with *Hind*III and *Bam*HI.
2. Separate the DNA by gel electrophoresis and purify each fragment using a gel extraction kit.
3. Clone the isolated cro fragment into pmCherry-C1 to create an N-terminal mCherry fused to cro (pmCherry-croC1).
4. Amplify a promoter-less fragment of mCherry-cro using the *Xho*I-mCherry-fwd and *Eco*RI-cro-rev primers (*see* Table 1) using pmCherry-croC1 as DNA template and typical thermocycler conditions of 94 °C for 30 s, 55 °C for 30 s, and 72 °C for 30 s for a total of 30 cycles.

5. Digest the promoter-less mCherry PCR product with *Xho*I and *Eco*RI.
6. Digest the pTM3 plasmid with *Xho*I and *Eco*RI.
7. Gel-purify both the PCR product and the digested plasmid on an agarose gel.
8. Ligate the PCR product and plasmid to create pTM3-mCherry-cro.

3.2 Generation of VACV-pE/L-mCherry(t) and VACV-mCherry-cro Recombinants by Infection/Transfection

1. Infect BSC-40 cells (60 mm dishes) with VACV (strain WR) at a multiplicity of infection (MOI) of 3 in serum-free MEM for 1 h.
2. Remove virus inoculum and replace with OPTI-MEM.
3. Transfect 2 μg of linearized plasmid DNA (pTM3-mCherry-cro or pTM3-pE/L-mCherry(t)) 2 h postinfection using Lipofectamine 2000™.
4. Twenty-four hours postinfection, scrape the infected cells and perform three freeze-thaw cycles to release recombinant virus.
5. Select recombinant VACV expressing GPT with MEM containing mycophenolic acid selection medium (*see* **Note 5**).
6. Confirm the purity of the recombinant VACV containing mCherry-cro or pE/L-mCherry(t) by PCR analysis (*see* Fig. 1).
7. Grow up and purify stocks of recombinant VACV by ultracentrifugation through a sucrose cushion (*see* **Note 6**).

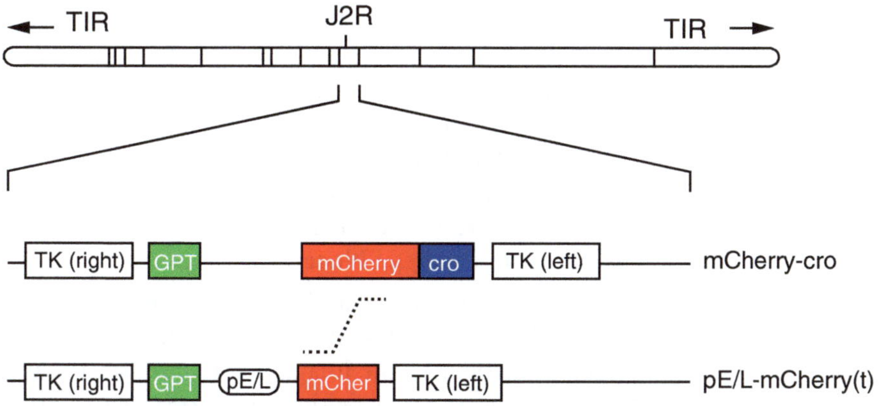

Fig. 1 Schematic representation of the recombinant VACV strains. Recombinant VACV was constructed bearing either a promoterless full-length mCherry-cro fusion protein (mCherry-cro) or a truncated mCherry gene under the control of a synthetic poxvirus early/late (pE/L) promoter (pE/L-mCherry(t)). These inserts also encoded a *gpt* cassette (GPT) for selection using mycophenolic acid. Both constructs are rescued into the VACV J2R locus. For simplicity, these constructs are shown in the conventional orientation, but the inserts are actually inverted relative to the virus genome. Coinfecting cells with these two viruses leads to mCherry expression following virus recombination

3.3 Time-Lapse Microscopy of Virus Recombination During VACV Infection

3.3.1 Infecting EGFP-cro BSC-40 Cells

1. One day before the experiment, seed 2.5×10^5 EGFP-cro BSC-40 cells onto a 35 mm glass-bottom dish.
2. Visualize the EGFP-cro BSC40 cells the following day under a fluorescent microscope to ensure that EGFP-cro signal is predominantly nuclear and that the cell monolayer is approximately 70% confluent.
3. Prior to infection, replace the medium with room-temperature complete MEM containing 10 mM HEPES and place cells at 4 °C for 30 min. During this time, prepare the virus inoculum.
4. In 0.5 mL of room-temperature serum-free MEM containing 10 mM HEPES, dilute VACV-pE/L-mCherry(t) and VACV-mCherry-cro so that each virus is at a MOI of 2.5 (total MOI = 5). Place inoculum on ice until the cells have cooled.
5. Synchronously infect the cells by adding the pre-cooled virus inoculum to the cells and return the cells to 4 °C for 1 h.
6. Following the 1 h of virus binding, remove the inoculum and gently wash the cells 2× with cooled PBS. Add 2 mL of pre-warmed complete FluoroBrite™ DMEM to the plate and incubate at 37 °C for 1 h.

3.3.2 Live-Cell Image Acquisition

1. Set up the spinning-disk confocal microscope for live-cell imaging. Strap the lens warmer around the 40× objective lens and warm the live-cell chamber to 37 °C with 5% CO_2 flow.
2. Seal the 35 mm glass-bottom dish with Parafilm™ and place on the preheated microscope stage. For an oil immersion lens, make sure to use extra oil as some will evaporate during a lengthy imaging experiment. Once in the chamber, carefully remove the Parafilm™ and place the cover onto the live-cell chamber.
3. Using the microscope eyepiece, quickly find EGFP-cro-expressing BSC-40 cells using the GFP eyepiece filter. When a desired cell has been found, switch over to the fluorescence channel. Center the cell of interest in the field of view (or if multiple cells are found, maneuver the field of view so that 2–3 cells can be imaged; *see* **Note** 7 and Fig. 2).
4. In the imaging software, add 8–10 "points" across the infected plate. Each point represents a different field of view. Using the Z-stack controller in the imaging software, set the center point (Z = 0 μm) on the first field of view to be acquired. Using the Z-stack controller, set the top and bottom points for the field of view (*see* **Note 8**).
5. To center the "zero points" of each field of view, center the first imaging point to Z = 0 μm. For each subsequent field of view, drag the Z-stack cursor to your desired center point. This will acquire images of each field of view with the same number of Z-stacks above and below Z = 0 μm.

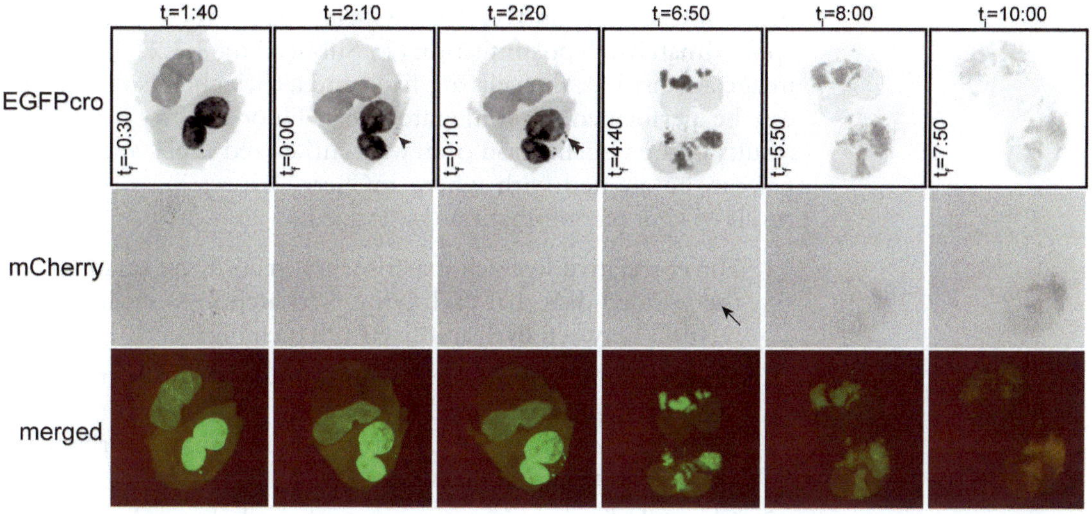

Fig. 2 Visualization of VACV recombination in BSC-40 cells. Live imaging of mCherry expression following coinfection with VACV-pE/L-mCherry(t) and VACV-mCherry-cro viruses. The images show VACV-infected EGFP-cro BSC-40 cells and start about 1 h 40 min post-virus infection (t_i). *Single arrowheads* depict factory formation (t_f). *Double arrowheads* track the time when two virus factories collide (t_f = 0:10). The *single arrow* marks the appearance of recombinant mCherry-cro protein (t_i = 6:50, t_f = 4:40)

6. In the "image acquisition" function of the imaging software, set the order of acquisition to start with the red confocal channel, followed by the green confocal channel. To help with image acquisition speed and sample protection, acquire all images within the Z-stack for channel 1 before switching to a second laser channel.
7. Collect 12 time points per hour (every 5 min) for 121 time points (10 h) for every image point. Z-stack intervals can be between 0.5 μm and 1.0 μm (*see* **Note 9**).
8. After checking all settings, click the "Start" button and watch the first 5-min interval of image acquisition. Make sure that all of the fields of view are finished acquiring before the end of the 5-min interval (*see* **Note 10**).

3.3.3 Detection of Newly Synthesized VACV DNA at Sites of Recombination

To investigate replication at sites of VACV recombination, cells can be pulsed with 5-ethynyl-2′-deoxyuridine (EdU) [17]. EdU is a thymidine analog that bears alkyl groups instead of methyl groups [14]. Fluorescent labels can be covalently attached to these alkyl groups by utilizing a copper-catalyzed click chemistry reaction that then fluorescently tags newly replicated DNA. EdU has multiple advantages compared to traditional BrdU labeling: the smaller molecules associated with click chemistry result in greater tissue penetration, the click reaction is rapid, and DNA does not need to be denatured to attach the fluorescent probe. The following section outlines how to visualize newly synthesized VACV DNA at sites of recombination following coinfection with VACV-pE/L-mCherry(t) and VACV-mCherry-cro; EdU is pulsed in at approximately 6 h postinfection (*see* Subheading 3.3.2). Following image acquisition, the cells are fixed and a click chemistry reaction can be performed to conjugate Alexa Fluor-647 to EdU. This results in the visualization of newly synthesized DNA and the signal can be aligned with where mCherry expression occurs as a result of viral recombination (*see* Fig. 3).

1. For correlative live-cell experiments, record the quadrant on the gridded dish that each point is located in to facilitate correlation between live- and fixed-cell imaging.
2. Remove the cover of the 35 mm gridded glass-bottom dish and place a piece of Parafilm™ over the opening to loosely seal the dish. This Parafilm™ lid introduces minimal disturbance of the dish in a subsequent step.
3. Channels for imaging are captured similarly to the previous section. The red confocal channel is captured prior to the green confocal channel. However, the DIC channel should also be used to capture a reference image point. Typically reference images are captured at the middle of the stack and aid in the correlation between live- and fixed-cell imaging.
4. Set the microscope to capture images every 10 min until 6 h postinfection. After 6 h, capture images every 5 min to facilitate visualization of recombination (*see* **Note 11**).
5. At approximately 4 h post-image acquisition (corresponds to 6 h postinfection), carefully remove the cover to the top-stage incubator system and add approximately 2 μL of 10 mM EdU solution directly to the cell culture dish.
6. Acquire images for an additional 1–2 h (*see* subheading 3.3.2) until mCherry expression is first detected.
7. Remove the live-cell tissue culture dish from the microscope and immediately fix the cells in PFA for at least 30 min at 4 °C (*see* **Note 12**).
8. Permeabilize cells and quench free aldehydes with 100 mM glycine in 0.1% Triton X-100 1× PBS for 20 min at room temperature.
9. Wash cells 3× with wash buffer and incubate cells in the click chemistry reaction (preparation outlined in Table 2) for 30 min at room temperature and protect from light (*see* **Note 13**).
10. Remove the click chemistry reaction. If additional staining is desired (*see* **Note 14**), continue to antibody and DNA labeling; however, EdU can be visualized at this point.
11. Wash the cells 3× with wash buffer.
12. Stain all DNA with DAPI (1 μg/mL) for an additional 10 min at room temperature in the dark.
13. Wash cells 3× with wash buffer and mount coverslips as desired.

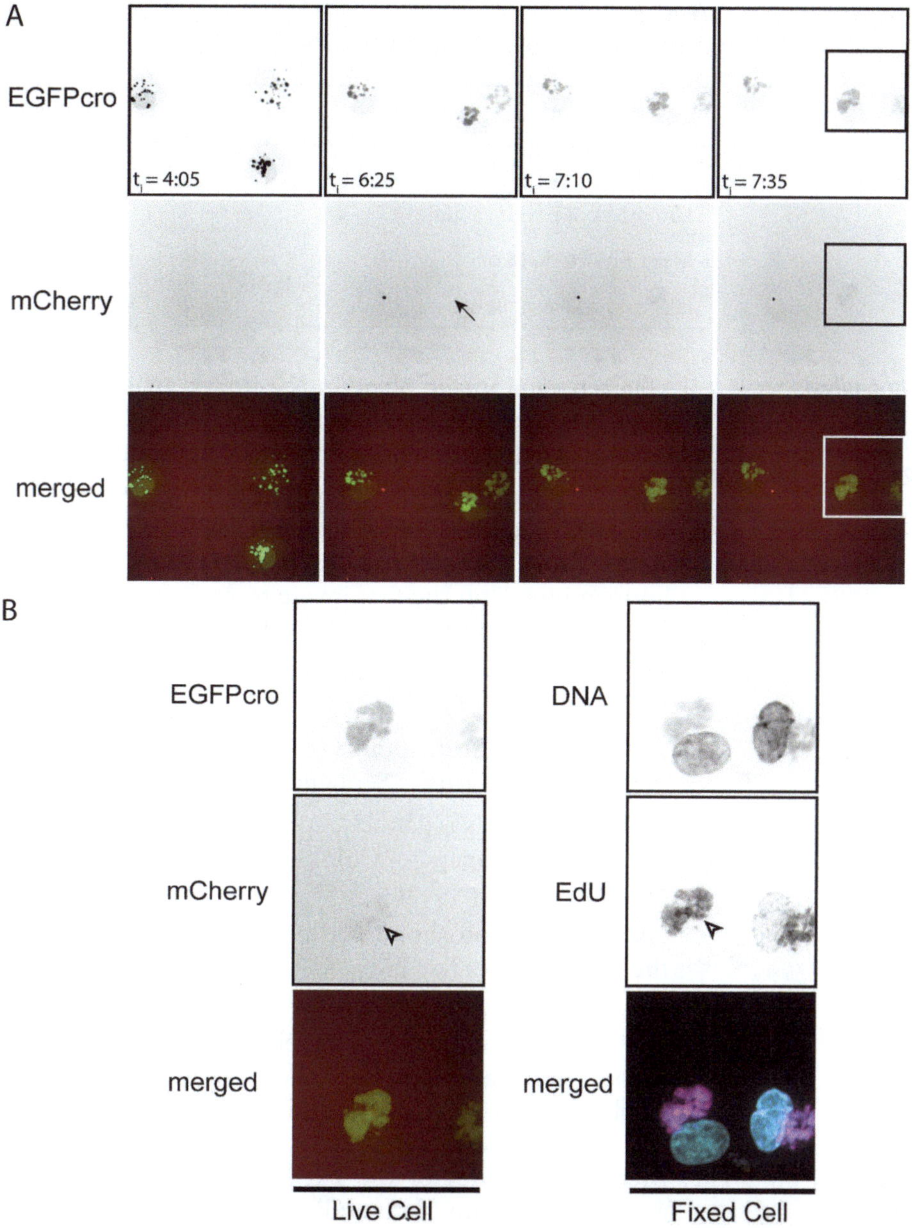

Fig. 3 Detection of newly synthesized DNA at sites of virus-by-virus recombination. (**a**) Live-cell time-lapse microscopy showing the appearance of recombinant mCherry protein following coinfection with VACV-pE/L-mCherry(t) and VACV-mCherry-cro viruses. The images show VACV-infected EGFP-cro BSC-40 cells and begin about 4 h 5 min postinfection (t_i). At t_i = 6:00, 10 μM EdU was added to the infected cells and further images were acquired until mCherry expression was detected at t_i = 7:35. The *single arrow* marks the first appearance of recombinant mCherry-cro protein (t_i = 6:25). (**b**) Correlative microscopy showing newly synthesized DNA near sites of recombinant mCherry expression. After the cells were fixed, click chemistry was used to conjugate Alexa Fluor 647 to the incorporated EdU and shows up in these images as magenta-colored structures. DAPI (cyan) was used to stain the rest of the DNA. The live-cell images shown (boxed) in **a** are enlarged in panel **b** before (at left) and after (at right) fluorescent labeling of the EdU

Table 2
Preparation of click chemistry reaction (per 500 μL)

Reaction component	Volume (μL)
100 mM tris pH 9.0	430
100 mM copper(II) sulfate	20
Alexa Fluor® azide	1.2
100 mM sodium ascorbate	50

3.4 Data Analysis

3.4.1 Live-Cell Microscopy

1. Flatten each image plane in the image analysis software (i.e., Volocity; Perkin Elmer) so that the infected cells can be continuously viewed throughout the image sequence.
2. Open the flattened imaging data in FIJI (*see* **Note 15**). Use the "Time Stamper" (Image → Stacks → Time Stamper) function to add the time to the image sequences. Set the time in hours:minutes on the first image to correspond to the approximate time of infection (*see* **Note 16**).
3. Save image as a merged multichannel video at this point using "Save As → AVI."
4. To extract serial time points from the complete image data, use the "Save As → Image Sequence."
5. To save channels separately, split the image data into the separate channels (Image → Color → Split Channels) and save file as described in **steps 3** and **4**.
6. Use Adobe Photoshop to add annotations, arrows, and etc. to the saved images (*see* Fig. 2).
7. Use Camtasia software (TechSmith) to create multi-panel videos using the separate channel videos and composite video tools. These can be put together side by side with transitions, labels, and arrows with Camtasia software.

3.4.2 Fixed-Cell Microscopy

1. Merge the planes of the fixed-cell images using Volocity to acquire projection (flattened) images.
2. Import the merged images into FIJI to add scale bars and choose appropriate image colors.
3. Use Adobe Photoshop (or similar software) to orient the images so that they align with the time-lapse images that were acquired during the live-cell image acquisition (*see* Fig. 3).

4 Notes

1. Protect from light and store up to several months at 4 °C.
2. Make up sodium L-ascorbate just prior to using it in the click chemistry reaction.

3. Generally, it takes about 4 days for the control BSC-40 cells to die following G418 selection.
4. It is important to confirm the uniformity of nuclear green fluorescence expression whenever a vial of BSC-40 cells expressing EGFP-cro is started from liquid nitrogen. We have observed a loss of EGFP-cro expression in cultures and recommend reselecting cells in G418 if this occurs.
5. In our laboratory we typically select recombinant GPT-expressing VACV under two rounds of liquid selection in MEM containing MPA followed by at least three rounds of selection in MEM containing MPA and 1.7% Noble Agar.
6. It is important that recombinant VACV stocks be purified by ultracentrifugation through a sucrose cushion for all subsequent experiments involving live-cell microscopy. In our laboratory we use poxvirus purification protocols that have been previously described [18, 19].
7. The cells should not be too close to the edges of the field of view, as they will move over time. If you would like to confirm that sites of virus recombination contain newly synthesized DNA, then it is important to take note of the quadrant on the gridded coverslip that each point is located in. This is necessary so that you can find your points after the click chemistry reaction has been carried out.
8. Add extra space on both the top and bottom of each point's Z-stack as the cells will move up and/or down during the imaging experiment.
9. Smaller Z-stack intervals result in larger data files, longer acquisition times, increased photobleaching, and reduced number of points that can be acquired within a 5-min interval.
10. Take note of the image acquisition start time relative to the time you initiated infection. This is important for future data analysis. Typically, we start image acquisition 1–2 h after we add warm medium to the virus-bound BSC-40 cells.
11. Add EdU to the culture dish during the 10-min intervals between image capture in the live-cell experiment. These 10-min intervals are designed to provide enough time to add EdU to the culture dish and realign the z-plane should any drift occur during the initial 4 h of image acquisition.
12. Make up fresh PFA for each experiment.
13. It is important to add the ingredients in the order listed and ensure that the solution is used within 15 min of preparation. Commercially available EdU labeling kits are also available from several commercial suppliers (i.e., Thermo Fisher Scientific).

14. If additional staining with antibodies is required, block the cells in 3% BSA for 20 min at room temperature before subsequent antibody staining.
15. FIJI software is useful for processing images post-acquisition across many different file source types.
16. If you start image acquisition approximately 2 h post-virus infection, then set the time to 2:00 in the Time Stamper.

Acknowledgments

This research was supported by a grant from the Natural Sciences and Engineering Research Council (NSERC). Quinten Kieser and Patrick Paszkowski were both supported by a province of Alberta Queen Elizabeth II graduate scholarship. Quinten Kieser was also supported by a NSERC summer student research award. We would like to thank Megan Desaulniers for critical reading of the manuscript and Greg Plummer in the University of Alberta faculty of Medicine and Dentistry Cell Imaging core facility for excellent technical assistance.

References

1. Moss B (1991) Vaccinia virus: a tool for research and vaccine development. Science 252(5013):1662–1667
2. Carroll MW, Moss B (1997) Poxviruses as expression vectors. Curr Opin Biotechnol 8(5):573–577. S0958-1669(97)80031-6
3. Sanchez-Sampedro L, Perdiguero B, Mejias-Perez E, Garcia-Arriaza J, Di Pilato M, Esteban M (2015) The evolution of poxvirus vaccines. Viruses 7(4):1726–1803. https://doi.org/10.3390/v7041726
4. Katsafanas GC, Moss B (2007) Colocalization of transcription and translation within cytoplasmic poxvirus factories coordinates viral expression and subjugates host functions. Cell Host Microbe 2(4):221–228. S1931-3128(07)00215-6
5. Mallardo M, Leithe E, Schleich S, Roos N, Doglio L, Krijnse Locker J (2002) Relationship between vaccinia virus intracellular cores, early mRNAs, and DNA replication sites. J Virol 76(10):5167–5183
6. Lin YC, Evans DH (2010) Vaccinia virus particles mix inefficiently, and in a way that would restrict viral recombination, in coinfected cells. J Virol 84(5):2432–2443. https://doi.org/10.1128/JVI.01998-09
7. Tolonen N, Doglio L, Schleich S, Krijnse Locker J (2001) Vaccinia virus DNA replication occurs in endoplasmic reticulum-enclosed cytoplasmic mini-nuclei. Mol Biol Cell 12(7):2031–2046
8. Paszkowski P, Noyce RS, Evans DH (2016) Live-cell imaging of vaccinia virus recombination. PLoS Pathog 12(8):e1005824. https://doi.org/10.1371/journal.ppat.1005824
9. Ptashne M, Jeffrey A, Johnson AD, Maurer R, Meyer BJ, Pabo CO et al (1980) How the lambda repressor and cro work. Cell 19(1):1–11
10. Evans DH, Stuart D, McFadden G (1988) High levels of genetic recombination among cotransfected plasmid DNAs in poxvirus-infected mammalian cells. J Virol 62(2):367–375
11. Merchlinsky M (1989) Intramolecular homologous recombination in cells infected with temperature-sensitive mutants of vaccinia virus. J Virol 63(5):2030–2035
12. Gammon DB, Evans DH (2009) The 3′-to-5′ exonuclease activity of vaccinia virus DNA polymerase is essential and plays a role in promoting virus genetic recombination. J Virol 83(9):4236–4250. https://doi.org/10.1128/JVI.02255-08
13. Willer DO, Mann MJ, Zhang W, Evans DH (1999) Vaccinia virus DNA polymerase promotes DNA pairing and strand-transfer reactions. Virology 257(2):511–523

14. Salic A, Mitchison TJ (2008) A chemical method for fast and sensitive detection of DNA synthesis in vivo. Proc Natl Acad Sci U S A 105(7):2415–2420. https://doi.org/10.1073/pnas.0712168105
15. Darzynkiewicz Z, Traganos F, Zhao H, Halicka HD, Li J (2011) Cytometry of DNA replication and RNA synthesis: historical perspective and recent advances based on "click chemistry". Cytometry A 79(5):328–337. https://doi.org/10.1002/cyto.a.21048
16. Moss B, Elroy-Stein O, Mizukami T, Alexander WA, Fuerst TR (1990) Product review. New mammalian expression vectors. Nature 348(6296):91–92. https://doi.org/10.1038/348091a0
17. Wang IH, Suomalainen M, Andriasyan V, Kilcher S, Mercer J, Neef A et al (2013) Tracking viral genomes in host cells at single-molecule resolution. Cell Host Microbe 14(4):468–480. https://doi.org/10.1016/j.chom.2013.09.004
18. Smallwood SE, Rahman MM, Smith DW, McFadden G (2010) Myxoma virus: propagation, purification, quantification, and storage. Curr Protoc Microbiol Chapter 14:Unit 14A11. https://doi.org/10.1002/9780471729259.mc14a01s17
19. Earl PL, Moss B (1998) Current protocols in molecular biology. John Wiley and Sons, Brooklyn, NY

Chapter 15

High-Content Analyses of Vaccinia Plaque Formation

Artur Yakimovich and Jason Mercer

Abstract

Vaccinia virus plaque assays are employed for quantification of virus titer through serial dilution of virus on a monolayer of cells. Once the virus titer is diluted enough to allow for only few cells of the monolayer to be infected, clonal spread of infection can be detected by observing the lesion in the cell monolayer or using virus-specific staining methods. Beyond simple titration, plaque formation bares priceless underlying information about subtle virus-host interactions and their impact on virus spread during multiple rounds of infection. These include virus infectivity, mode of virus spread, virus replication rate, and spatiotemporal spread efficacy. How this underlying information can be harnessed using a high-content imaging setup is discussed here.

Key words Vaccinia, Virus plaque assay, Virus spread, High-content imaging

1 Introduction

The plaque assay is one of the earliest quantitative methods developed in virology. Originally developed for bacteriophages, the assay was later adapted for mammalian viruses by Renato Dulbecco in 1953 [1, 2]. In a monolayer of producer cells cultured in a dish, viruses form plaques by first infecting only a few cells of the monolayer. These initial infected cells replicate and spread infection to their neighbors by cell-to-cell [3, 4] or cell-free transmission in a lytic or non-lytic fashion [5–7] eventually forming clonal populations of infected cells within the monolayer. Ultimately, cytopathic effect, infected cell migration [8], and cell lysis [5] form a lesion—an empty spot in the cultured cell monolayer. This empty spot is termed a plaque. Plaques can sometimes be seen by the naked eye but are more commonly visualized upon fixation and staining of the cell monolayer with crystal violet. The number of plaques corresponds directly to the number of fully infectious plaque-forming units (PFUs) in the virus inoculum.

In a broader sense, a viral plaque may refer to the resulting lesion—the endpoint of plaque formation—but also to a distin-

Jason Mercer (ed.), *Vaccinia Virus: Methods and Protocols*, Methods in Molecular Biology, vol. 2023,
https://doi.org/10.1007/978-1-4939-9593-6_15,

guishable clonal population of infected cells. The latter can be visualized using a virus that expresses a fluorescent transgene or immunocytochemistry combined with fluorescence microscopy of the infected monolayer [5, 9, 10]. While the majority of replication competent viruses form plaques in favorable conditions, the mechanisms of plaque formation vary significantly, depending on the mode of virus transmission, time postinfection, and state of the host cell and the culture medium [11].

In the aftermath of this complexity, viral plaques are often of different sizes and shapes, carrying a footprint of the underlying host-pathogen interactions that have occurred [5, 9–11]. In a liquid-culturing medium, viruses that spread via cell-cell contact form round plaques, whereas cell-free spread of viruses results in plaques that are typically elongated (comet shaped) as a result cell-free virus particles being subject to passive mass transfer in the liquid medium [5, 11]. The formation of comet-shaped plaques can be prevented by the addition of low-melt agarose, carboxymethylcellulose [12], or other gelling agents to the culture medium. This may, however, conceal important host-pathogen interactions manifesting in these phenotypes, including the amount of cell-free virus egressing from the initially infected cell, cell-free virus infectivity, relationship between cell-cell and cell-free spread, etc. [10, 11, 13]. Instead of artificially preventing these phenotypes from occurring it is possible to apply computer vision-based quantitative analysis combined with model-based fluid dynamics analysis to untangle the complex shapes that viral plaques take. To understand virus spread dynamics this approach can be further combined with live-cell time-lapse imaging and cell tracking [10, 11, 14, 15].

VACV spreads through a monolayer of cells using a combination of cell-cell and cell-free mechanisms. Cell-cell spread is mediated by cell-associated enveloped virions (CEVs), which after egress remain tethered to the cell surface. Cell-free spread is directed by extracellular enveloped virions (EEVs) which are formed when CEVs are released from the infected cell surface [10, 16, 17]. Cell-free VACV spread can also be mediated through intracellular mature virions (IMVs) released upon cell lysis. VACV plaque formation is additionally influenced by the production rate of CEVs [16, 18, 19], formation of actin tails [16, 20–24], virus-induced cell motility [8, 25], and superinfection repulsion [9, 26]. In light of this complexity, classical plaque assays yield limited mechanistic information regarding specific plaque phenotypes. High-content functional (e.g., fluorescence microscopy) and label-free (e.g., phase-contrast microscopy) microscopy-based plaque assays, in turn, can unveil information about the mode of spread (cell-cell or cell-free, lytic or pre-lytic [10, 13]), stage of infection (early or late VACV promoter-based transgenes [27]), host cell state, or monolayer dynamics employing time-lapse microscopy

[9]. This readily quantifiable imaging data allows for measurement of plaque features like signal intensity, size, shape, directionality, and growth dynamics which can then be correlated to important biological defining host-pathogen interactions. In this chapter we describe how to set up, perform, and analyze VACV plaque formation under live-cell conditions in a high-content fashion.

2 Materials

2.1 Endpoint High-content Plaque Assay

1. Depending on the scale: 96-Well or 384-well imaging-grade microtiter plates (*see* **Notes 1–3**).
2. Tissue culture cells (*see* **Note 4**).
3. Culture medium: Dulbecco's minimal essential medium (DMEM) containing 10% v/v fetal calf serum (FCS) (*see* **Note 5**).
4. Phosphate-buffered saline (PBS).
5. 0.25% Trypsin/EDTA.
6. Wild-type VACV stock (*see* **Note 6**).
7. Infection medium: DMEM without supplements.
8. Fixation solution: 4% w/v Paraformaldehyde (PFA) solution in PBS.
9. Quenching solution: 50 mM NH_4Cl in PBS.
10. Permeabilization solution: 0.1% Triton® X-100 in PBS.
11. Blocking buffer: 5% w/v Bovine serum albumin (BSA) in PBS.
12. Hoechst 33342 nuclear dye and other stains (*see* **Note 7**).
13. Primary antibody against viral protein, diluted in 5% BSA in PBS at an appropriate concentration.
14. Secondary antibody conjugated to a fluorescent dye (e.g., Alexa488) raised to recognize the species of the primary antibody.

2.2 Time-Lapse High-Content Plaque Assay

1. Depending on the scale: 96-Well or 384-well imaging-grade microtiter plates (*see* **Notes 1–3**).
2. Tissue culture cells (*see* **Note 4**).
3. Culture medium: Dulbecco's minimal essential medium (DMEM) containing 10% v/v fetal calf serum (FCS) (*see* **Note 5**).
4. Infection medium: DMEM without supplements.
5. Live-imaging medium (*see* **Note 8**).
6. Phosphate-buffered saline (PBS).
7. 0.25% Trypsin/EDTA.
8. Fluorescent recombinant VACV stock (*see* **Note 9**).
9. Deionized water for humidity control.

3 Methods

3.1 High-Content Plaque Assay Imaging

1. Move the confluent indicator cell culture dish into the laminar safety cabinet.
2. Wash the cells with PBS.
3. Detach the cells by incubating in 0.25% trypsin/EDTA for 3–5 min at 37 °C.
4. Count cells using either automated cell counter or manual cell counting chamber.
5. Dilute the cell suspension to the desired concentration with cell culture medium in multichannel pipette reservoir (*see* **Note 10**).
6. Resuspend cells in the reservoir and dispense them into the microtiter plate at desired volume using either automated or manual multichannel pipette (*see* **Note 11**).
7. To ensure an even distribution of the cells throughout the growth surface leave cells standing at room temp for 20–40 min before transferring them into the incubator.
8. Culture cells overnight at 37 ° C in a humidified cell culture incubator.
9. One day after cell seeding verify homogeneous monolayer formation using inverted transmission light microscope.
10. Prepare warm virus stock and infection medium to 37 ° C (*see* **Note 12**).
11. If serial pre-dilution prior to inoculum dispensing into the plate is required perform it as outlined in Fig. 1a or perform the serial dilution of the virus inoculum directly in the plate as outlined in Fig. 1b and **step 12** of this protocol (*see* **Notes 13** and **14**).
12. To perform a serial dilution in the plate using multi-compartment reservoir add an appropriate amount of fresh infection medium into the compartments or reservoirs, making sure that medium volume in the wells of the multi-titer plate corresponds to the desired dilution factor (*see* Fig. 1a, b and **Note 13**).
13. Using a multichannel pipette dilute the first step in the group of wells of the plate designated for the first step (e.g., column 1, Fig. 1b).
14. Mix the added inoculum with the medium in the well by aspirating and dispensing 3×.
15. Exchange the pipette tips (*see* **Note 14**).

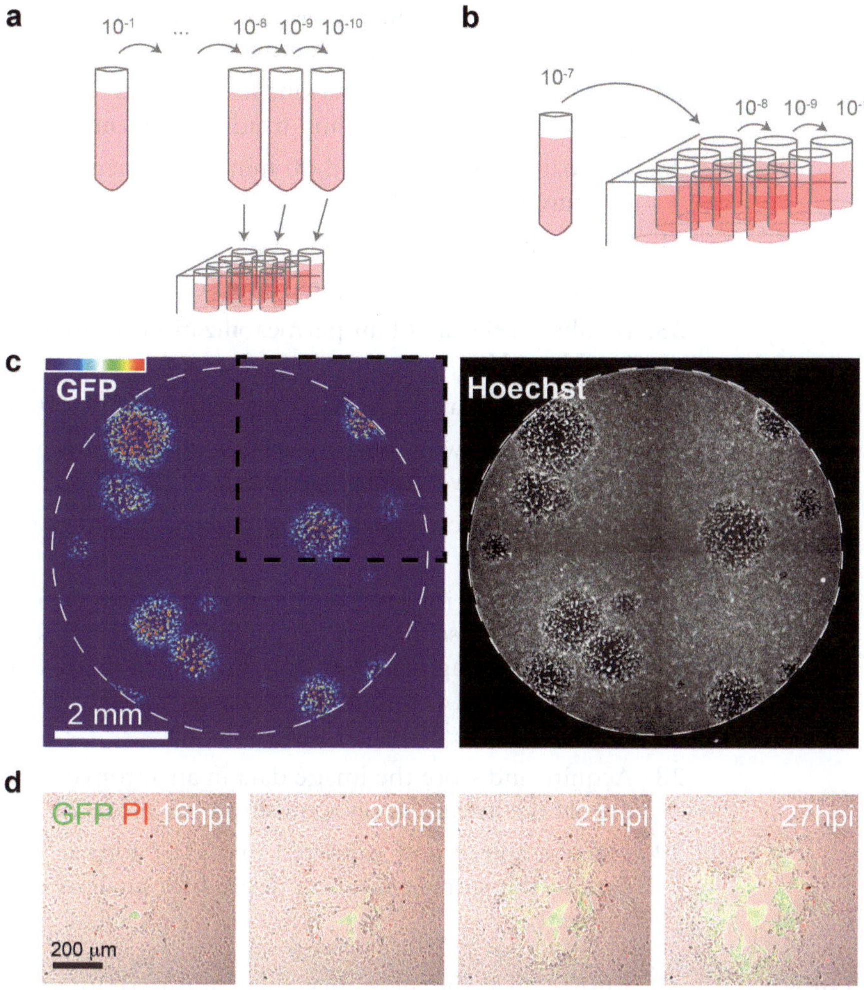

Fig. 1 Principles of the high-content VACV plaque assay. (**a**) Schematic representation of the serial pre-dilution prior to inoculum dispensing into the plate. (**b**) Schematic representation of serial dilution performed directly in the plate. (**c**) A typical example of high-content fluorescence microscopy-based plaque assay (Plaque2.0) performed as an endpoint assay using VACV WR E/L EGFP virus. On the left-hand side, image shows that virus EGFP signal intensity is color-coded violet (minimum) to red (maximum). On the right-hand side, image shows Hoechst signal from the monolayer nuclei. (**d**) Typical images from high-content fluorescence microscopy-based plaque assay performed in live-cell imaging setting. Still frames of merged transmission light, propidium iodide (PI, red), and VACV IHD-J E/L EGFP (GFP, green) signals are arranged in a time-dependent manner showing dynamics of the growing VACV plaque. Here PI signal is used to indicate monolayer cell death. Time points are indicated as hours postinfection (hpi)

16. From the first step column using a multichannel pipette transfer the amount of liquid appropriate to the dilution factor to the next column and so on (*see* **Notes 15** and **16** and Fig. 1b).
17. Incubate for the desired time at 37 °C or other temperature required by experimental conditions.

18. For endpoint imaging proceed to **step 19**. For time-lapse imaging proceed to Subheading 3.2.
19. After 12–24 h of incubation, fix the plate by removing the culture medium and adding fixation solution (*see* **Note 17**).
20. Incubate between 30 and 60 min on the bench at room temperature (*see* **Note 18**).
21. Wash cells 3× with PBS.
22. Incubate cells at RT in quenching solution for 5 min.
23. Incubate cells at RT in permeabilization solution for 10 min (*see* **Note 19**).
24. Incubate cells at RT in blocking buffer for 30–60 min.
25. Stain samples with primary and secondary antibodies according to respective staining protocols.
26. Stain for host cell nuclei using Hoechst or other nuclear dyes (*see* **Note 7**).
27. Upon staining imaging of the plate using an automated high-content microscope can be performed. For this choose the appropriate magnification and number of fields of view, wells desired, as well as fluorescence channels in the microscope settings (*see* **Note 20**).
28. Acquire and store the image data in an automated fashion and export the images upon acquisition (*see* **Note 21** and Fig. 1c).
29. Once acquired and exported in "TIF" format, move image data to the storage location desired for analysis by Plaque2.0 software.

3.2 Time-Lapse High-Content Plaque Assay

1. Ensure that the necessary incubation parameters (required temperature, CO_2, and humidity) are possible and switched on at the microscope at least 1 h before imaging.
2. Proceeding from Subheading 3.1, **step 17**, live-cell image the plate immediately or after a desired period of time using an automated high-content microscope (*see* **Notes 20**, **22**, and **23** and Fig. 1d).
3. After the desired period of incubation (typically 24 h), fix cells for additional endpoint analyses (proceed to Subheading 3.1, **step 19**), or dispose of virus-infected cells according to established safety protocols.
4. Images should be exported and stored as single time point per folder (a single time point is inclusive of multiple field-of-view positions), where each position and channel is stored as an individual "TIF" file (*see* **Note 24**).

3.3 High-Content Analysis Using Plaque 2.0

1. Download Plaque2.0 from http://plaque2.github.io/ website.
2. Once downloaded, follow the installation instructions on the website (*see* **Note 25**).

3. Once installed, start the software and set up the "Main Parameters" (*see* Fig. 2) according to the location of your data and the preferred location and name of analysis results.
4. In case the full well picture has been obtained in tiled acquisition of several sites (fields of view) per well, stitching of the image tiles is required. To perform stitching, check the box next to the "Stitch" pane and set up respective parameters (*see* Fig. 2a and Table 1).
5. Use "Test Settings" button, to ensure that parameters are set correctly. It may make sense to run this module before running the rest of the analysis.
6. In case the stitched outside-of-the-well area (circular wells) is present in the square field of view of the image, to improve the precision of the analysis these parts of the image can be masked. For this, activate the check box in the "Mask" pane and set up the masking parameters (*see* Fig. 2b and Table 1).
7. Use "Test Settings" button, to ensure that parameters are set correctly.
8. If nuclear stain was used in the assay to visualize total cells in the well, activate "Monolayer" pane and set up the nuclear detection parameters (including thresholding strategy, etc.) in the respective window (*see* Fig. 3a and Table 1).
9. Use "Test Settings" button, to ensure that parameters are set correctly.
10. To detect VACV plaques in the HCI data, activate the "Plaque" pane. Set relevant parameters, including plaque signal threshold (*see* Fig. 3b and Table 1; **Note 26**).
11. Prior to starting the analysis use "Test Settings" button, to ensure that parameters are set correctly.
12. Once parameters are set and tested press "Run" button to run the analysis on the whole folder of images. Once complete, results will be saved as comma-separated value (CSV) files in the folder selected.
13. For time-lapse acquisition data, repeat the analysis on each folder containing single time point data per folder (*see* Subheading 3.2, **step 4**).
14. After analyses, locate results consisting of two CSV files. The first CSV file contains data relevant to the image-based feature readouts (i.e., number of plaques, number of cells in the monolayer; *see* Table 2, for the full list of image-based features and their detailed explanation). The second CSV file contains individual plaque-based features (i.e., plaque size and shape; number of infected cells in the plaque; *see* Table 2 for the full list of plaque-based readouts and their detailed explanation).
15. Correlate the plaque objects using their coordinates (*see* **Note 24**).

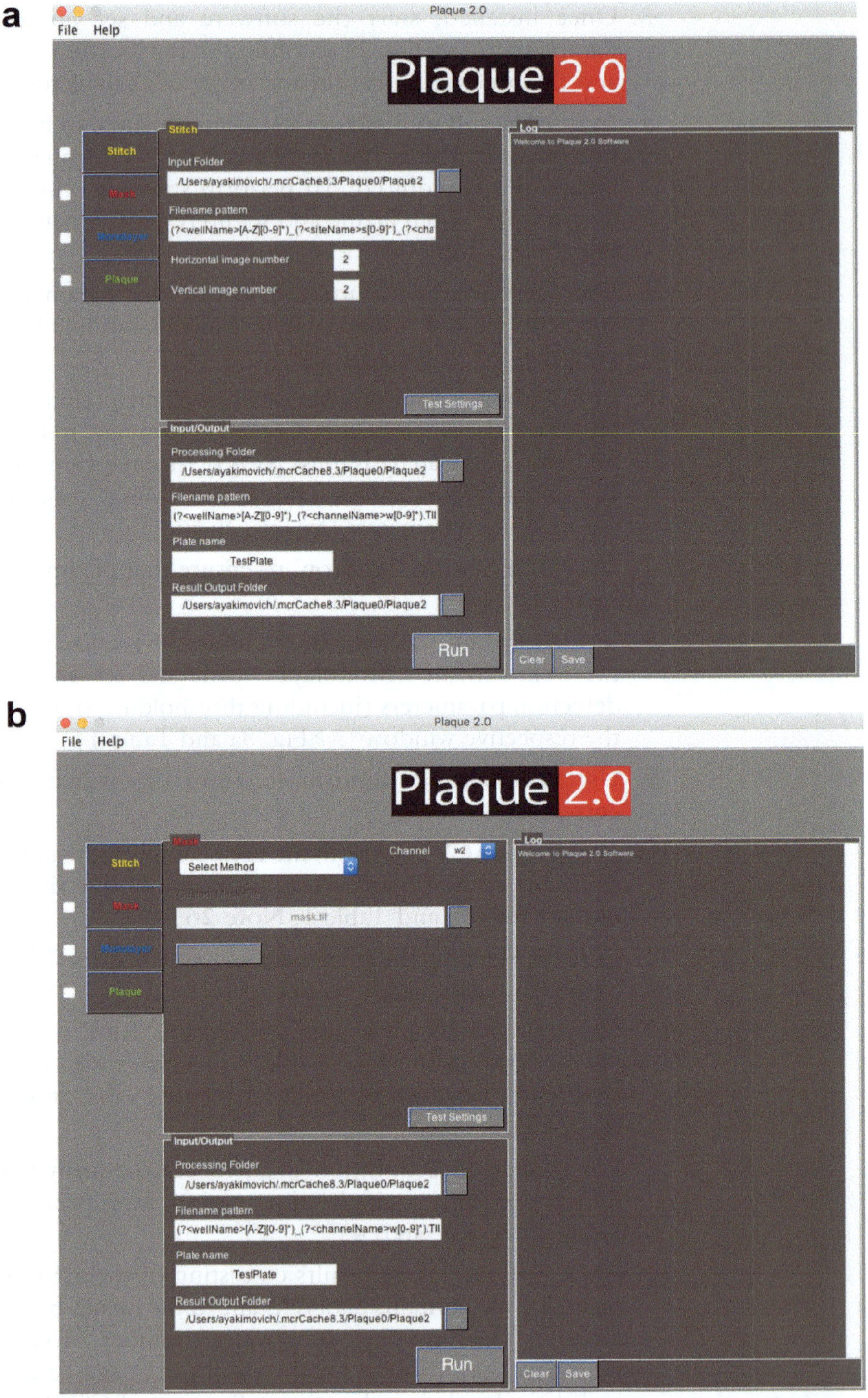

Fig. 2 Plaque2.0 software interface view showing the main window and the respective activated pane. (**a**) View of the user-defined parameters in the "Stitch" pane. (**b**) View of the user-defined parameters in the "Mask" pane

Table 1
Detailed overview of all user-defined parameters in the Plaque2.0 software

Parameter	Description
Main parameters (input/output)	
Processing folder	Used to indicate the location of microscopy images that do not require stitching or have been processed by the "Stitch" module
Filename pattern	This input defines the regular expression (RegEx) that is used to parse the metadata from the image filenames (well rows, well columns, channels). Metadata will be used to formulate readouts
Plate name	This input defines a name for the current plate analyzed. The parameter is used as a prefix for output files
Result output folder	Defines location of the output files
Stitch window	
Input folder	Used to indicate the location of the microscopy images to be stitched by this module
Filename pattern	This input defines the RegEx used to parse the metadata (well rows, well columns, channels, sites) from the filenames of the images
Horizontal image number, vertical image number	This input specifies the number of well sites for full well stitching in horizontal and vertical direction accordingly
Mask window	
Load custom mask	If this mask definition method is selected the well mask is loaded from the file provided by the input path
Manual mask definition	If this mask definition method is selected the "Define Mask" button becomes enabled. Upon pressing the latter, a new window where the user can specify the mask for the plate wells is opened. This is done by drag resizing an oval outline over the first stitched image from the "Processing Folder." The mask determined is then saved in the current working folder by double-clicking the image
Automatic mask definition	If this mask definition method is selected, the mask is determined automatically using the first stitched image from the "Processing Folder"
Monolayer window	
Artifact threshold	This setting defines an upper threshold (grayscale values stretched between 0 and 1) used to filter out very bright imaging artifacts

(continued)

Table 1
(continued)

Parameter	Description
Manual thresholding	If method is selected for thresholding a manual user-defined value will be used to perform foreground segmentation
Otsu global thresholding	Selecting this method will perform automatic Otsu global thresholding-based segmentation of foreground pixels
Otsu local thresholding	Selecting this method will perform automatic Otsu local thresholding-based segmentation of foreground pixels. Here the image is split into smaller blocks. Otsu thresholding is performed separately for each of the blocks allowing local threshold values to be used
Threshold	This value specifies the manual threshold (defined between 0 and 1) for manual thresholding
Block size	This input provides one size (one dimension) of a square block side (pixels) to be used for the "Otsu local thresholding" method. Note that width or height of the image should be divisible by the "block size"
Minimal threshold	This value (between 0 and 1) defines the lowest threshold possible for the "Otsu global thresholding"
Correction factor	Here a threshold correction factor (between 0 and 1—to decrease, higher than 1 to increase) can be defined to fine-tune the results of the "Otsu global thresholding"
Min/max cell area	These values are used for cell number calculation. Specify minimum and maximum nucleus area
Correction "ball" radius	If used, this value provides the radius of a, so-called, rolling ball, used in the "Rolling Ball" algorithm for the illumination correction. Radius should be lower than the typical size of an object in the image
Plaque window	
Fixed threshold	This value (between 0 and 1) defines a fixed manual threshold for virus image foreground detection
Minimal plaque area	Here the minimum area of a detected object to be considered a plaque is specified
Connectivity	This input defines the maximum distance plaque regions (pixels)
Min/max cell area	This value defines minimum and maximum cellular area in the virus signal image for cell number calculation. This parameter is used only when the cell number estimation based on the nuclear segmentation is disabled
Gaussian filter size	This input defines the "blurring bell" (Gaussian filter) for virus image convolution (pixels). It is used for detecting plaques

(continued)

Table 1 (continued)

Parameter	Description
Gaussian filter sigma	Here the width of the "blurring bell" (standard deviation of the Gaussian filter) for virus image convolution is defined. It is used for detecting plaques together with the previous parameter
Peak region size	This input defines the size of a plaque region (maximum) in pixels. It is assumed that only one intensity maximum can be detected per plaque

Parameter column represents the parameter name in the software. Description software represents details about the specific parameter. Section headers correspond to the respective name of the pane

4 Notes

1. In case the assay is performed in fluorescence microscopy setting, to lower the background coming from the reflection of the stray excitation light black-walled imaging plates are used. Imaging quality microtiter plates have flat clear bottom, with thickness ranging between 0.17 mm (so-called #1.5 thickness) and 0.64 mm (#4 thickness).
2. Material and coating of the plate's bottom play an important role for cell adhesion; typically plates sold for tissue culture are coated. Material, evenness, and thickness of the plate's bottom may have an impact on the quality of the images obtained in the automated microscopy, especially if autofocusing is used. Altogether, microscopic imaging plates should be chosen based on experimental conditions, experimental scale, imaging hardware, and other specific requirements. As a rule of thumb, thin mineral coverslip glass-bottom plates give a better signal-noise in fluorescence imaging, as well as autofocusing performance; however they require a better coating for cell adhesion and are significantly more expensive.
3. Plate bottom coating is of key importance for good cell adhesion and, depending on the nature of the indicator cells used (concerning their properties like contact inhibition, readiness to grow in a monolayer), may be crucial in the formation of a homogenous monolayer. A homogeneous monolayer is, in turn, crucial to obtain reproducible results in plaque assay. Assure that the bottom of the plates is made of either organic (polystyrene or cyclic olefin copolymer) or mineral coverslip glass.
4. Plaque assay indicator cell line often refers to the cell the virus is typically grown in, as they need to be able to support virus

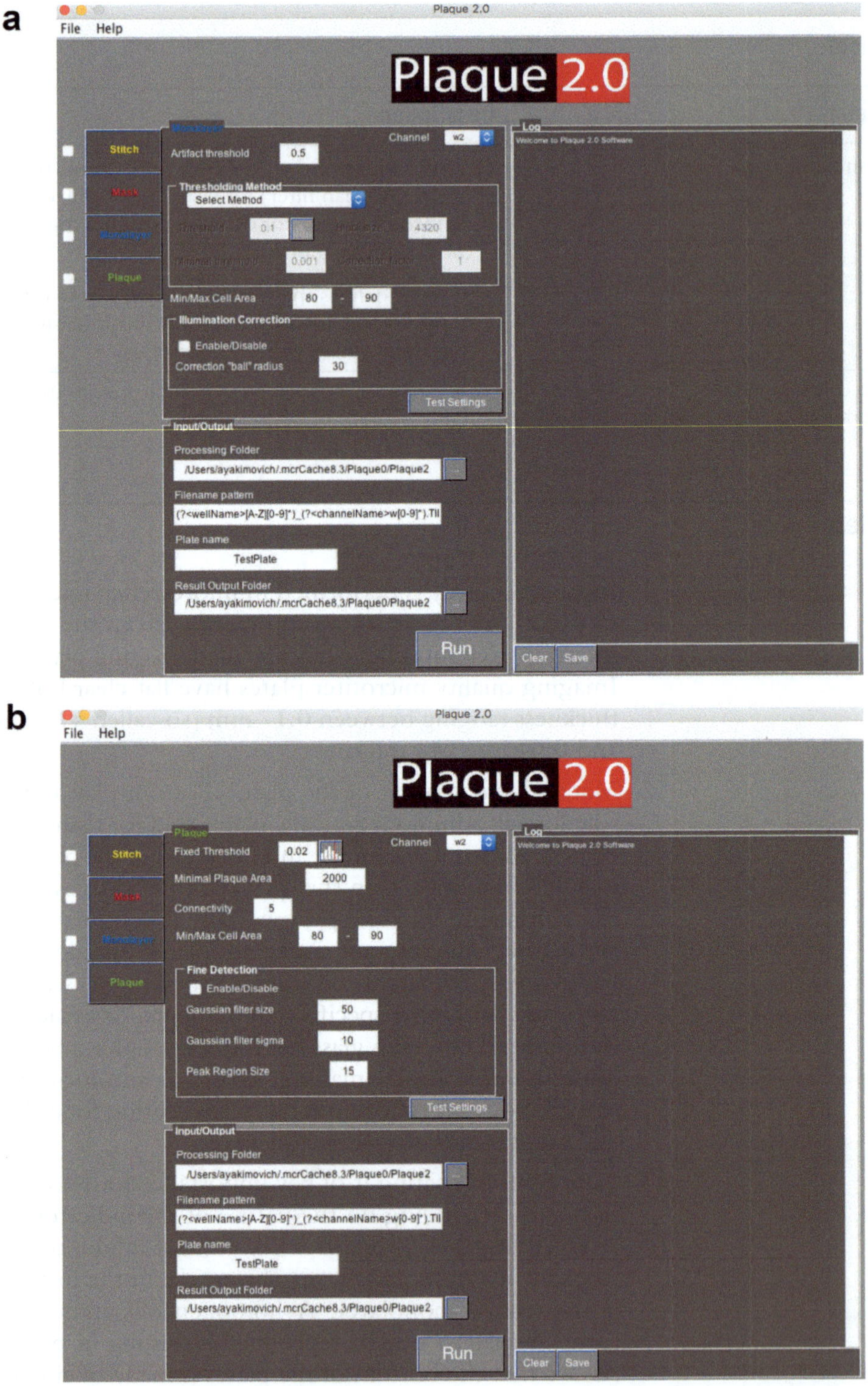

Fig. 3 Plaque2.0 software interface view showing the main window and the respective activated pane. (**a**) View of the user-defined parameters in the "Monolayer" pane. (**b**) View of the user-defined parameters in the "Plaque" pane

Table 2
Detailed overview of all readouts provided by the Plaque2.0 software in the resulting comma-separated value files

Feature	Description
Image-based features	
NucleiImageName	Measured file name (nuclei signal image)
wellRow	Measured plate row
wellCollumn	Measured plate column
maxNucleiIntensity	Nuclear intensity maximum in the image measured
totalNucleiIntensity	Nuclear intensity sum in the image measured (after masking)
meanNucleiIntensity	Nuclear intensity mean in the image measured (after masking)
numberOfNuclei	Number of nuclei estimated per image
VirusImageName	Measured file name (virus signal image)
maxVirusIntensity	Virus intensity maximum in the image measured (after masking)
totalVirusIntensity	Virus intensity sum in the image measured (after masking)
meanVirusIntensity	Virus intensity mean in the image measured (after masking)
numberOfPlaques	Number of detected plaques per image
numberOfInfectedNuclei	Number of infected nuclei detected after overlaying the nuclear image (if applicable) with thresholder virus image. If no nuclear image is provided, virus image is used for the estimation
Plaque-based features	
Area	Area of the plaque region (pixel)
Centroid	Centroid position of the plaque region
BoundingBox	Coordinates of the plaque region in boundaries provided in the following format: *x y width height*
MajorAxisLength	Size of the major axis of the fitted ellipsoid to the measured plaque region
MinorAxisLength	Size of the minor axis of the fitted ellipsoid to the measured plaque region
Eccentricity	Ellipsoid eccentricity—i.e., how much the ellipse fitted to the measured plaque region varies from being circular
ConvexArea	Convex area of the measured plaque region
numberOfNucleiInPlaque	Nuclei number estimated in the plaque
numberOfInfectedNucleiInPlaque	Infected nuclei number estimated in the plaque
wellRow	Measured plate row
wellCollumn	Measured plate column
maxIntensityGFP	Virus intensity maximum in the image measured (after masking)
totalIntensityGFP	Virus intensity sum in the image measured (after masking)
meanIntensity	Virus intensity mean in the image measured (after masking)

Section headers correspond to the respective readout group (image based or object based) saved as a single file per group

replication, as well as be highly susceptible to the virus infection. We routinely use BSC40 or HeLa cells, but one can use any cell line as long as they can be grown to confluency and infected by VACV. Depending on the experimental setup dictated by the goal of the study or visualization modality different cells may be used.

5. To prevent the spread of cell-free viral particles in the cell culture medium through advection a gelling agent may be added. For example this can be ultra-low-melting cell culture-grade agarose cell at a final concentration of 0.5–3% [13].
6. Visualization strategies for VACV include (but are not limited to) use of immunofluorescent (IF) staining against viral proteins and recombinant viruses expressing fluorescent proteins. Since IF visualization is possible only in an endpoint assay upon fixation this protocol focuses on IF and wild-type VACV (VACV). However, endpoint plaque assays can also be performed with fluorescent recombinant VACVs. While VACV strain Western Reserve produces circular plaques, e.g., VACV strain, International Health Department J (VACV IHD-J) produces comet-shaped plaques [27]. Both strains may be used in this assay.
7. While nuclear staining is not strictly necessary for the high-content plaque assay, it may be very informative to evaluate, e.g., total cell count and other parameters of the system. Since some nuclear dyes may inhibit VACV infection [27], nuclear staining here is only included in the endpoint assay postfixation.
8. Live imaging-compatible medium typically does not contain phenol red, since the latter may significantly increase background fluorescence and lead to uneven background throughout the duration of the longer time-lapse acquisition, due to pH changes. Additionally, it is important to choose a suitable buffering agent for live imaging culture medium. For example HEPES is often used to buffer medium for short-term incubations.
9. Recombinant VACV-expressing fluorescent proteins (e.g., VACV encoding GFP under early promoter) or chimeric fluorescent proteins (conjugated with viral proteins) may be used here. The main criteria of selection here are signal-to-noise (the higher the better) in the imaging system used and signal uniformity in the infected cell (the more uniform the better).
10. It is advisable to seed the right number of cells to ensure highest possible confluence. This way a close tight and homogeneous monolayer may be achieved on the morning after seeding. For example a good range to get a monolayer of BSC40 cells in a 96-well plate overnight is between 5000 and 50,000 cells per well (depending on the cultivation condi-

tions, cell line passage, etc.). At low dispensing volumes and while working with hydrophobic surfaces, it is important to ensure medium contact with the plate bottom during dispensing.

11. It is important to have sufficient medium volume in the well to ensure homogenous cell seeding. For example for a typical 96-well plate a minimum recommended volume is 50 μL. However, higher volume is preferred for better reproducibility.
12. Upon thawing the frozen VACV aliquot it is important to sonicate it to ensure precise titration results.
13. Serial dilution of the virus can be performed either prior to dispensing the inoculum into the plate (pre-dilution) or directly in the plate (Fig. 1a, b). The choice between the two depends on the viral titer of the stock virus solution and area of the well in the plate used.
14. As virus particles may become trapped even by low-adhesive plastic surfaces, it is important to plan liquid handling to ensure as little as possible change of vessels. At the same time, since adsorption of virus particles on plastic surfaces is reversible, it is important to exchange tips between dilution steps, to ensure titration precision.
15. Add the inoculum into the first dilution and proceed transferring volume (according to the chosen dilution factor, e.g., 1/10 or 1/2 of total) from one dilution into subsequent in a stepwise manner. Make sure to exchange the disposable pipettes or tips upon each dilution (Fig. 1a).
16. For example in a 96-well plate for 1/10 or 1/2 factors it is convenient to have 90 or 50 μL, respectively.
17. Alternatively, fixation can be performed by diluting a higher percentage PFA solution directly in the culture/infection medium to arrive at 4% final PFA concentration. In such case the inoculum medium is not removed.
18. Longer fixation times are important to minimize the loss of cells in the monolayer upon washing.
19. Alternative permeabilization agents such as methanol or saponin can be used.
20. Depending on the average plaque size and time of fixation, typically 2× to 10× magnification is used to image VACV plaques. To obtain a representative titer typically imaging of a complete well is desired. To date sCMOS chip sizes allow imaging a full 96-well plate well in one image at 2× magnification. For example using 4× magnification on many systems a

full 96-well plate well may be acquired in four separate images (fields of view), which can be stitched together in the image processing step. Increasing magnification and number of the fields of view increases acquisition time and requirements on storage and analysis.

21. If the data is saved in a proprietary format, make sure that individual channels and fields of view are exported in "TIF" format for further analysis. This can be done using either the proprietary software of the high-content microscope or the open-source Fiji/ImageJ software.
22. Live-cell imaging of VACV plaques typically requires a fluorescent reporter protein to be expressed by VACV assayed. However other live staining options may be possible.
23. Make sure that the time it takes to acquire all required wells in the plate is not longer than the required time interval.
24. Correlation of the individual plaques may be performed in a downstream analysis, by associating respective positions of the plaque centers. This can be performed outside of the Plaque2.0 software (e.g., using Python or KNIME software [28]). A good starting point for such analysis is to correlate the plaque center coordinates using the nearest-neighbor algorithm within a restricted neighborhood comparable to the size of a plaque [10].
25. To use the latest version of the Plaque2.0 software we suggest using the MATLAB™ (Mathworks) source code. For this download the source code folder. Download and install the MATLAB™ (Mathworks) integrated development environment (IDE). Start the IDE and open the source code folder. In the IDE run the PlaqueGUI.m file appropriate for your platform.
26. To ensure that the plaque's related signal is treated equally throughout the dataset global manual threshold is used. This parameter can be set interactively, by opening the threshold selection tool button next to the value input. Apart from thresholding and detection of putative plaque centers, the algorithm of Plaque2.0 software performs plaque area detection separating merging plaques using parameters defined in the optional "Fine detection" section (Fig. 3b and Table 1).

References

1. D'Herelle F (1926) The bacteriophage and its behavior. The Williams & Wilkins Company, Baltimore, MD
2. Dulbecco R, Vogt M (1953) Some problems of animal virology as studied by the plaque technique. Cold Spring Harb Symp Quant Biol 18:273–279
3. Mothes W, Sherer NM, Jin J, Zhong P (2010) Virus cell-to-cell transmission. J Virol 84:8360–8368

4. Sattentau Q (2008) Avoiding the void: cell-to-cell spread of human viruses. Nat Rev Microbiol 6:815–826
5. Yakimovich A, Gumpert H, Burckhardt CJ, Lütschg VA, Jurgeit A, Sbalzarini IF et al (2012) Cell-free transmission of human adenovirus by passive mass transfer in cell culture simulated in a computer model. J Virol 86:10123–10137
6. Burckhardt CJ, Greber UF (2009) Virus movements on the plasma membrane support infection and transmission between cells. PLoS Pathog 5(11):e1000621
7. Bär S, Daeffler L, Rommelaere J, Nüesch JP (2008) Vesicular egress of non-enveloped lytic parvoviruses depends on gelsolin functioning. PLoS Pathog 4(8):e1000126
8. Sanderson CM, Way M, Smith GL (1998) Virus-induced cell motility. J Virol 72(2):1235–1243
9. Doceul V, Hollinshead M, van der Linden L, Smith GL (2010) Repulsion of superinfecting virions: a mechanism for rapid virus spread. Science (New York, NY) 327:873–876
10. Yakimovich A, Andriasyan V, Witte R, Wang I-H, Prasad V, Suomalainen M, Greber UF (2015) Plaque2. 0—a high-throughput analysis framework to score virus-cell transmission and clonal cell expansion. PLoS One 10(9):e0138760
11. Yakimovich A, Yakimovich Y, Schmid M, Mercer J, Sbalzarini IF, Greber UF (2016) Infectio: a generic framework for computational simulation of virus transmission between cells. mSphere 1(1):e00078–e00015
12. Russell WC (1962) A sensitive and precise plaque assay for herpes virus. Nature 195(4845):1028–1029
13. Yakimovich A, Gumpert H, Burckhardt CJ, Lutschg VA, Jurgeit A, Sbalzarini IF, Greber UF (2012) Cell-free transmission of human adenovirus by passive mass transfer in cell culture simulated in a computer model. J Virol 86(18):10123–10137
14. Sbalzarini IF, Koumoutsakos P (2005) Feature point tracking and trajectory analysis for video imaging in cell biology. J Struct Biol 151:182–195
15. Tinevez J-Y, Perry N, Schindelin J, Hoopes GM, Reynolds GD, Laplantine E, Bednarek SY, Shorte SL, Eliceiri KW (2017) TrackMate: an open and extensible platform for single-particle tracking. Methods 115:80–90
16. Smith GL, Vanderplasschen A, Law M (2002) The formation and function of extracellular enveloped vaccinia virus. J Gen Virol 83(12):2915–2931
17. Moss B (2013) Poxviridae. In: Fields BN, Knipe DM, Howley PM et al (eds) Fields virology, vol 1, 6th edn. Lippincott Williams & Wilkins, a Wolters Kluwer business, Philadelphia, PA, p 2664
18. Condit RC, Moussatche N, Traktman P (2006) In a nutshell: structure and assembly of the vaccinia virion. Adv Virus Res 66:31–124
19. Roberts KL, Smith GL (2008) Vaccinia virus morphogenesis and dissemination. Trends Microbiol 16(10):472–479
20. Blasco R, Sisler J, Moss B (1993) Dissociation of progeny vaccinia virus from the cell membrane is regulated by a viral envelope glycoprotein: effect of a point mutation in the lectin homology domain of the A34R gene. J Virol 67(6):3319–3325
21. Cudmore S, Cossart P, Griffiths G, Way M (1995) Actin-based motility of vaccinia virus. Nature 378(6557):636–638
22. Frischknecht F, Moreau V, Röttger S, Gonfloni S, Reckmann I, Superti-Furga G, Way M (1999) Actin-based motility of vaccinia virus mimics receptor tyrosine kinase signalling. Nature 401(6756):926–929
23. McIntosh A, Smith GL (1996) Vaccinia virus glycoprotein A34R is required for infectivity of extracellular enveloped virus. J Virol 70(1):272–281
24. Wolffe EJ, Weisberg AS, Moss B (1998) Role for the vaccinia virus A36R outer envelope protein in the formation of virus-tipped actin-containing microvilli and cell-to-cell virus spread. Virology 244(1):20–26
25. Valderrama F, Cordeiro JV, Schleich S, Frischknecht F, Way M (2006) Vaccinia virus-induced cell motility requires F11L-mediated inhibition of RhoA signaling. Science 311(5759):377–381
26. Doceul V, Hollinshead M, Breiman A, Laval K, Smith GL (2012) Protein B5 is required on extracellular enveloped vaccinia virus for repulsion of superinfecting virions. J Gen Virol 93(Pt 9):1876–1886
27. Yakimovich A, Huttunen M, Zehnder B, Coulter LJ, Gould V, Schneider C, Kopf M, McInnes CJ, Greber UF, Mercer J (2017) Inhibition of poxvirus gene expression and genome replication by bisbenzimide derivatives. J Virol 91(18):e00838–e00817
28. Fillbrunn A, Dietz C, Pfeuffer J, Rahn R, Landrum GA, Berthold MR (2017) KNIME for reproducible cross-domain analysis of life science data. J Biotechnol 261:149–156

Chapter 16

Super-resolution Microscopy of Vaccinia Virus Particles

Robert Gray and David Albrecht

Abstract

Super-resolution microscopy enables the study of vaccinia architecture at subviral resolution with molecular specificity. Here, we outline how to use structured illumination microscopy (SIM) and stochastic optical reconstruction microscopy (STORM) to detect fluorescently tagged or immunolabeled viral proteins on purified virions. Tens to hundreds of individual virions can be imaged in a single field of view providing data for single-particle averaging or quantitative analysis of viral protein spatial organization.

Key words Vaccinia virus, Super-resolution microscopy, Immunofluorescence, SIM, STORM

1 Introduction

Imaging-based approaches have a long-standing tradition in virus research [1, 2]. The broad structure of vaccinia virus (VACV) particles was determined decades ago using electron microscopy [3]. In recent years technological advances in this field have led to ultrahigh-resolution reconstructions of the virus architecture [4, 5]. However, electron microscopy does not provide the necessary molecular specificity to localize the 80 plus different viral proteins within the particle. Mapping viral proteins can provide vital information both on their function and their relationship to other viral or cellular components. For example, despite being a hallmark structure of poxviruses, it was only in 2013 that the first lateral body proteins were unequivocally localized [6].

Conventional fluorescence microscopy provides molecular specificity, but diffraction of light limits the resolution to around 200 nm, meaning it is unsuitable for determining the exact location or distribution of proteins in VACV. Super-resolution microscopy (SRM) is a new field that goes beyond the diffraction limit

Robert Gray and David Albrecht contributed equally to this work.

Jason Mercer (ed.), *Vaccinia Virus: Methods and Protocols*, Methods in Molecular Biology, vol. 2023,
https://doi.org/10.1007/978-1-4939-9593-6_16, © Springer Science+Business Media, LLC, part of Springer Nature 2019

providing the potential to bridge the gap between ultrastructural EM imaging and molecular specific fluorescence microscopy. SRM techniques exploit spectral properties and photophysics to reach nanometer resolutions. This makes them suitable for determining the location of labeled proteins within individual virions. While there are numerous ways to circumvent the diffraction limit we will focus on structured illumination microscopy (SIM, 5) and the single-molecule localization microscopy (SMLM) methods: photoactivated localization microscopy (PALM, 6), stochastic optical reconstruction microscopy (STORM, 7), and stimulated emission depletion microscopy (STED, 8). The three approaches differ in their achievable resolution, sample preparation, optical setup requirements, time resolution, and potential for live imaging. For in-depth comparisons and explanations of the techniques, *see* recent reviews [9]. Briefly, SMLM methods can achieve the best resolutions with moderate requirements for the optical setup but require long imaging times and elaborate sample preparation. SIM and STED require little in terms of sample preparation and acquisition speed is much faster than SMLM. However, both require specialist microscopes and provide lower resolution. The differences are summarized in Table 1.

SIM, STORM, and STED of isolated vaccinia particles [10–12] and SIM of vaccinia-infected cells [13] have been demonstrated. Examples of SIM and STORM images of vaccinia virus cores and lateral bodies are shown in Fig. 1. The best method to

Table 1
Comparison of some properties of three modalities of super-resolution microscopy

Method	SIM	STED	SMLM
Lateral resolution	100–130 nm	20–70 nm	10–30 nm
Temporal resolution	Seconds	Seconds	Minutes—tens of minutes
Microscope required	Specialized	Specialized	General wide-field, high NA objective
Post-processing	Reconstruction with microscope software	Deconvolution with microscope software	Reconstruction with SMLM software; many choices available
Possible probes	Most dyes and FPs	Specialized dyes and FPs	Specialized photoswitchable dyes and FPs
Live-cell compatible	Yes	(Yes)	No
Laser power/ photodamage	Low	High	High

For more detailed information *see* ref. 9. Resolutions are in the best case

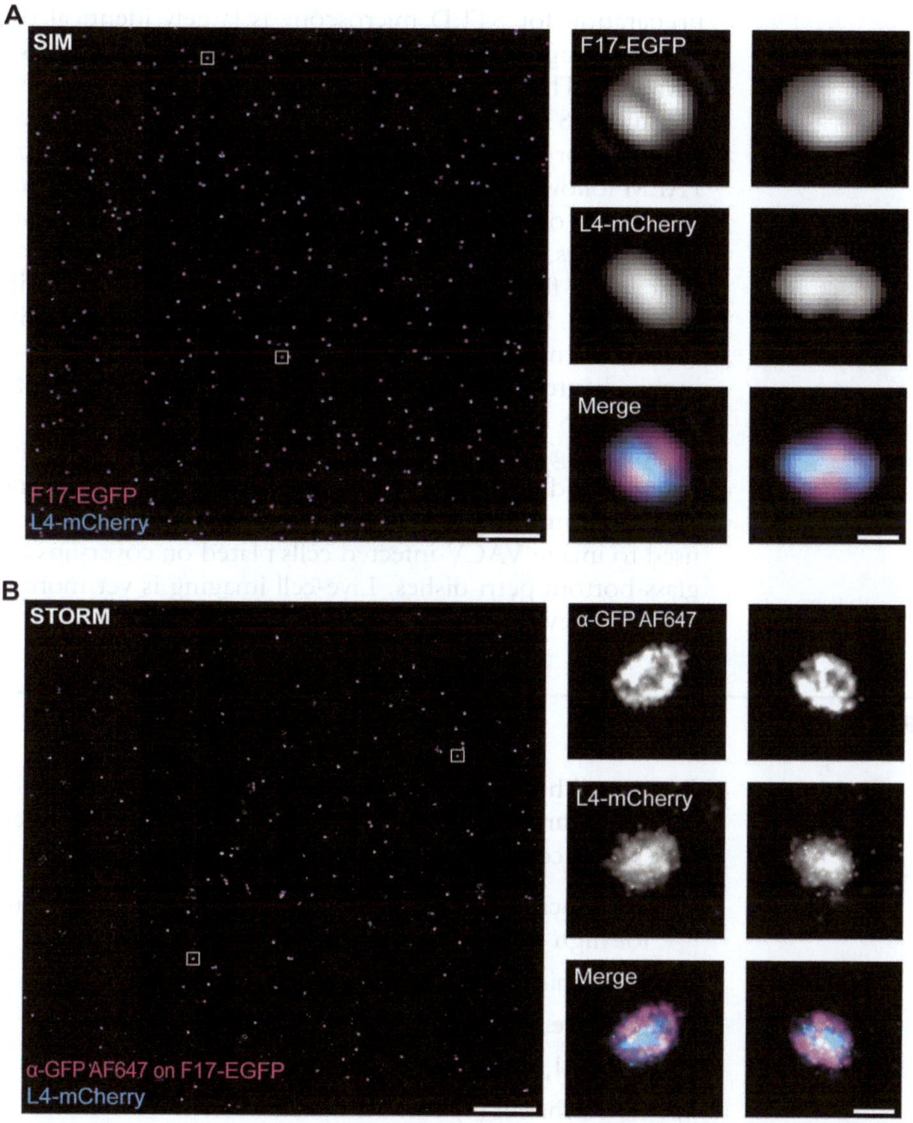

Fig. 1 Super-resolution microscopy images of vaccinia virus. (**a**) Dual-color SIM image of lateral body protein F17-EGFP and core protein L4-mCherry. Close-up of individual virions show the two lateral bodies flanking the elongated core in the sagittal orientation. (**b**) Dual-color STORM image of F17-EGFP labeled with AF647-conjugated anti-GFP nanobodies and L4-mCherry

use will depend on the requirements of the experiment. In any case, the protein of interest will require labeling with a fluorescent probe. There are various methods to achieve this [14] although the most common are immunofluorescence with antibodies directed against the protein of interest, or incorporation of a fluorescent fusion protein tagged to the protein of interest.

In this chapter we describe how to prepare isolated VACV particles for imaging with SIM and STORM modalities. Sample preparation for STED microscopy is largely identical to that for SIM, although requirements for the secondary antibodies are more stringent. The manual of the user's STED microscope should be consulted. STORM can be used with an activator and reporter dye pair [15] or in reducing buffer with blinking organic dyes [16]. PALM follows the same principles as STORM but relies on photoswitchable or photoconvertible fluorescent proteins rather than organic dyes. A method for the construction of recombinant VACV containing fluorescent fusion proteins has been described elsewhere [17]. Fluorescent protein tags can be combined with immunofluorescence for multichannel imaging. The methods described in this chapter can be combined for multicolor imaging of the same sample with different SRM modalities.

Imaging of infected cells and intracellular virus is more challenging, and the exact methodology required will depend greatly on the nature of the experiment. However, this protocol can be used to image VACV-infected cells plated on coverslips or suitable glass-bottom petri dishes. Live-cell imaging is yet more challenging and relevant reviews should be consulted [18].

2 Materials

Prepare all buffers with deionized water. Be sure to sterile filter all solutions through 0.2 μm filters prior to use to remove particles that might contribute to background during imaging.

1. High-performance objective matched coverslips (usually #1.5 for high-NA objectives) (*see* **Note 1**).
2. Glass microscopy slides.
3. Acetone, ultrapure.
4. Ethanol, ultrapure.
5. 70% Ethanol (v/v in deionized water).
6. 1 mM Tris, pH 9.0.
7. 1 M KOH solution.
8. 0.1 M KOH solution.
9. Deionized water.
10. Phosphate-buffered saline (PBS).
11. Fixation buffer: 4% Paraformaldehyde in PBS, pH 7.4 (*see* **Note 8**).
12. Quenching buffer: 0.25% NH_4Cl in PBS.
13. Blocking buffer: 5% Bovine serum albumin (BSA) in PBS.
14. Permeabilization buffer: 0.2% TritonX-100 in PBS.

15. Permeabilization/blocking buffer: 0.2% TritonX-100, 5% BSA, 1% serum in PBS.
16. STORM buffer (oxyrase-MEA): 20% (v/v) Sodium DL-lactate in PBS, adjusted to pH 8.0. Add 3% (v/v) OxyFluor and 50–100 mM cysteamine (MEA) freshly before imaging [19].
17. Parafilm.
18. Mounting medium: Vecta Shield, ProLong Gold, or other mounting medium.
19. Lint-free paper tissues.
20. Nail polish.
21. Tabletop sonication water bath.
22. Orbital shaker.
23. Scalpel.
24. Tweezers.
25. 12-Well cell culture plates.
26. Silicone spacer for STORM buffer reservoir (e.g., by EM Sciences).

3 Methods

3.1 Preparation of Coverslips

1. Wash coverslips with ultrapure ethanol, acetone, and deionized water sequentially (*see* **Note 1**).
2. Repeat **step 1** three times.
3. Check coverslips are clean and hydrophobic by placing a small drop of deionized water on the coverslip. Droplet should retain its shape and run off easily.
4. Sonicate coverslips in 1 M KOH solution for 20 min (*see* **Note 2**).
5. Wash coverslips three times with deionized water.
6. Ensure that no dust particles or smears are visible on the coverslips.
7. Cleaned coverslips can be stored in a sealed container submerged in deionized water.

3.2 Preparing Isolated Virus Particles on Coverslips

1. Place coverslips into a suitable container, e.g., round 18 mm coverslip in 12-well plate.
2. Defrost virus stock solution.
3. Vortex virus stock solution for 30 s.
4. Sonicate virus stock solution for 30 s.
5. Repeat **steps 8** and **9** two times (*see* **Note 3**).

6. For each 18 mm coverslip dilute the virus stock solution in 100 µL 1 mM Tris to a final concentration of ~2×10^7 pfu/mL (*see* **Note 4**).
7. Sonicate the diluted solution of virus particles two times for 30 s.
8. Place 100 µL virus solution on each clean coverslip. The solution will spread on the hydrophilic surface of the coverslip and coat it entirely.
9. Leave at room temperature for 30–60 min for the virus to bind to the coverslip (*see* **Notes 5–7**). Virus particles will stay bound to the coverslips during subsequent steps.
10. Carefully remove the virus solution with a pipette (*see* **Note 8**).
11. Fix the virus by careful addition of fixation buffer (*see* **Note 9**). Ensure that the coverslip is covered completely and does not float on top of the PFA solution.
12. Incubate at RT for 15 min. After PFA fixation the virus is inactivated and coverslips can be handled outside of biological safety cabinets.
13. Remove PFA solution.
14. Wash coverslips with PBS.
15. Remove PBS and incubate in quenching buffer for 5 min at RT to reduce PFA autofluorescence.
16. Place coverslips in PBS.
17. Proceed to Subheading 3.3 for staining or samples can be imaged directly or stored in PBS at 4 °C for several days.

3.3 Immunofluorescence Staining for Super-resolution Microscopy

The following steps are for immunofluorescence staining of one or more specific target proteins. This can be done in combination with fluorescent fusion proteins, such as a GFP-tagged viral protein incorporated into the VACV particle. To reduce photobleaching, perform all incubation steps in the dark, especially avoiding direct sunlight, e.g., by placing the dish with coverslips in a small cardboard box. Ideally, perform all washing steps on an orbital shaker at 5–10 rpm to ensure good mixing while avoiding washing off the virus particles bound to the coverslip.

1. Incubate virus-coated coverslips from Subheading 3.2, **step 17**, in permeabilization buffer for 10 min (*see* **Note 10**).
2. Incubate coverslips with blocking buffer for 30 min at RT (*see* **Notes 11** and **12**).
3. Stain with primary antibody diluted in staining buffer for 1 h at RT (*see* **Notes 13** and **14**).
4. If the primary antibodies are directly conjugated to fluorophores proceed to **step 9**.

5. Wash for 5 min with PBS.
6. Move coverslips to a fresh container to reduce background (e.g., move coverslip to a new well in a 12-well plate).
7. Wash two more times for 5 min with PBS.
8. Stain with secondary antibody in staining buffer for 1 h at RT (*see* **Notes 13–15**).
9. Wash one time for 5 min with PBS.
10. Move coverslips to a fresh container to reduce background.
11. Wash two more times for 5 min with PBS.
12. Proceed to Subheading 3.4 for structured illumination or STED microscopy, Subheading 3.5, for STORM, or store coverslips in PBS until mounting.

3.4 Coverslip Mounting for Structured Illumination and STED Microscopy

1. Wearing clean gloves wash glass slides (one per coverslip) with ethanol and lint-free tissue.
2. Pipette 10 μL of mounting medium onto the center of the glass slide (*see* **Note 16**).
3. Remove the coverslip from PBS with tweezers.
4. Hold the coverslip vertically and remove excess PBS by carefully touching the edge of the coverslip onto lint-free tissue.
5. Set the coverslip, with the bound virus particles facing down, onto the mounting medium on the glass slide.
6. Gently press down onto the center of the coverslip with a pair of tweezers to remove any small air bubbles, and push them outwards until they escape at the side. No air bubbles should be trapped between glass slide and coverslip.
7. Seal the coverslip with nail polish if using a non-self-curing mounting medium (*see* **Note 17**).
8. Dry the slide at RT for a minimum of 10 min.
9. Carefully clean the top of the coverslip using lint-free tissue and 70% ethanol being careful not to damage the nail polish seal (*see* **Note 18**).
10. Samples are now ready for imaging and post-acquisition image analysis (*see* **Note 19**).

3.5 Sample Preparation for Localization Microscopy (STORM)

The method described here is based on a cysteamine (MEA)-based buffer although alternative buffer recipes can be used [19, 20]. We use commercially available silicone spacers as a buffer reservoir (illustrated in Fig. 2). A buffer reservoir is highly recommended as it greatly improves STORM performance. There are several non-commercial options available (*see* **Note 20**). Be sure to wear clean gloves for all subsequent steps.

1. Proceed with samples from Subheading 3.3, **step 12**.

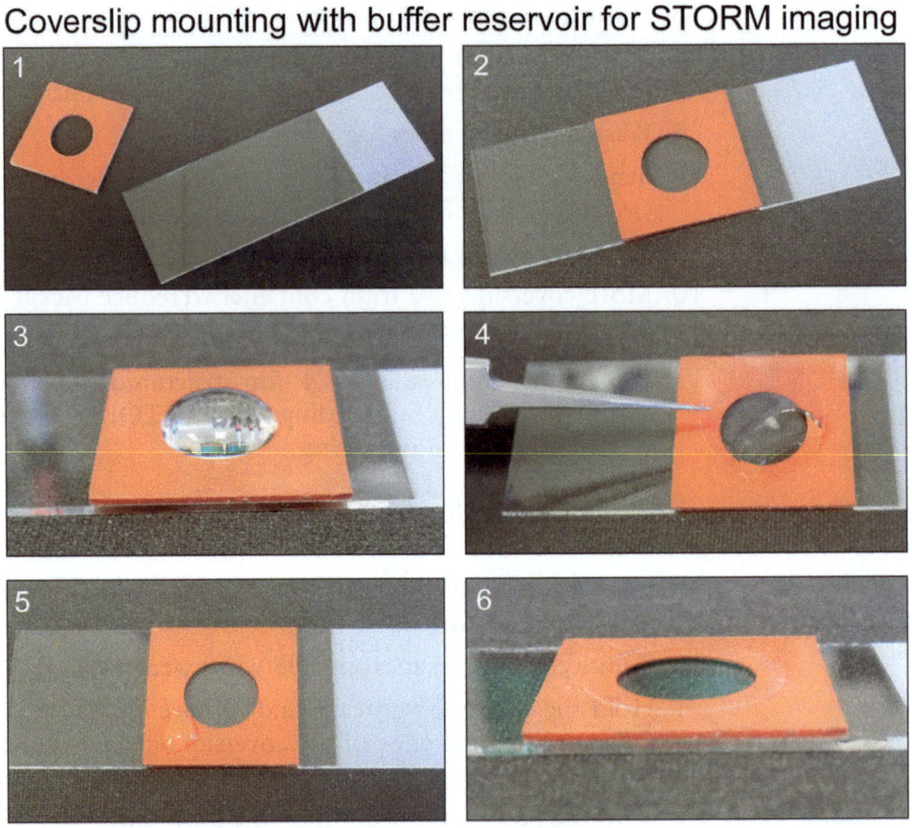

Fig. 2 Slide preparation for STORM with buffer reservoir. Silicone spacers (1) with an adhesive (e.g., from *EM Sciences*) are glued to a clean microscopy slide (2). The vacancy in the center serves as a reservoir for the STORM buffer (3) onto which the coverslip is placed with the sample facing downwards (4). Gently pressing down the coverslip extrudes excess buffer (5) and forms an airtight seal between the coverslip and the silicone (6)

2. Clean glass microscopy slide with ultrapure ethanol and lint-free tissue.
3. Remove protective cover from silicone spacer and attach firmly to the clean slide.
4. Assure that no air bubbles are trapped between the adhesive and the slide.
5. Clean the exposed center of the glass slide inside the silicone spacer and surrounding silicone using ultrapure ethanol and lint-free tissue (*see* **Note 21**).
6. Prepare fresh STORM buffer (*see* **Note 22**). Ensure that buffer is mixed well.
7. Briefly sonicate STORM buffer to degas it. Do not vortex.
8. Pipette 50–100 μL of buffer into the chamber created on the slide by the silicone spacer.

9. Remove the coverslip from PBS with tweezers.
10. Hold the coverslip vertically and remove excess PBS by carefully touching the edge of the coverslip onto lint-free tissue.
11. Set the coverslip with the bound virus particles facing down onto the chamber containing the STORM buffer. Avoid trapping air bubbles by slowly setting the coverslip down at a tilted angle.
12. Gently press down on the coverslip using tweezers or a finger to squeeze excess buffer out of the side. If the coverslip slips easily, apply more pressure.
13. Remove excess buffer with a lint-free tissue without shifting the position of the coverslip. The surface of the coverslip should be dry and the coverslip firmly attached to the silicone, sealing the buffer in the chamber (*see* **Note 23**).
14. Clean the coverslip using lint-free tissue and 70% ethanol without sliding it off the silicone.
15. Sample is ready for imaging. Images should be acquired within the next few hours.
16. Acquire STORM image series with TIRF or HiLo illumination (*see* **Note 24**).
17. After imaging, the coverslip may be recovered by carefully lifting it off the silicone with tweezers.
18. Store sample in PBS at 4 °C for up to a few days for subsequent experiments (*see* **Note 25**).
19. Analyze raw data with any single-molecule localization software, e.g., the free ThunderSTORM ImageJ plug-in [21] or QuickPALM [22].
20. Further data mining such as single-particle analysis with VirusMapper [10], or cluster analysis with SR-Tesseler [23] can be performed on reconstructed images or localization tables.

4 Notes

1. Round 13 or 18 mm coverslips fit into standard 12-well cell culture plates and require approx. 500 μL to be covered completely. Square 18 mm coverslips and larger coverslips fit into 6-well plates and require at least 1 mL per well. 60 mm Dishes can be used for incubation/washing of coverslips; however, they require even higher volumes. Choose the right coverslip and container according to the application and adjust volumes appropriately during all subsequent steps.

2. Coverslips can alternatively be made hydrophobic by incubating in 0.1 M KOH solution for 16 h at RT.
3. Sonication is most important to disperse aggregated virions. If aggregates persist increase sonication time (e.g., 5 × 5 min). Take care that the solution does not heat up as that damages the virus (e.g., add ice to sonicator). Improve purity by banding the virus in a sucrose gradient.
4. For different size coverslips adjust the volume of virus used per coverslip accordingly. Some experimentation may be required to achieve optimal density for each virus.
5. Virus binding to the coverslip is concentration dependent and incubating for more than 60 min will hardly increase the density. Experiment with the concentration to get the desired number of viruses per field of view. Particle number per PFU can differ depending on the virus/mutant used; also *see* ref. [24]. It is possible to reuse the same solution on multiple coverslips with nearly identical densities. Precious samples can be recovered and stored at −80 °C for later use. If it is difficult to achieve sufficiently high virus densities, dilute virus in smaller volumes (e.g., 1 μL in 20 μL 1 mM Tris or use stock virus solution). Use hydrophobic coverslips (Subheading 3.1, **step 3**) and place a single drop into the center of the coverslip. *See* Fig. 3 for examples of different virus densities.
6. To prevent the solution containing virus particles from running off the coverslip, place the coverslip on a small patch of hydrophobic parafilm which will ensure that the solution stays on the coverslip.
7. If evaporation of the virus solution during binding on the coverslip occurs, use a humidified chamber. For example, place coverslip in an enclosed box together with a wet tissue paper.
8. Remove and add solutions carefully to avoid washing virus particles off the coverslip during all steps. Never pipette directly onto the center of the coverslip that will be imaged. Always keep coverslips submerged in solution as drying up will decrease sample quality. Use appropriate containers and volumes (*see* **Note 1**).
9. For best performance, prepare fixation buffer fresh from EM-grade methanol-free 16% PFA. Fixation buffer can be stored at 4 °C for 2 weeks or aliquoted and frozen. Alternatively, prepare fixation buffer by dissolving paraformaldehyde (PFA) powder in PBS.
10. If immunofluorescence labeling is poor, try permeabilizing for 20 min in 1% Triton X-100 in PBS. Poor labeling could be due to low penetration depth of antibodies which could be circum-

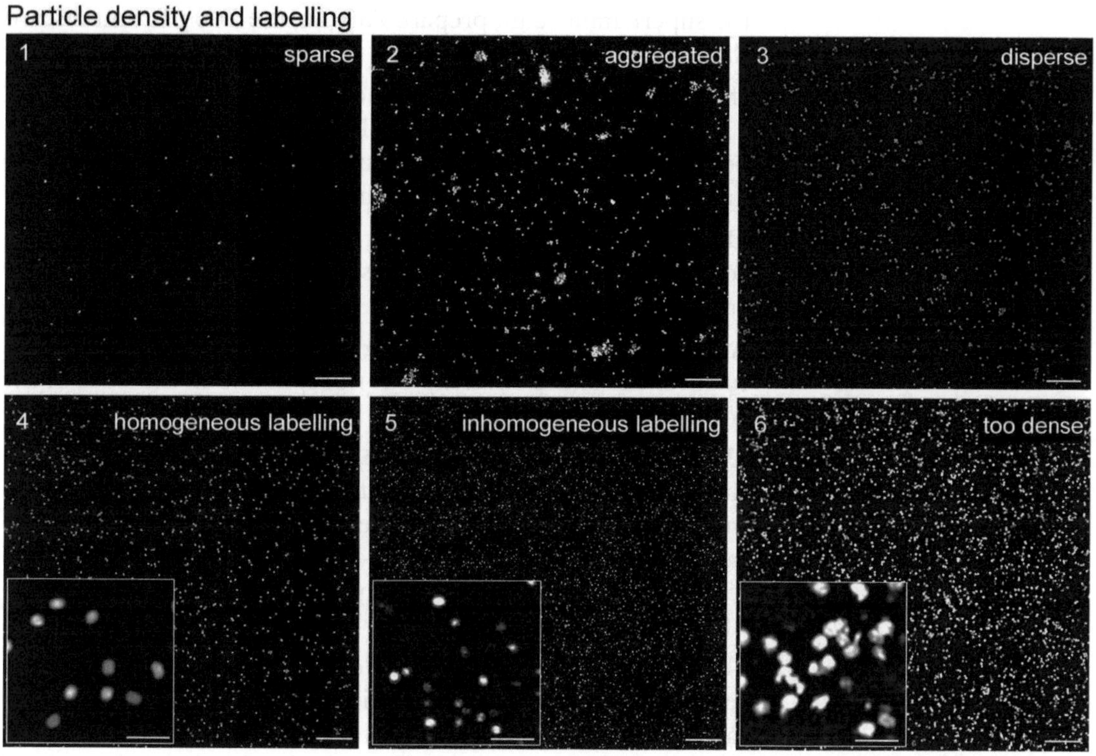

Fig. 3 Virus particle density and labeling. Purified virus particles in a typical 50 × 50 μm field of view. (1) Low virus concentrations result in a sparse sample that yields few particles per field of view. (2) Poor virus quality or insufficient sonication results in aggregates. (3–4) Ideally, virions are monodisperse at high densities with little background. (5) Inhomogeneous labeling of virions may be the result of insufficient permeabilization or low-affinity antibodies. (6) Too high particle densities obstruct the analysis of single virions. Scale bars: 5 μm and 1 μm insets

vented by using smaller labels (e.g., Fab fragments or nanobodies).

11. Blocking can be performed at RT for 1 h but also can be performed at 4 °C overnight.
12. If the background is very high, add 1% heat-inactivated serum (goat, bovine, horse, etc.) to the blocking buffer to reduce unspecific binding of antibodies.
13. Use dilutions typically employed for immunofluorescence microscopy and increase the concentration if signal is too low or reduce the concentration if the background is too high. A good starting point is 1:1000 for affinity-purified antibodies and 1:100 for serum. Incubation time with secondary antibodies can be increased but usually staining for more than 2 h negligibly improves labeling. To avoid bright aggregates on the sample, spin down the diluted primary and secondary antibody at max speed on a tabletop centrifuge for 10 min and use only

the supernatant (e.g., prepare 750 μL, use 700 μL, and discard the last ~50 μL).

14. If extremely high antibody concentrations are required (e.g., 1:20) place a drop of 30–50 μL containing the diluted antibody on a patch of parafilm and invert the coverslip onto the drop. Place in a humidified chamber (*see* **Note** 7).
15. Optionally add DAPI, or other fluorescent probes for multicolor imaging to the staining solution containing secondary antibodies.
16. Adjust the amount of mounting medium so that it is sufficient to cover the entire coverslip but not spill out on the sides of the coverslip.
17. Use clear nail polish to seal the edges of the mounted coverslip. If no good seal is achieved test different brands. If chosen mounting medium cures (hardens) then it may not be necessary to seal with nail polish; however, it usually takes much longer for the sample to be ready for imaging (follow the manufacturer's instructions).
18. To ensure that the coverslip is clean for example hold it up and reflect the light from a lamp off it. Any residues (e.g., inorganic salts from buffers, smeared nail polish, fingerprints, dust particles) are easily detectable. Repeat cleaning procedures until no dirt is visible by eye.
19. The acquired structured illumination images may provide good information about the position of the proteins that have been labeled in the virus, but further image analysis can be important. Use an image analysis software package such as the free Fiji [25]. There are numerous plug-ins to this software that can be used to further analyze images of vaccinia particles, such as the plug-in VirusMapper [10, 26], which is designed for averaging images of viruses together to improve precision and signal-to-noise ratio.
20. With a suitable microscope, samples prepared for STORM can also be used for SIM or STED, imaging the same or a different target structure. SIM images should be acquired first as the method induces less photobleaching (*see* Table 1). There are multiple ways to mount samples on a sealed buffer reservoir for STORM. Alternatively, parafilm gaskets can be self-made from minimal materials [27] and other approaches are equally feasible (magnetic holder, coverslip sandwich). Each should be adapted to the microscope stage requirements.
21. Slides with silicone spacers attached can be cleaned with 70% ethanol and reused.
22. STORM buffer containing a reducing agent should always be prepared fresh directly before imaging. The concentration of

MEA (50–150 mM) can be adjusted for different fluorophores. 50 mM yields good results with Alexa Fluor 647. Alternative STORM buffer (BME based): 150 mM Tris, 10 mM NaCl, 1% glucose, 1% glycerol, pH 8. Add 1% (v/v) β-mercaptoethanol (BME) and oxygen-scavenging system (0.5 mg/mL glucose oxidase, 40 mg/mL catalase) freshly before imaging. For dual-color STORM imaging ensure that the buffer is compatible with all fluorophores.

23. If any air bubbles are visible in the chamber carefully remove coverslip and go back to Subheading 3.5, **step 8**. If the silicone is dirty or not perfectly flat the coverslip will not form a good seal. If no good seal is formed, carefully recover the coverslip and clean the slide or prepare a new slide (Subheading 3.5, **step 1**).
24. Typically acquire 30,000 frames at 25-ms exposure time. Acquire wide-field/confocal/SIM images first if imaging with multiple modalities due to high photodamage during STORM imaging. If activated fluorophore density is too high, change to a different buffer system, use a microscope with higher laser intensity, or illuminate at maximum power until a suitable steady state is reached (can be several minutes). If the background from floating fluorophores is too high, remount using fresh buffer.
25. Do not store samples in reducing buffers as the samples will decay.

Acknowledgments

R.G. is funded by the Engineering and Physical Sciences Research Council (EP/M506448/1). D.A. received funding from the European Union's Horizon 2020 research and innovation programme under the Marie Skłodowska-Curie grant agreement No 750673.

References

1. Kausche GA, Pfankuch E, Ruska H (1939) Die Sichtbarmachung von pflanzlichem Virus im Übermikroskop. Naturwissenschaften 27:292–299
2. Green RH, Anderson TF, Smadel JE (1942) Morphological structure of the virus of vaccinia. J Exp Med 75:651–656
3. Dales S (1963) The uptake and development of vaccinia virus in strain L cells followed with labeled viral deoxyribonucleic acid. J Cell Biol 18:51–72
4. Cyrklaff M, Risco C, Fernández JJ et al (2005) Cryo-electron tomography of vaccinia virus. Proc Natl Acad Sci U S A 102:2772–2777
5. Grünewald K, Cyrklaff M (2006) Structure of complex viruses and virus-infected cells by electron cryo tomography. Curr Opin Microbiol 9:437–442
6. Schmidt FI, Bleck CKE, Reh L et al (2013) Vaccinia virus entry is followed by core activation and proteasome-mediated release of the Immunomodulatory effector VH1 from lateral bodies. Cell Rep 4:464–476

7. Gustafsson MGL (2000) Surpassing the lateral resolution limit by a factor of two using structured illumination microscopy. J Microsc 198:82–87
8. Betzig E, Patterson GH, Sougrat R et al (2006) Imaging intracellular fluorescent proteins at nanometer resolution. Science 313:1642–1645
9. Sydor AM, Czymmek KJ, Puchner EM, Mennella V (2015) Super-resolution microscopy: from single molecules to supramolecular assemblies. Trends Cell Biol 25:730–748
10. Gray RDM, Beerli C, Pereira PM et al (2016) VirusMapper: open-source nanoscale mapping of viral architecture through super-resolution microscopy. Sci Rep 6:29132
11. Horsington J, Turnbull L, Whitchurch CB, Newsome TP (2012) Sub-viral imaging of vaccinia virus using super-resolution microscopy. J Virol Methods 186:132–136
12. Culley S, Albrecht D, Jacobs C et al (2018) Quantitative mapping and minimization of super-resolution optical imaging artifacts. Nat Methods 15(4):263–266
13. Horsington J, Lynn H, Turnbull L et al (2013) A36-dependent actin filament nucleation promotes release of vaccinia virus. PLoS Pathog 9:e1003239
14. Sakin V, Paci G, Lemke EA, Müller B (2016) Labeling of virus components for advanced, quantitative imaging analyses. FEBS Lett 590:1896–1914
15. Rust MJ, Bates M, Zhuang X (2006) Sub-diffraction-limit imaging by stochastic optical reconstruction microscopy (STORM). Nat Methods 3:793–796
16. Heilemann M, van de Linde S, Schüttpelz M et al (2008) Subdiffraction-resolution fluorescence imaging with conventional fluorescent probes. Angew Chem Int Ed Engl 47:6172–6176
17. Marzook NB, Procter DJ, Lynn H, et al (2014) Methodology for the efficient generation of fluorescently tagged vaccinia virus proteins. J Vis Exp e51151. https://doi.org/10.3791/51151
18. Henriques R, Griffiths C, Rego EH, Mhlanga MM (2011) PALM and STORM: unlocking live-cell super-resolution. Biopolymers 95:322–331
19. Nahidiazar L, Agronskaia AV, Broertjes J et al (2016) Optimizing imaging conditions for demanding multi-color super resolution localization microscopy. PLoS One 11:1–18
20. Olivier N, Keller D, Gönczy P, Manley S (2013) Resolution doubling in 3D-STORM imaging through improved buffers. PLoS One 8:1–9
21. Ovesný M, Křížek P, Borkovec J et al (2014) ThunderSTORM: a comprehensive ImageJ plug-in for PALM and STORM data analysis and super-resolution imaging. Bioinformatics 30:2389–2390
22. Henriques R, Lelek M, Fornasiero EF et al (2010) QuickPALM: 3D real-time photoactivation nanoscopy image processing in ImageJ. Nat Methods 7:339–340
23. Levet F, Hosy E, Kechkar A et al (2015) SR-Tesseler: a method to segment and quantify localization-based super-resolution microscopy data. Nat Methods 12:1065–1071
24. Stiefel P, Schmidt FI, Dörig P et al (2012) Cooperative vaccinia infection demonstrated at the single-cell level using FluidFM. Nano Lett 12:4219–4227
25. Schindelin J, Arganda-Carreras I, Frise E et al (2012) Fiji: an open-source platform for biological-image analysis. Nat Methods 9:676–682
26. Gray RDM, Mercer J, Henriques R (2017) Open-source single-particle analysis for super-resolution microscopy with VirusMapper. J Vis Exp e55471–e55471. https://doi.org/10.3791/55471
27. Pereira PM, Almada P, Henriques R (2015) High-content 3D multicolor super-resolution localization microscopy. Methods Cell Biol 125:95–117
28. Hell SW, Wichmann J (1994) Breaking the diffraction resolution limit by stimulated emission: stimulated-emission-depletion fluorescence microscopy. Opt Lett 19:780

Chapter 17

Bioluminescence Imaging as a Tool for Poxvirus Biology

Beatriz Perdiguero, Carmen Elena Gómez, and Mariano Esteban

Abstract

Bioluminescence imaging, with luciferase as a reporter-encoding gene, has been successfully and widely used for studies to follow viral infection in an organism and to measure therapeutic efficacy of antiviral agents in small animal models. Bioluminescence is produced by the reaction of a luciferase enzyme stably inserted into the viral genome with a defined substrate systemically delivered into the animal. The light emitted is captured allowing the detection of viral infection sites and the quantification of viral replication in the context of tissues of a living animal. The goal of this chapter is to provide a technical background for the evaluation of poxvirus infection in cells and animals through bioluminescence imaging technology using luciferase-expressing recombinant poxviruses.

Key words Bioluminescence, In vivo imaging, Luciferase, Poxvirus, Infection, Biodistribution, Pathogenesis

1 Introduction

Bioluminescence imaging (BLI) is a powerful and versatile technology to study viral pathogenesis [1], host immune responses to infection, and efficacy of vaccines and antiviral therapies in living animals, especially in mice [2]. Unlike conventional methods that analyze viral infection by monitoring survival, euthanizing animals at multiple time points, dissecting tissues, and quantifying plaque-forming units (p.f.u) in those tissues, bioluminescence technology allows the study of the course of infection over time in the same animal minimizing the effects of biological variation and reducing the number of animals required to generate statistically meaningful data [1]. Additionally, BLI can identify unexpected sites of infection and/or patterns of host response that could be easily missed by analyzing only selected tissues at predetermined time points [3].

BLI studies of viral infection are typically performed with recombinant viruses engineered to express luciferase gene under the control of a viral promoter. The introduction of the firefly luciferase gene into the genome of vaccinia virus (VACV) was the

Jason Mercer (ed.), *Vaccinia Virus: Methods and Protocols*, Methods in Molecular Biology, vol. 2023,
https://doi.org/10.1007/978-1-4939-9593-6_17,

first example of an animal virus being used for measurements of emitted light as an alternative approach to traditional plaque assays to follow replication of the virus in cell cultures and tissues of infected animals [1]. VACV is a large DNA-containing virus, which allows for insertion of the luciferase cassette gene under a VACV promoter without affecting viral replication or virulence in animal models. Recombinant VACV produces ~10,000 particles/cell and thus a significant pool of luciferase molecules are present in infected organs [1].

Luciferase activity could be detected as early as 1 h after in vitro infection of cells. Bioluminescence signal increased in direct proportion to viral replication [4]. The limits of detection were one infected cell in a background of a million noninfected cells [1]. Both the minimum time to detection and minimal detectable input titers of virus are dependent upon surface area occupied by infected cells, multiplicity of infection, and promoter used to drive luciferase. However, these data provide a general estimate of the sensitivity of BLI for quantifying infection with VACV recombinants expressing luciferase reporter gene in vitro.

In vivo studies using mice infected with luciferase-expressing VACV have also shown that measurements of photon fluxes correlated in linear fashion with viral loads measured in organs isolated from VACV-infected mice at different days post-infection by plaque formation assay, thus supporting the notion that bioluminescence provides a direct measure of viral dissemination [5, 6]. Luciferase-expressing VACV was still pathogenic compared with wild-type Western Reserve strain in vitro or in vivo, which validates the use of this reporter virus for studies of poxvirus pathogenesis. BLI can also be used to predict lethality in challenge models and test novel anti-smallpox vaccines and antiviral treatments [4, 5, 7].

Multiple recombinant viruses expressing luciferase derived from RNA- or DNA-containing viruses were subsequently generated, like equine encephalitis virus [8], influenza virus [9], herpes simplex virus type I [10–15], Sendai virus [16], Sindbis virus [17, 18], dengue virus [19], varicella zoster virus [20], and different poxvirus vectors [4, 6, 7, 21–29].

Until BLI was described, the detection of luciferase activity in a mice infected with a recombinant virus expressing luciferase requires sacrifice, excision of different tissues, and homogenization to measure enzyme activity in a conventional luminometer. Light emitted from the homogenized tissues after the exposure of luciferase to luciferin substrate in the presence of ATP can be detected by a luminometer equipped with a highly sensitive computer and a software platform for data analysis. A single-injector luminometer contains a light-tight box in which the sample is added in a tube along with luciferin substrate just prior to the measurement of luciferase activity and the sliding door is closed. After reading is completed, the software provides complete sample documentation for further analysis. The luciferase activity is normalized to the protein content of the sample. A plate-reading luminometer uses a 96-well plate for the measurement of the luciferase activity instead of the single tube allowing the reading of multiple samples at the same time [30].

Since bioluminescence is the emission of light from biochemical reactions that occurs within a living organism, BLI requires the availability of reporter luciferase enzymes expressed in living cells, administration of luciferase substrate, and very sensitive cooled charge-coupled device (CCD) cameras that operate in the visible to near-infrared regions of the spectrum to detect the low levels of light emitted from the body of an animal [31, 32]. These detectors convert the photons per unit area into electrons [33] and the imaging software converts electron signals into a two-dimensional image. The software is also able to quantify the intensity of the emitted light (number of photons striking the detectors) and to convert these numerical values into a pseudocolor graphic. Therefore, BLI generates planar projection data displayed as pseudo-colored images to represent signal intensity that is localized over grayscale reference images of the subjects [34]. Since anatomic resolution is relatively poor in whole-body images, a high-magnification lens can be directed at sites in the body where labeled cells have been localized by whole-body imaging to produce high-resolution images that complement the lower resolution images taken of the whole animal [35–38]. In fact, BLI data are typically quantified using region of interest (ROI) analysis to measure photons of light emitted from a defined anatomic site.

The main advantages of BLI are the following: (1) highly sensitive approach, allowing detection of as few as 1×10^2 pfu of luciferase-expressing viruses in vivo [10]; (2) clearance of the substrate; (3) photon emission is relatively rapid allowing for repeated testing in a given animal over a reasonable time period [39]; (4) luciferase is a good optical indicator in mammalian cells and tissues since it is nearly absent in mammals; (5) noninvasive, virtually free of any background signal, does not need external light excitation, and allows for semiquantitative real-time detection of biological processes; and (6) relatively economical.

However, although BLI has several advantages for small-animal imaging studies as described above, there are also several factors that can influence the quantification and sensitivity of this technique (deeply described at [2, 40]), together with certain limitations that must be taken into consideration when performing this type of imaging analysis: (1) BLI signal intensity is correlated with the amount of luciferase present; therefore it is important to consider the half-lives of the luciferase transcript and protein when analyzing and interpreting data; (2) signal attenuation due to the absorption and scattering of light by mammalian tissues limits spatial resolution of bioluminescence imaging to 1–3 mm. Since light is attenuated approximately ten-fold per centimeter of overlying tissue, light from superficial sites is detected to a greater extent than light emitted from deeper organs and tissues [31]. Therefore, a direct comparison of bioluminescence signals between organs would not be accurate without correlation with viral measurements in isolated tissues; (3) attenuation of light in tissues is also relatively greater for luciferase enzymes that emit blue-green light, such as *Gaussia* and *Renilla* luciferases, than for those that emit red and infrared light, such as firefly luciferase and click beetle red luciferase (see below); (4) while tomographic techniques generate three-dimensional images, BLI typically produces a single two-dimensional image of the whole animal. Although there are three-dimensional optical imaging systems and/or reconstruction techniques commercially available, it still may be difficult to discriminate photons emitted by infected cells in two immediately adjacent areas; (5) it is unlikely to be used in humans because of the limitations in detecting light produced in deep tissues.

BLI studies have been performed using different luciferase enzymes [40–42]. The substrates for these proteins are oxidized and chemically consumed by the luciferases in reactions that require oxygen and ATP and generate an excited-state molecule that emits light (photons) detected by external and sensitive CCD cameras.

In the intact animal, absorption of light by tissue, and particularly absorption by hemoglobin and other pigmented molecules such as melanin, attenuates the bioluminescent signal generated by cells of interest. Red and infrared light (wavelengths >600 nm) suffers less signal attenuation than does blue-green light with shorter wavelengths (<600 nm). This is an advantage to firefly luciferase (Fluc; derived from the firefly *Photinus pyralis*) and click beetle red luciferase (CBRluc; derived from *Pyrophorus plagiophthalamus*) over Renilla luciferase (Rluc; derived from the sea pansy *Renilla reniformis*) and Gaussia luciferase (Gluc; derived from the copepod *Gaussia princeps*) [43]. Firefly luciferase is the most commonly used reporter gene for BLI because of the sustained kinetics of light emission, favorable pharmacokinetics of the luciferin substrate, and relatively red-shifted emission spectrum [44]. Specific mutations of this gene can result in increased enzymatic activity with subsequently enhanced bioluminescent signal [31, 45]. Both Fluc and CBRluc exhibit emission peaks at approximately 560 nm, shifting to a peak of 610 nm at body temperature (37 °C), whereas the emission peak of Rluc and Gluc is at approximately 480 nm, in a region where absorption by hemoglobin is relatively high [46]. However, Renilla and Gaussia luciferases also have several advantages for BLI: (1) they do not require ATP as a cofactor during bioluminescent reaction and thus can be used for imaging cells independently of their metabolic state; (2) they are smaller than Fluc, which may be beneficial for integrating into viral genomes and constructing genetic fusions with other proteins of interest

[47]; (3) their optical properties and substrate allow discrimination from firefly luciferase. In vivo applications of these marine luciferases will be greatly enhanced by generating longer wavelength variants of these enzymes, as it has been achieved for *Renilla* luciferase [48]. Beyond isolating luciferase enzymes from different organisms and optimizing emission wavelengths, there are also ongoing efforts to modify the stability of luciferase enzymes to improve the temporal resolution or increase the levels of bioluminescence [49].

The substrate for Fluc and CBRluc, *D-luciferin*, is relatively stable in the body and has a relatively long circulation time. Its delivery is accomplished through intraperitoneal, intravenous, or subcutaneous injection and imaging begins 8–10 min after injection with a peak at 15–20 min and a bioluminescence signal relatively stable for another 15–20 min before decreasing as a result of substrate clearance [5, 46]. Synthetic analogs based on new luciferins, caged luciferins, or luciferin precursors with mutant and latent luciferases that replace the natural *D-luciferin* substrate to broad the scope of BLI have been reported (reviewed in [50]). In contrast, the substrate for Rluc and Gluc, *coelenterazine*, is rapidly cleared from the body, displays a more limited biodistribution, and auto-oxidizes in serum, increasing background signal [46, 51]. Therefore, it should be administered by intravenous or intracardiac routes [52] and data acquisition must be completed within a few seconds to a minute after injection, since the peak of light emission occurs at about 1–2 min after injection of substrate and decreases rapidly over 10–15 min [5, 53].

Therefore, by using luciferase enzymes with different emission spectra and/or substrates, it should be possible to monitor two different viral processes in the same animal, such as replication and sites of tissue injury. The combination of imaging reporter viruses with genetically engineered reporter mice is also expected to substantially enhance the power of bioluminescence imaging for quantitative studies of viral and host factors that control disease outcome and effects of established and new antiviral therapeutic agents.

2 Materials

2.1 Equipment

1. Balance and container to weigh mice.
2. Plexiglas anesthesia system.
3. In vivo bioluminescence imaging system equipped with a light-tight chamber in which the animals are placed for imaging, a sensitive cooled CCD camera to detect the emitted light (CCD cameras that can reconstruct 3D views are helpful), and a computer equipped with the Living Image software. A broad variety of imaging systems are commercially available [2] but IVIS instruments are the most widely used. Each model is equipped with its specific software.
4. Luminometer, single injector or plate reader.

2.2 Reagents and Buffers

1. Appropriate mouse strain.
2. Luciferase-expressing VACV viral stock for infections.
3. Sterile Dulbecco's phosphate-buffered saline (DPBS).
4. Luciferin stock solution: 30 mg/mL D-Luciferin firefly, potassium salt in DPBS, and filtered through a 0.2 μm filter (*see* **Notes 1–3**).
5. 1 ml syringes and 23 gauge (23G) needles.
6. Tissue lysis buffer: 25 mM glycylglycine pH 7.8, 15 mM $MgSO_4$, 4 mM EGTA, 1% Triton X-100, 1 mM dithiothreitol.
7. Luciferase assay buffer: 25 mM glycylglycine pH 7.8, 15 mM $MgSO_4$, 4 mM EGTA, 15 mM KPO_4, 3 mM dithiothreitol, 3 mM ATP.
8. Small-animal electric hair clipper.
9. Depilatory cream of choice.
10. Isoflurane.
11. Charcoal paper.
12. Illuminated magnifying lamp.
13. Bifurcated vaccination needles.
14. 0.2 μm filters.
15. Tissue homogenizer.
16. 70% ethanol.
17. TexPure detergent.

3 Methods

To obtain reproducible BLI data, a standardized imaging protocol, including viral dose, route of administration of recombinant virus and luciferase substrate, amount of injected substrate per animal weight, delay between injection, and imaging and positioning of mice, is required.

3.1 Considerations Prior to Infection (See Notes 4 and 5)

1. One week before imaging studies, mice should be fed with alfalfa-free rodent chow to reduce the alfalfa-related luminescent background in the gut.
2. Mice must be shaved 3–4 days prior to imaging analysis since removal of fur increases the sensitivity of detection of bioluminescent signal.

3.2 Infection of Mice with Luciferase-Expressing Recombinant VACVs (See Notes 6–8)

Luciferase-expressing VACV recombinant viruses can be administered to mice by different routes including intraperitoneal, intramuscular, intrarectal, intragastric, intradermal, intranasal, intracraneal, intravenous, intraocular, or tail scarification. Depending on the immunization route, mice should be anesthetized prior to the infection. According to the experiment, different viral doses and routes of administration will be used and bioluminescence imaging will be performed at different times post-infection (hours, days, or until lethality).

1. For intranasal infection [7, 29]: Once mice are weighed and anesthetized, hold them by scruff and using an illuminated magnifying lamp infect them by slowly dispensing 10 μL of the virus stock into one nostril; be sure that the entire droplet is delivered to nasal cavity. Leave mice resting for at least 24 h before proceeding with whole-body imaging.
2. For tail scarification [29]: 10 μL of virus inoculum containing the appropriate viral dose was pipetted onto the skin covering cartilaginous area proximal to the base of the tail. A bifurcated vaccination needle was used to deliver 25 punctures through the inoculum solution. Leave the inoculum to be absorbed.

3.3 Substrate Administration and Anesthesia (See Notes 9–20)

1. Weigh mice.
2. Thaw an aliquot of luciferin solution at room temperature.
3. Prepare IVIS imaging equipment switching on the apparatus, waiting for the unit to self-calibrate, and ensuring that the heating pad within the imaging station is warm via both temperature readout and tactile confirmation. Place a sheet of charcoal paper at the stage of the imaging chamber.
4. Briefly shake the vial of luciferin, load a syringe, inject mice intraperitoneally with the warmed luciferin solution at a dose of 150 mg/kg body weight, and place them into a clear Plexiglas anesthesia box connected to an isoflurane vaporizer (2% isoflurane provided in 1 l/min oxygen) for about 10–15 min before imaging. 1 mL syringe with a 23G needle is recommended.
5. Observe mice for signs of general anesthesia, notably cessation of movement and decreased respiratory rate.

3.4 Imaging (See Notes 21–28)

1. Once mice are fully anesthetized (2 min after the last mouse has ceased moving), they should be immediately transferred from the anesthesia box to the nose cones attached to the manifold in the imaging chamber previously protected with charcoal paper. Adjust the heads to allow free delivery of the anesthesia to the nostrils through nose cones inserted in the openings of the tube. To facilitate efficient transfer, arrange animals within the anesthesia box in the order that they will be placed into the heated imaging chamber. Mice should be placed with the desired side up.
2. Close the door, acquire a grayscale photographic reference image of mice under weak illumination, and then take the initial bioluminescence image in complete darkness. The luciferase activity will be represented using a pseudo-color scale,

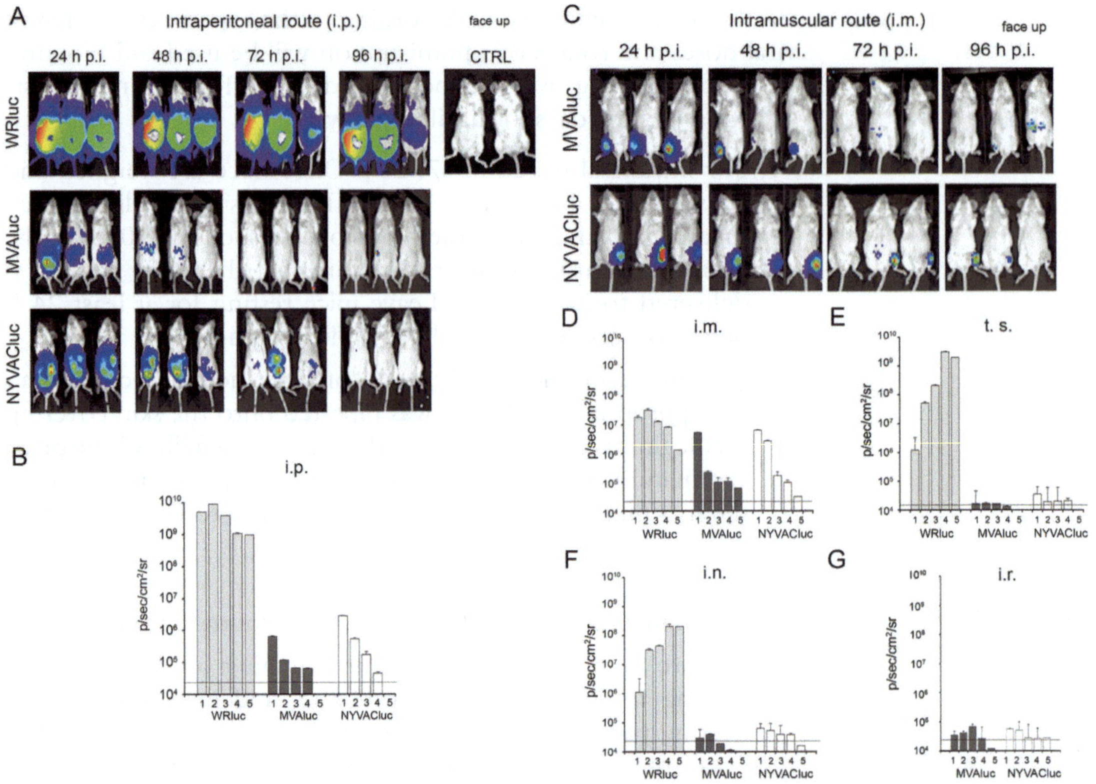

Fig. 1 (**a**, **c**) Bioluminescence distribution of different strains of VACV recombinant viruses expressing luciferase. WRLuc, MVALuc, and NYVACLuc recombinants were inoculated in BALB/c mice by intraperitoneal (**a**) or intramuscular (**c**) route. In the right-hand panel of (**a**), noninfected mice (CTRL) are shown. (**b**, **d–g**) Quantification of the luciferase signal in the ROI during intraperitoneal (**b**), intramuscular (**d**), tail scarification (**e**), intranasal (**f**), or intrarectal (**g**) infections. Mean ± SD values for photon fluxes over time are represented. The solid line represents the background level of bioluminescence (with permission from [23])

using red as the highest and blue as the lowest photon flux. This pseudo-color image is superimposed on the grayscale reference image showing the spatial distribution of detected photons (Fig. 1).

3. Mice should be imaged in multiple positions to confirm and localize bioluminescent signals. Conventional views include ventral, dorsal, and two lateral images. Maintain consistency in the order of orientation for imaging between sequential animals. When the mice are turned from one side to the other for imaging, they should be observed for any signs of distress or changes in vitality. The mice are again imaged and the procedure is completed.

4. After the desired time-course bioluminescence images are acquired, the anesthesia is ceased and mice are returned to their cages where they are monitored and given thermal sup-

port until they recover from anesthesia (mice usually awake within 15 min) or may be euthanized and specific tissues harvested, weighed, and homogenized for the measurement of viral loads by plaque assay and/or by in vitro luciferase assay.

5. All equipment should be disinfected at the completion of imaging.

3.5 Image Analysis and Data Presentation (See Notes 29–32)

After imaging is completed, the images should be saved and analyzed to quantify the light detected from specific regions of interest from the mice at each time point. For each Living Image software, follow the manufacturer's recommendations but the general steps that should be performed to quantify bioluminescence signals are the following:

1. Regions of interest (ROI) need to be designated for each animal.
2. Click the first image and select the desired ROI shape within the organ that exhibits bioluminescent signal to determine the photon/s value for each organ for each day. Adjust the size and position of the ROI to include the region(s) that you want to analyze.
3. Once all ROIs are drawn for the first image, apply them to the other images from the series. Position each ROI in the desired location but do not change the size of the ROI.
4. A selected sequence of images is then displayed automatically in a table that can be saved in a *.csv format.
5. Statistical analysis is performed using Excel software by exporting the *.csv table with the ROI data from Living Image into Excel.
6. Light emission is measured as photons per second per centimeter squared per steradian (p/s/cm^2/sr; photon flux) and represented as mean values ± SEM. Normalize this data by subtracting the photon flux from the noninfected time point.

3.6 Luciferase Assays

1. Tissues/organs from individual mice are homogenized in 1 mL of tissue lysis buffer.
2. Approximately 300 μL of the homogenate is removed and centrifuged at 15,000 × *g* for 5 min and 200 μL of the supernatant is assayed for luciferase activity.
3. Samples (200 μL) are mixed with 1× luciferin solution and 100 μL of luciferase assay buffer and luciferase activity is measured in a luminometer and expressed as luciferase reference units (LRU; [23]). These relative light units can be normalized for sample size by correcting for total protein content.

4 Notes

4.1 Substrate Preparation

1. The D-luciferin firefly powder is stored at −20 °C. It is recommended to allow powder to warm up at room temperature before dissolving it in DPBS.
2. Luciferin should be protected from light and purchased and stored in the smallest quantity possible to prevent conversion of luciferin to dehydroluciferin, a competitive inhibitor of the luciferase/luciferin reaction.
3. Unused reconstituted luciferin substrate can be stored as aliquots at −20 °C until use. Long-term storage as a frozen solution is not recommended for sensitive applications such as whole-animal studies. The solution should be thawed and warmed to room temperature immediately prior to application.

4.2 Considerations Prior to Infection

4. The majority of the studies of poxvirus infection using BLI involve mice as the host species since their small size minimizes attenuation of the bioluminescence signal. However, the technique is not limited to mice and has been applied to other species such as bats [28], rats ([27]; luciferin administration at 50 mg/kg body weight), dormice [54], prairie dogs ([55]; luciferin administration at 125 mg/kg body weight), and squirrels [56].
5. Depending on the coat color of the mice and the area where bioluminescence is to be measured, hair may be removed from the skin over the area of interest by shaving using a veterinary clipper for small animals or by using a depilatory cream. If shaving is used, mice are briefly anesthetized with isoflurane. While either depilation method may be effective alone, it may be necessary to combine both methods, with mechanical shaving followed by chemical depilation, for optimal results. If this method is employed, the mechanical shaving must be carried out meticulously to avoid skin abrasion and subsequent severe irritation by the cream [57]. White mice usually do not require removal of hair since the absorption of light is minimal; however, a substantial decrease in signal due to light absorption by melanin in the skin and fur can occur in black mice if hair is not removed. Hairless mice are also available but may not represent suitable host models because they tend to be associated with immunodeficiencies. Wipe with alcohol to sterilize the skin after shaving.

4.3 Infection of Mice with Luciferase-Expressing Recombinant VACVs

6. Mice can be scanned prior to infection to confirm not to have autoluminescence beyond background levels.
7. Poxviruses used in bioluminescence imaging analysis include raccoon poxvirus [28], monkeypox virus [54–56, 58], cowpox

virus [59], and different VACV strains such as Western Reserve (WR; [6, 7, 23, 24, 59]), International Health Department-J strain (IHD-J; [24, 26, 29]), Wyeth [59], Tian Tan strain [27], and the highly attenuated strains MVA [23, 28] and NYVAC [23].

8. Most of the recombinants generated express the firefly luciferase gen and the doses used depend on the mouse strain, the attenuation level of the poxvirus assayed, and the route of administration. For the sake of orientation, the viral doses for BALB/c mice could range from 10^4 to 10^6 pfu/mouse for intranasal route, from 10^6 to 10^7 pfu/mouse for intraperitoneal route, and from 10^5 to 10^6 pfu/mouse for tail scarification.

4.4 Substrate Administration and Anesthesia

9. If more than one mouse is imaged simultaneously, luciferin injection for each mouse should be performed as fast as possible so that all mice are administered with the substrate at about the same time.
10. To maximize detection of bioluminescence signal and achieve the least variation in experimental data, it is crucial to standardize the dose of injected luciferin and the time between injection and the start of imaging. BLI studies typically use 150 mg/kg of luciferin in a solution of sterile phosphate-buffered saline. However, it is important to note that intraperitoneal injection of this concentration of luciferin does not produce maximal bioluminescence from firefly luciferase in tissues. In fact, luciferin administration at 150 mg/kg versus 450 mg/kg body weight results in a dose-dependent increase in luminescence emitted [60].
11. For the intraperitoneal luciferin injection, mice should be manually restrained, abdomen side up, with cranial end of the animal pointed down. Needles should be level side up and slightly angled when entering the abdominal cavity. Penetrate just through the abdominal wall (about 4–5 mm) of the animal's left lower abdominal quadrant. Be careful not to go too deep or you can damage internal organs.
12. Occasionally, a droplet of luciferin solution can be exuded from the injection site indicating that the injection was not peritoneal but perhaps intra-adipose, intramuscular, intradermal, or even subcutaneous. In these situations, it is unlikely that a valuable signal will be obtained but there is little damage in attempting to image the mouse regardless. Do not try to reinject when this occurs.
13. The biodistribution of luciferin in the mouse is not homogenous, and access to some tissues (e.g., the brain) is relatively low [44]. In these cases, an intravenous luciferin administra-

tion is helpful for reaching anatomic sites like the brain, but clearance of the substrate will occur much more rapidly than when substrate is delivered intraperitoneally. For intravenous injection, mice should be anesthetized prior to luciferin administration.

14. A luciferin kinetic study should be performed for each animal model to determine peak signal time after substrate administration.
15. For the testing of engineered VACVs as potential oncolytic agents (reviewed in [61]), Renilla luciferase has also been inserted into the viral genome instead of firefly luciferase and, therefore, coelenterazine is injected as luciferase substrate immediately before imaging [62–65].
16. The exact order of delivery of substrate and anesthesia will depend on the system being used.
17. Do not over-anesthetize the mice for prolonged periods of time.
18. A maximum of five mice should be anesthetized simultaneously.
19. Inhalational anesthesia by using induction chambers significantly reduces the risk of trauma associated with injectable anesthetics, more easily controls and standardizes depth and duration, is more cost effective, typically results in fewer adverse events, and is rapidly reversible.
20. Isoflurane is the preferred anesthetic agent for both induction and maintenance due to the reduced stress on the liver.

4.5 Imaging

21. Remember to place charcoal paper in the imaging box to help absorb any light not emitted by the infected mice. For intranasal poxvirus inoculations, cover the mouse head with another sheet of charcoal paper to avoid saturation of the CCD camera due to high signal from the nasal cavity.
22. The body weight artifact may be controlled by including in the imaging subject an internal control such as luminescent beads [66].
23. Careful and consistent positioning of mice every time during imaging is required in BLI analyses.
24. Younger mice with lower body weight are preferred for BLI since photons emitted from internal organs or tissues can penetrate more efficiently.
25. When imaging, the field of view should be adjusted to fit the number of mice in the image. It should be at the highest setting that allows visualization of all mice to be imaged.

26. If the signal is expected to be strong shorter exposure times (e.g., 1 s) are appropriate. Longer times (e.g., 5–10 min) are required if the signal is predicted to be weak. If the signal strength is unknown, a 1-min initial exposure time is a reasonable place to start. Once the signal exposure time has been optimized, begin taking images every minute until the signal reaches a maximum. By monitoring the signal intensity at each ROI you can easily determine when the signal begins to decrease, indicating that the maximum signal intensity has been reached. The maximum signal should be used as your final data.
27. "Saturated" images are nonquantitative since the saturated pixels cannot collect more signals. The previously described parameters should be adjusted to collect an image that reflects a signal strength that is adequate for anatomic localization but not strong enough to saturate the camera.
28. At the end of the imaging session, it is necessary to perform a decontamination of both the IVIS imaging chamber and the induction chamber using 70% ethanol and TexPure detergent, respectively.

4.6 Image Analysis and Data Presentation

29. Background measurements should be taken and subtracted from experimental data. Subtractions are automatically calculated by the Living Image software.
30. For presentation purposes, each luminescence image needs to be standardized to a light intensity scale.
31. The region of interest for calculation of photonic counts has to be selected such that the area is uniform with each measurement.
32. It is recommended to add the ROI to the last time point of the series since this image will likely have the largest signal area. Once this ROI is drawn, it is copied to the other images in the series maintaining the same size. This allows for direct comparison of ROI signal over time.

Acknowledgments

This investigation was supported by a Spanish grant funded by MINECO/FEDER 2013-45232-R. We thank Victoria Jimenez for expert technical assistance with virus growth and members of the Poxvirus and Vaccine group for their help and contribution to studies in the biology of vaccinia virus.

References

1. Rodriguez JF, Rodriguez D, Rodriguez JR, McGowan EB, Esteban M (1988) Expression of the firefly luciferase gene in vaccinia virus: a highly sensitive gene marker to follow virus dissemination in tissues of infected animals. Proc Natl Acad Sci U S A 85(5):1667–1671
2. Andreu N, Zelmer A, Wiles S (2011) Noninvasive biophotonic imaging for studies of infectious disease. FEMS Microbiol Rev 35(2):360–394. https://doi.org/10.1111/j.1574-6976.2010.00252.x
3. Hutchens M, Luker GD (2007) Applications of bioluminescence imaging to the study of infectious diseases. Cell Microbiol 9(10):2315–2322. https://doi.org/10.1111/j.1462-5822.2007.00995.x
4. Luker KE, Hutchens M, Schultz T, Pekosz A, Luker GD (2005) Bioluminescence imaging of vaccinia virus: effects of interferon on viral replication and spread. Virology 341(2):284–300. https://doi.org/10.1016/j.virol.2005.06.049
5. Luker KE, Luker GD (2008) Applications of bioluminescence imaging to antiviral research and therapy: multiple luciferase enzymes and quantitation. Antivir Res 78(3):179–187. https://doi.org/10.1016/j.antiviral.2008.01.158
6. Zaitseva M, Kapnick SM, Scott J, King LR, Manischewitz J, Sirota L, Kodihalli S, Golding H (2009) Application of bioluminescence imaging to the prediction of lethality in vaccinia virus-infected mice. J Virol 83(20):10437–10447. https://doi.org/10.1128/JVI.01296-09
7. Zaitseva M, Kapnick S, Golding H (2012) Measurements of vaccinia virus dissemination using whole body imaging: approaches for predicting of lethality in challenge models and testing of vaccines and antiviral treatments. Methods Mol Biol 890:161–176. https://doi.org/10.1007/978-1-61779-876-4_10
8. Gardner CL, Burke CW, Tesfay MZ, Glass PJ, Klimstra WB, Ryman KD (2008) Eastern and Venezuelan equine encephalitis viruses differ in their ability to infect dendritic cells and macrophages: impact of altered cell tropism on pathogenesis. J Virol 82(21):10634–10646. https://doi.org/10.1128/JVI.01323-08
9. Pan W, Dong Z, Li F, Meng W, Feng L, Niu X, Li C, Luo Q, Li Z, Sun C, Chen L (2013) Visualizing influenza virus infection in living mice. Nat Commun 4:2369. https://doi.org/10.1038/ncomms3369
10. Luker GD, Bardill JP, Prior JL, Pica CM, Piwnica-Worms D, Leib DA (2002) Noninvasive bioluminescence imaging of herpes simplex virus type 1 infection and therapy in living mice. J Virol 76(23):12149–12161
11. Luker GD, Prior JL, Song J, Pica CM, Leib DA (2003) Bioluminescence imaging reveals systemic dissemination of herpes simplex virus type 1 in the absence of interferon receptors. J Virol 77(20):11082–11093
12. Krug A, Luker GD, Barchet W, Leib DA, Akira S, Colonna M (2004) Herpes simplex virus type 1 activates murine natural interferon-producing cells through toll-like receptor 9. Blood 103(4):1433–1437. https://doi.org/10.1182/blood-2003-08-2674
13. Luker GD, Leib DA (2005) Luciferase real-time bioluminescence imaging for the study of viral pathogenesis. Methods Mol Biol 292:285–296
14. Burgos JS, Guzman-Sanchez F, Sastre I, Fillat C, Valdivieso F (2006) Non-invasive bioluminescence imaging for monitoring herpes simplex virus type 1 hematogenous infection. Microbes Infect 8(5):1330–1338. https://doi.org/10.1016/j.micinf.2005.12.021
15. Hwang S, Wu TT, Tong LM, Kim KS, Martinez-Guzman D, Colantonio AD, Uittenbogaart CH, Sun R (2008) Persistent gammaherpesvirus replication and dynamic interaction with the host in vivo. J Virol 82(24):12498–12509. https://doi.org/10.1128/JVI.01152-08
16. Miyahira AK, Shahangian A, Hwang S, Sun R, Cheng G (2009) TANK-binding kinase-1 plays an important role during in vitro and in vivo type I IFN responses to DNA virus infections. J Immunol 182(4):2248–2257. https://doi.org/10.4049/jimmunol.0802466
17. Cook SH, Griffin DE (2003) Luciferase imaging of a neurotropic viral infection in intact animals. J Virol 77(9):5333–5338
18. Tseng JC, Levin B, Hurtado A, Yee H, Perez de Castro I, Jimenez M, Shamamian P, Jin R, Novick RP, Pellicer A, Meruelo D (2004) Systemic tumor targeting and killing by Sindbis viral vectors. Nat Biotechnol 22(1):70–77. https://doi.org/10.1038/nbt917
19. Schoggins JW, Dorner M, Feulner M, Imanaka N, Murphy MY, Ploss A, Rice CM (2012) Dengue reporter viruses reveal viral dynamics in interferon receptor-deficient mice and sensitivity to interferon effectors in vitro. Proc Natl Acad Sci U S A 109(36):14610–14615. https://doi.org/10.1073/pnas.1212379109
20. Zhang Z, Rowe J, Wang W, Sommer M, Arvin A, Moffat J, Zhu H (2007) Genetic analysis of varicella-zoster virus ORF0 to ORF4 by use of

a novel luciferase bacterial artificial chromosome system. J Virol 81(17):9024–9033. https://doi.org/10.1128/JVI.02666-06

21. Rivera R, Hutchens M, Luker KE, Sonstein J, Curtis JL, Luker GD (2007) Murine alveolar macrophages limit replication of vaccinia virus. Virology 363(1):48–58. https://doi.org/10.1016/j.virol.2007.01.033
22. Hung CF, Tsai YC, He L, Coukos G, Fodor I, Qin L, Levitsky H, Wu TC (2007) Vaccinia virus preferentially infects and controls human and murine ovarian tumors in mice. Gene Ther 14(1):20–29. https://doi.org/10.1038/sj.gt.3302840
23. Gomez CE, Najera JL, Domingo-Gil E, Ochoa-Callejero L, Gonzalez-Aseguinolaza G, Esteban M (2007) Virus distribution of the attenuated MVA and NYVAC poxvirus strains in mice. J Gen Virol 88(Pt 9):2473–2478. https://doi.org/10.1099/vir.0.83018-0
24. Zaitseva M, Kapnick SM, Meseda CA, Shotwell E, King LR, Manischewitz J, Scott J, Kodihalli S, Merchlinsky M, Nielsen H, Lantto J, Weir JP, Golding H (2011) Passive immunotherapies protect WRvFire and IHD-J-Luc vaccinia virus-infected mice from lethality by reducing viral loads in the upper respiratory tract and internal organs. J Virol 85(17):9147–9158. https://doi.org/10.1128/JVI.00121-11
25. Zaitseva M, Shotwell E, Scott J, Cruz S, King LR, Manischewitz J, Diaz CG, Jordan RA, Grosenbach DW, Golding H (2013) Effects of postchallenge administration of ST-246 on dissemination of IHD-J-Luc vaccinia virus in normal mice and in immune-deficient mice reconstituted with T cells. J Virol 87(10):5564–5576. https://doi.org/10.1128/JVI.03426-12
26. Zaitseva M, McCullough KT, Cruz S, Thomas A, Diaz CG, Keilholz L, Grossi IM, Trost LC, Golding H (2015) Postchallenge administration of brincidofovir protects healthy and immune-deficient mice reconstituted with limited numbers of T cells from lethal challenge with IHD-J-Luc vaccinia virus. J Virol 89(6):3295–3307. https://doi.org/10.1128/JVI.03340-14
27. Liu Q, Fan C, Zhou S, Guo Y, Zuo Q, Ma J, Liu S, Wu X, Peng Z, Fan T, Guo C, Shen Y, Huang W, Li B, He Z, Wang Y (2015) Bioluminescent imaging of vaccinia virus infection in immunocompetent and immunodeficient rats as a model for human smallpox. Sci Rep 5:11397. https://doi.org/10.1038/srep11397
28. Stading BR, Osorio JE, Velasco-Villa A, Smotherman M, Kingstad-Bakke B, Rocke TE (2016) Infectivity of attenuated poxvirus vaccine vectors and immunogenicity of a raccoonpox vectored rabies vaccine in the Brazilian Free-tailed bat (*Tadarida brasiliensis*). Vaccine 34(44):5352–5358. https://doi.org/10.1016/j.vaccine.2016.08.088
29. Zaitseva M, Thomas A, Meseda CA, Cheung CYK, Diaz CG, Xiang Y, Crotty S, Golding H (2017) Development of an animal model of progressive vaccinia in nu/nu mice and the use of bioluminescence imaging for assessment of the efficacy of monoclonal antibodies against vaccinial B5 and L1 proteins. Antivir Res 144:8–20. https://doi.org/10.1016/j.antiviral.2017.05.002
30. Sadikot RT, Blackwell TS (2008) Bioluminescence: imaging modality for in vitro and in vivo gene expression. In: Armstrong D (ed) Advanced protocols in oxidative stress I. Humana Press, Totowa, NJ, pp 383–394. https://doi.org/10.1007/978-1-60327-517-0_29
31. Contag CH, Bachmann MH (2002) Advances in in vivo bioluminescence imaging of gene expression. Annu Rev Biomed Eng 4:235–260. https://doi.org/10.1146/annurev.bioeng.4.111901.093336
32. Sadikot RT, Blackwell TS (2005) Bioluminescence imaging. Proc Am Thorac Soc 2(6):537–540. https://doi.org/10.1513/pats.200507-067DS. 511–532
33. Spibey CA, Jackson P, Herick K (2001) A unique charge-coupled device/xenon arc lamp based imaging system for the accurate detection and quantitation of multicolour fluorescence. Electrophoresis 22(5):829–836. https://doi.org/10.1002/1522-2683()22:5<829::AID-ELPS829>3.0.CO;2-U
34. Kusy S, Contag CH (2014) Reporter gene technologies for imaging cell fates in hematopoiesis. Methods Mol Biol 1109:1–22. https://doi.org/10.1007/978-1-4614-9437-9_1
35. Olson JA, Zeiser R, Beilhack A, Goldman JJ, Negrin RS (2009) Tissue-specific homing and expansion of donor NK cells in allogeneic bone marrow transplantation. J Immunol 183(5):3219–3228. https://doi.org/10.4049/jimmunol.0804268
36. Reichardt W, Durr C, von Elverfeldt D, Juttner E, Gerlach UV, Yamada M, Smith B, Negrin RS, Zeiser R (2008) Impact of mammalian target of rapamycin inhibition on lymphoid homing and tolerogenic function of nanoparticle-labeled dendritic cells following allogeneic hematopoietic cell transplantation. J Immunol 181(7):4770–4779
37. Fujisaki J, Wu J, Carlson AL, Silberstein L, Putheti P, Larocca R, Gao W, Saito TI, Lo

Celso C, Tsuyuzaki H, Sato T, Cote D, Sykes M, Strom TB, Scadden DT, Lin CP (2011) In vivo imaging of Treg cells providing immune privilege to the haematopoietic stem-cell niche. Nature 474(7350):216–219. https://doi.org/10.1038/nature10160

38. Park D, Spencer JA, Koh BI, Kobayashi T, Fujisaki J, Clemens TL, Lin CP, Kronenberg HM, Scadden DT (2012) Endogenous bone marrow MSCs are dynamic, fate-restricted participants in bone maintenance and regeneration. Cell Stem Cell 10(3):259–272. https://doi.org/10.1016/j.stem.2012.02.003
39. Chen X, Larson CS, West J, Zhang X, Kaufman DB (2010) In vivo detection of extrapancreatic insulin gene expression in diabetic mice by bioluminescence imaging. PLoS One 5(2):e9397. https://doi.org/10.1371/journal.pone.0009397
40. Keyaerts M, Caveliers V, Lahoutte T (2012) Bioluminescence imaging: looking beyond the light. Trends Mol Med 18(3):164–172. https://doi.org/10.1016/j.molmed.2012.01.005
41. Thorne N, Inglese J, Auld DS (2010) Illuminating insights into firefly luciferase and other bioluminescent reporters used in chemical biology. Chem Biol 17(6):646–657. https://doi.org/10.1016/j.chembiol.2010.05.012
42. Prescher JA, Contag CH (2010) Guided by the light: visualizing biomolecular processes in living animals with bioluminescence. Curr Opin Chem Biol 14(1):80–89. https://doi.org/10.1016/j.cbpa.2009.11.001
43. Sato A, Klaunberg B, Tolwani R (2004) In vivo bioluminescence imaging. Comp Med 54(6):631–634
44. Berger F, Paulmurugan R, Bhaumik S, Gambhir SS (2008) Uptake kinetics and biodistribution of 14C-D-luciferin – a radiolabeled substrate for the firefly luciferase catalyzed bioluminescence reaction: impact on bioluminescence based reporter gene imaging. Eur J Nucl Med Mol Imaging 35(12):2275–2285. https://doi.org/10.1007/s00259-008-0870-6
45. Harwood KR, Mofford DM, Reddy GR, Miller SC (2011) Identification of mutant firefly luciferases that efficiently utilize aminoluciferins. Chem Biol 18(12):1649–1657. https://doi.org/10.1016/j.chembiol.2011.09.019
46. Zhao H, Doyle TC, Coquoz O, Kalish F, Rice BW, Contag CH (2005) Emission spectra of bioluminescent reporters and interaction with mammalian tissue determine the sensitivity of detection in vivo. J Biomed Opt 10(4):41210. https://doi.org/10.1117/1.2032388
47. Venisnik KM, Olafsen T, Gambhir SS, Wu AM (2007) Fusion of Gaussia luciferase to an engineered anti-carcinoembryonic antigen (CEA) antibody for in vivo optical imaging. Mol Imaging Biol 9(5):267–277. https://doi.org/10.1007/s11307-007-0101-8
48. Loening AM, Wu AM, Gambhir SS (2007) Red-shifted *Renilla reniformis* luciferase variants for imaging in living subjects. Nat Methods 4(8):641–643. https://doi.org/10.1038/nmeth1070
49. Loening AM, Fenn TD, Wu AM, Gambhir SS (2006) Consensus guided mutagenesis of Renilla luciferase yields enhanced stability and light output. Protein Eng Des Sel 19(9):391–400. https://doi.org/10.1093/protein/gzl023
50. Adams ST Jr, Miller SC (2014) Beyond D-luciferin: expanding the scope of bioluminescence imaging in vivo. Curr Opin Chem Biol 21:112–120. https://doi.org/10.1016/j.cbpa.2014.07.003
51. Pichler A, Prior JL, Piwnica-Worms D (2004) Imaging reversal of multidrug resistance in living mice with bioluminescence: MDR1 P-glycoprotein transports coelenterazine. Proc Natl Acad Sci U S A 101(6):1702–1707. https://doi.org/10.1073/pnas.0304326101
52. Tannous BA, Kim DE, Fernandez JL, Weissleder R, Breakefield XO (2005) Codon-optimized Gaussia luciferase cDNA for mammalian gene expression in culture and in vivo. Mol Ther 11(3):435–443. https://doi.org/10.1016/j.ymthe.2004.10.016
53. Bhaumik S, Lewis XZ, Gambhir SS (2004) Optical imaging of Renilla luciferase, synthetic Renilla luciferase, and firefly luciferase reporter gene expression in living mice. J Biomed Opt 9(3):578–586. https://doi.org/10.1117/1.1647546
54. Earl PL, Americo JL, Cotter CA, Moss B (2015) Comparative live bioluminescence imaging of monkeypox virus dissemination in a wild-derived inbred mouse (*Mus musculus* castaneus) and outbred African dormouse (*Graphiurus kelleni*). Virology 475:150–158. https://doi.org/10.1016/j.virol.2014.11.015
55. Falendysz EA, Londono-Navas AM, Meteyer CU, Pussini N, Lopera JG, Osorio JE, Rocke TE (2014) Evaluation of monkeypox virus infection of black-tailed prairie dogs (*Cynomys ludovicianus*) using in vivo bioluminescent imaging. J Wildl Dis 50(3):524–536. https://doi.org/10.7589/2013-07-171
56. Falendysz EA, Lopera JG, Doty JB, Nakazawa Y, Crill C, Lorenzsonn F, Kalemba LN, Ronderos MD, Mejia A, Malekani JM, Karem

K, Carroll DS, Osorio JE, Rocke TE (2017) Characterization of monkeypox virus infection in African rope squirrels (Funisciurus sp.). PLoS Negl Trop Dis 11(8):e0005809. https://doi.org/10.1371/journal.pntd.0005809

57. Goldman SJ, Jin S (2014) The bioluminescent imaging of spontaneously occurring tumors in immunocompetent ODD-luciferase bearing transgenic mice. Methods Mol Biol 1098:129–143. https://doi.org/10.1007/978-1-62703-718-1_11
58. Lopera JG, Falendysz EA, Rocke TE, Osorio JE (2015) Attenuation of monkeypox virus by deletion of genomic regions. Virology 475:129–138. https://doi.org/10.1016/j.virol.2014.11.009
59. Americo JL, Sood CL, Cotter CA, Vogel JL, Kristie TM, Moss B, Earl PL (2014) Susceptibility of the wild-derived inbred CAST/Ei mouse to infection by orthopoxviruses analyzed by live bioluminescence imaging. Virology 449:120–132. https://doi.org/10.1016/j.virol.2013.11.017
60. Paroo Z, Bollinger RA, Braasch DA, Richer E, Corey DR, Antich PP, Mason RP (2004) Validating bioluminescence imaging as a high-throughput, quantitative modality for assessing tumor burden. Mol Imaging 3(2):117–124. https://doi.org/10.1162/1535350041464865
61. Haddad D (2017) Genetically engineered vaccinia viruses as agents for cancer treatment, imaging, and transgene delivery. Front Oncol 7:96. https://doi.org/10.3389/fonc.2017.00096
62. Thorne SH (2009) Design and testing of novel oncolytic vaccinia strains. Methods Mol Biol 542:635–647. https://doi.org/10.1007/978-1-59745-561-9_32
63. Kelly KJ, Brader P, Woo Y, Li S, Chen N, Yu YA, Szalay AA, Fong Y (2009) Real-time intraoperative detection of melanoma lymph node metastases using recombinant vaccinia virus GLV-1h68 in an immunocompetent animal model. Int J Cancer 124(4):911–918. https://doi.org/10.1002/ijc.24037
64. Haddad D, Chen N, Zhang Q, Chen CH, Yu YA, Gonzalez L, Aguilar J, Li P, Wong J, Szalay AA, Fong Y (2012) A novel genetically modified oncolytic vaccinia virus in experimental models is effective against a wide range of human cancers. Ann Surg Oncol 19(Suppl 3):S665–S674. https://doi.org/10.1245/s10434-011-2198-x
65. Mansfield DC, Kyula JN, Rosenfelder N, Chao-Chu J, Kramer-Marek G, Khan AA, Roulstone V, McLaughlin M, Melcher AA, Vile RG, Pandha HS, Khoo V, Harrington KJ (2016) Oncolytic vaccinia virus as a vector for therapeutic sodium iodide symporter gene therapy in prostate cancer. Gene Ther 23(4):357–368. https://doi.org/10.1038/gt.2016.5
66. Virostko J, Chen Z, Fowler M, Poffenberger G, Powers AC, Jansen ED (2004) Factors influencing quantification of in vivo bioluminescence imaging: application to assessment of pancreatic islet transplants. Mol Imaging 3(4):333–342. https://doi.org/10.1162/1535350042973508

Chapter 18

Growth and Purification of Vaccinia Virus Stocks for MPM Imaging

Glennys V. Reynoso, John P. Shannon, Jeffrey L. Americo, James Gibbs, and Heather D. Hickman

Abstract

This chapter provides methods for the propagation, purification, and titration of vaccinia virus (VACV) and the highly attenuated strain-modified vaccinia Ankara (MVA). Additionally, we provide information on VACV recombinants we have used for intravital imaging with multiphoton excitation.

Key words Vaccinia virus, Modified vaccinia Ankara, Virus purification

1 Introduction

In this chapter, we describe how to cultivate permissive cells, and grow and purify viral stocks of both vaccinia virus (VACV) and modified vaccinia Ankara (MVA). While VACV can be propagated on a large number of cells due to its large host range [1], we find good yields after growth in the human thymidine kinase-negative (TK^-) 143B cells we describe in this chapter. In contrast, the restricted host range of MVA [2] allows replication in a limited number of cells, including primary chicken embryo fibroblast (CEF) cells, described here, or BHK-21 cells [3].

2 Materials

2.1 Materials for the Preparation of VACV Stocks

All solutions should be prepared in a biological safety cabinet using aseptic techniques. Reagents should be warmed to 37 °C prior to use unless otherwise stated and should be stored at 4 °C. We do not add any antibiotics to cell culture media.

1. Human thymidine kinase-negative (TK^-) 143B cells (ATCC, CRL-8303).

Jason Mercer (ed.), *Vaccinia Virus: Methods and Protocols*, Methods in Molecular Biology, vol. 2023, https://doi.org/10.1007/978-1-4939-9593-6_18,

2. Cell culture medium: Dulbecco's minimum essential medium (DMEM) containing high glucose (GlutaMAX), pyruvate, and 7.5% fetal bovine serum (FBS).
3. Trypsin/EDTA 0.05% with phenol red.
4. 0.1% Bovine serum albumin (BSA) solution: 10% BSA (Sigma, cat# 03117332001) added to a measured volume of ultrapure water and filtered prior to use.
5. 10 mM HEPES.
6. Infection medium: Add 10 mL of the 0.1% BSA solution and 5 mL of 10 mM HEPES to 500 mL of DMEM GlutaMAX supplemented with pyruvate.
7. Balanced salt solution (BSS) + 0.1% bovine serum albumin (BSA): Add 86.71 g of NaCl to a 20 L Nalgene bottle containing 8 L of ultrapure water. Place the bottle on a large stir plate along with a large stir bar, and stir at a moderate speed to begin dissolving the NaCl. In a 1000 mL beaker, mix 29.2 mL of stock solution B, 14.6 mL of stock solution D, 145.98 mL of stock solution H, and 107.66 mL of HEPES media (*see* **Note 1** and Table 1). Add mixture of the stock solutions to the 8 L of ultrapure water containing the NaCl; let mixture stir. In a separate beaker, add 100 mL of ultrapure water, slowly drip in 22 mL of stock solution C (*see* **Note 1** and Table 1), and check for any calcium precipitates. Reduce stirring speed, slowly add the calcium mixture to the 20 L Nalgene bottle, and check for any precipitation. If no precipitation is observed, add ultrapure water in order to bring up the solution to a final volume of 10 L. Add 10 g of BSA to 10 L of BSS solution. Stir for 1–2 h, or until all BSA is dissolved. Sterilize the solution by filtering through a 0.22 μM EMD Millipore Millipak filter. We typically aliquot the solution into 500 mL sterile Nalgene glass bottles. Rinse the 20 L Nalgene bottle

Table 1
Stock solution composition

Stock solution	Chemical	Molarity	Amount in 500 mL
B	KCl	1.68	62.63 g
D	$MgSO_4 \cdot 7HO_2$	0.84	103.51 g
E	KH_2PO_4	0.84	57.15 g
F	K_2HPO_4	1.12	97.55 g
G	HEPES	1 M pH 7.3	106.15 g
C	$CaCl_2$ $2H_2O$	1.12	82.33 g

and hoses with deionized water and 100% ethanol, and allow to air-dry overnight.

8. Filter-sterilize 1 mM and 10 mM Tris-Cl, pH 9.0.
9. Filter-sterilize 36% sucrose solution in 10 mM Tris-Cl, pH 9.0.
10. Filter-sterilize 25% and 40% sucrose solutions in 1 mM Tris-Cl, pH 9.0.
11. Crystal violet stain: Mix 50 mL of 95% ethanol, 300 mL of 37% formaldehyde, and 1.5 g of crystal violet powder (*see* **Note 2**). Store at room temperature.
12. Corning 162-cm^2 (T-150) tissue culture flasks.
13. Beckman ultracentrifuge tubes, polyallomer 25 × 89 mm or 14 × 89 mm, autoclave sterilized.
14. 70% Ethanol.
15. Humidified, 37 °C, 9% CO_2 incubator.
16. Thermo Scientific Sorvall ST 40 benchtop centrifuge.
17. Ultracentrifuge with Beckman SW28 and SW41 rotors.
18. Biocomp Gradient Master.
19. 15 mL and 40 mL Pyrex Dounce tissue grinder.
20. Sonics Vibracell cup horn sonicator.

2.2 Materials for the Preparation of MVA Stocks

1. Primary chick embryo fibroblasts (CEFs) prepared from 10-day-old, specific pathogen-free embryonated chicken eggs (commercially purchased from Charles River Laboratories).
2. Cell culture medium: Eagle's minimal essential medium (EMEM) supplemented with 10% fetal bovine serum (FBS), glutamine, and penicillin-streptomycin.
3. Infection medium: EMEM supplemented with 2% fetal bovine serum (FBS), glutamine, and penicillin-streptomycin.
4. Virus-overlay medium: Methylcellulose in EMEM supplemented with 2% fetal bovine serum (FBS), glutamine, and penicillin-streptomycin.
5. Sterile microdissecting scissors and forceps.
6. 70% and 100% ethanol.
7. Deionized H_2O.
8. 100 cm^2 Sterile petri dishes.
9. Trypsin/EDTA: 0.25% (w/v) Trypsin, 0.02% (w/v) EDTA.
10. Trypsinization flask.
11. Humidified incubators set to 37 °C, 5% CO_2, or 31 °C, 5% CO_2.
12. 37 °C Water bath.
13. Vortex.

14. 10, 25, and 50 mL serological pipettes.
15. Autoclaved cheesecloth.
16. Sterile 500 mL glass beaker.
17. Sterile 50 mL and 250 mL centrifuge tubes/bottles.
18. Low-speed refrigerated centrifuge with 50 mL and 250 mL adaptors.
19. Ultra-speed refrigerated centrifuge with SW28 and SW41 rotors and buckets.
20. Corning 162 cm^2 (T-150) tissue culture flasks.
21. Beckman ultracentrifuge tubes, polyallomer 25 × 89 mm or 14 × 89 mm, autoclave sterilized.
22. Corning 6-well tissue culture plates.
23. Sonics Vibracell cup horn sonicator.
24. Wet ice.
25. 40 cm Cell scrapers.
26. Sterile 1.8 mL Sarstedt sample tubes.
27. Filter sterilized 1 mM and 10 mM Tris-Cl, pH 9.
28. 40 mL Pyrex Dounce tissue grinder.
29. Biocomp Gradient Master.
30. Filter sterilize 36% sucrose solution in 10 mM Tris-Cl, pH 9.
31. Filter sterilize 25% and 40% sucrose solution in 1 mM Tris-Cl, pH 9.
32. Luria-Bertani (LB) broth.
33. Shaking incubators set to 37 °C.
34. 1:1 Acetone:methanol fixative solution.
35. Phosphate-buffered saline (PBS).
36. Diluted rabbit anti-WR VACV-IgG (Quality Biological, Lot# 304680, January 2006): 1:500 in PBS supplemented with 3% FBS.
37. Diluted horseradish peroxidase (HRP)-conjugated secondary antibody (Thermo Fisher, Cat# 32400): 1:4000 in PBS supplemented with 3% FBS.
38. Saturated O-dianisidine (Sigma, Cat# D913) solution: Add 0.5 mL 100% ethanol to a small bead of substrate, vortex thoroughly, incubate for 30 min at 37 °C, and clarify by centrifugation at 15,000 × *g* for 30 s.
39. 30% H_2O_2 (Sigma Aldrich, Cat# H1009—500 mL).
40. Working substrate solution: 0.2 mL Clarified o-dianisidine and 10 μL 30% H_2O_2 to every 10 mL sterile PBS.

3 Methods for the Preparation of VACV and MVA Stocks

While working with VACV, follow all Biosafety Level 2 (BSL-2) procedures, including the use of the appropriate personal protective equipment (PPE). Tissue culture should be performed using aseptic technique to avoid contamination of cells or viral stocks.

3.1 Preparation of VACV Stocks

3.1.1 Cell Preparation for VACV Growth

1. Aspirate the medium from a confluent monolayer of HuTK$^-$ 143B cells in a T-150 tissue culture flask.
2. Wash cells with trypsin/EDTA 0.05% or PBS in order to remove the remaining serum.
3. Overlay cells with 5 mL of trypsin/EDTA. Place flask in a humidified, 37 °C, 9% CO_2, incubator for 5 min. Cells should start detaching from flasks. Gently tap the flasks to completely detach the cells.
4. Add 10 mL of cell culture medium. Pipette up and down several times to break up any cell clumps.
5. Add 2 mL of the cell suspension to a new T-150 tissue culture flask containing 30 mL of fresh maintenance medium. Swirl the flasks around in order to cover the entire flask area and to evenly distribute the cells. Place the flasks in a humidified, 37 °C, 9% CO_2, incubator until the cells are confluent. At this dilution, cells will be confluent in about 2 days. For the expansion of VACV, add 2 mL cell suspension to 15 T-150 tissue culture flasks.

3.1.2 Growth of VACV Stocks

1. Add 150 mL of infection medium and 750 μL of the desired VACV stock to a sterile 250 mL Nalgene media bottle. The amount of virus propagated will vary based on the plaque-forming units (PFU)/mL of the seed virus and how many cells are infected. Preferably, we infect with a 0.1–0.5 multiplicity of infection (MOI) and estimate approximately 2.5×10^7 cells per T-150 tissue culture flask.
2. Aspirate medium from each flask and add 10 mL of infection medium + virus mixture. Swirl flasks around to ensure that the entire surface of the flask is covered.
3. Place flasks in a humidified, 37 °C, 9% CO_2, incubator. Gently rock flasks about every 30 min. Incubate flasks for 1.5–2 h.
4. Add 30 mL of cell culture medium to each flask. Allow flasks to incubate for 48–72 h before harvesting. Examine the flasks after 24 h to determine how the infection is proceeding, judging by cell death (or virus-induced cytopathic effects (CPE)). If by 48 h there is minimal CPE, allow cells to incubate for an additional 24 h.

3.1.3 Harvesting VACV Stocks

1. After incubation, note CPE in the flasks to determine the appropriate time of harvest. The cells should be rounded up and no longer strongly adherent to the flasks before harvest.
2. Carefully tap the flasks to detach the loosely adherent cells.
3. Transfer contents of five T-150 tissue culture flasks to a conical-bottom 225 mL centrifuge tube. Repeat for the next ten flasks.
4. Spin the centrifuge tubes at 874 × *g* for 30 min, aspirate media, and discard.
5. Resuspend the cell pellet from each 225 mL centrifuge tube (from five flasks) in 10 mL of BSS + 0.1% BSA, 4 °C. Transfer the pellet to the second centrifuge tube, and repeat for the third centrifuge bottle. Place the resuspended pellets into a sterile 50 mL Falcon tube.
6. To minimize the loss of virus, rinse the centrifuge tubes with 5 mL of BSS + 0.1% BSA, 4 °C. Carefully transfer the 5 mL of BSS + 0.1% BSA between the centrifuge tubes, as described above. The final volume after the rinse step will be ~20 mL.
7. Freeze resuspended pellets at −80 °C until ready to aliquot.

3.1.4 Preparing and Storing VACV Stocks

1. Freeze/thaw cell pellet three times on dry ice.
2. Carefully pour pellet mixture into a sterile 15 mL Pyrex Dounce tissue grinder.
3. Slowly homogenize mixture in order to minimize the formation of bubbles or aerosols. Homogenize pellet by driving the pestle up and down (~20 times), or until a uniform solution is observed.
4. Pipette mixture back into the 50 mL Falcon tube.
5. Wash the homogenizer with 10 mL of ice-cold BSS + 0.1% BSA; volume will be ~30 mL.
6. Sonicate for 30 s at 40% power; repeat three times.
7. Add ~20 mL of ice-cold BSS + 0.1% BSA, bringing the volume up to 50 mL.
8. Quick spin the 50 mL Falcon tube, and allow the centrifuge to come up to 874 × *g*. Immediately stop the centrifuge once that speed is reached.
9. Carefully pour out the supernatant into a sterile 125 mL media bottle.
10. Wash and break up the pellet with 10 mL of ice-cold BSS + 0.1% BSA, and add another 20 mL of BSS + 0.1% BSA, bringing the total volume to 30 mL.
11. Repeat the sonification process, as described in **step 6**.
12. Repeat the quick spin process, as described in **step 8**.

13. Pour supernatant into the media bottle containing the previous supernatant.
14. Wash and break up the pellet again, as described in **step 10**. Bring up to a final volume of 30 mL. Repeat the sonification and quick spin steps. Add supernatant to the 125 mL media bottle. A final volume of ~110 mL of virus will yield a VACV stock of 1–3 × 10^8 plaque-forming units (PFU)/mL.
15. Mix virus well and aliquot into sterile conical-bottom free-standing Sarstedt tubes. Store aliquoted virus at −80 °C. Alternatively, proceed to purification (Subheading 3.1.5).

3.1.5 Purification of VACV Stocks

1. Thaw frozen VACV stocks or proceed directly from Subheading 3.1.4. For larger yields of purified virus, begin with 40 T-150 tissue culture flasks for VACV growth.
2. Centrifuge harvested VACV stock at 1365 × *g* for 10 min, 4 °C. Discard the supernatant.
3. Resuspend pellets in 1.5 mL of 10 mM Tris-Cl per T-150 tissue culture flask. Final volume for 40 T-150 tissue culture flasks will be ~60 mL.
4. In a 40 mL Pyrex Dounce tissue grinder, slowly homogenize cell suspension in order to minimize the formation of bubbles or aerosols. Homogenize pellet by driving the pestle up and down (~40 times), or until a uniform solution is observed.
5. Transfer homogenate to 50 mL conical-bottom tubes and centrifuge at 874 × *g* for 5 min in order to remove the cell nuclei. Save the supernatant.
6. Pool supernatants and sonicate at 30–40% power for 60 s, keeping the lysate on ice. Repeat the sonification step three to four times, allowing the supernatant to rest for 30 s in between each sonification.
7. Layer the supernatant onto a cushion of 36% sucrose in sterile ultracentrifuge tubes. Centrifuge at 32,900 × *g* for 80 min, 4 °C.
8. Aspirate and discard supernatant. Resuspend virus pellet in 1 mL of 1 mM Tris-Cl, pH 9.0, per 20 T-150 flasks. Virus can be stored and aliquoted at this point if optimal purity is not required.
9. If further purification is needed, prepare a 25–40% sucrose gradient either manually or through the use of an automatic gradient maker, such as the Biocomp Gradient Master. For the Gradient Master, layer 25% sucrose over 40% sucrose. Carefully cap tubes avoiding any air bubbles. After running the Gradient Master, check that the interface between the two different sucrose concentration is no longer visible.

10. Sonicate virus for 30 s, two times at 30–40% power, 4 °C.
11. Carefully discard 1–3 mL off the top of the gradient in order to overlay the virus pellet on top of the gradient. Layer 1–3 mL of purified virus onto the 25–40% sucrose gradient and centrifuge for 1 h at 26,000 × *g*, 4 °C.
12. Observe the milky band near the middle of the gradient and carefully aspirate sucrose above the virus band.
13. Collect the virus band with a 10 mL serological pipette, transfer to a 50 mL conical tube, and store the sample on ice.
14. Aspirate the remaining sucrose from the gradient and resuspend the pellet in 1 mL of 1 mM Tris-Cl, pH 9.0.
15. Sonicate the resuspended pellet in a cup horn filled with wet ice for 1 min at 40% power.
16. Layer 1 mL of the resuspended pellet onto a new 25–40% sucrose gradient in a centrifuge tube and centrifuge at 26,000 × *g* for 1 h, 4 °C.
17. Observe the milky band near the middle of the gradient and carefully aspirate sucrose above the virus band.
18. Collect the virus band with a 10 mL serological pipette and pool with the previous virus band stored on ice in a 50 mL conical tube.
19. Add 2–3 volumes of 1 mM Tris-Cl, pH 9.0, to the 50 mL conical tube and mix by pipetting up and down several times.
20. Transfer the diluted virus to a centrifuge tube and centrifuge for 1 h at 32,000 × *g*, 4 °C.
21. Aspirate and discard the supernatant and resuspend the virus pellet in 1 mL of 1 mM Tris-Cl, pH 9.0, per 20 T-150 flasks.
22. Sonicate the purified virus in a cup horn filled with wet ice for 30 s, two times at 40% power with 30-s rest between each sonication.
23. Aliquot and store at −80 °C.

3.1.6 Titer Determination of VACV Stocks

1. Seed HuTK$^-$ 143B cells onto sterile 6-well plates until a confluent monolayer is achieved. Generally, one confluent T-150 tissue culture flask of HuTK$^-$ 143B will yield 8–10 6-well plates for use 24 h later.
2. Serially dilute VACV tenfold in infection medium in a 24-well plate.
3. Aspirate the medium from the confluent 6-well plates. Add 500 μL of each virus dilution per well, performing duplicates for each dilution. Usually 10^5, 10^6, and 10^7 dilutions are sufficient to determine titer.

4. Incubate the 6-well plates with the 500 µL of virus dilutions for 2 h. Gently rock the plates every 20 min.
5. After the 2-h incubation period, add 5 mL of cell culture media and incubate in a humidified, 37 °C, 9% CO_2, incubator for 48 h. If possible, avoid moving the plates to minimize the potential for satellite colonies.
6. Once the 48-h incubation is complete, carefully aspirate the media from the wells. Avoid disturbing the cell monolayer.
7. Add 500 µL of crystal violet stain. Rotate plates to cover the whole well. Aspirate the excess stain and allow the plates to air-dry.
8. Quantitate the number of plaques per well at the dilution of virus allowing clear discrimination of plaques.

3.2 Preparation of MVA Stocks

3.2.1 Preparation of Primary Chick Embryo Fibroblast Cells (CEFs) for MVA Growth

1. Place commercially purchased 10-day fertilized eggs in a biosafety cabinet and position eggs with blunt end up; this ensures that the internal air sac remains at the top of the egg.
2. Spray all eggs with 70% ethanol.
3. Using sterile scissors and forceps, crack the top part of the egg and remove the upper portion of the shell to expose the inner membrane.
4. Remove the inner membrane with sterile forceps to expose the viable embryo.
5. Using two pairs of sterile forceps, gently grasp the legs and neck and transfer the embryo to a sterile 100 cm^2 petri dish.
6. Remove the wings, head, and feet from each isolated embryo and place the remaining portion of the body into a sterile 100 cm^2 petri dish containing 10 mL EMEM.
7. Load a 6 mL syringe with chick embryos (maximum of five embryos/syringe) and plunge tissue through the syringe barrel into a sterile trypsinization flask.
8. Add 50 mL of trypsin/EDTA to the trypsinization flask and incubate in a 37 °C, 5% CO_2, incubator for 10 min while stirring.
9. Gently decant cells through sterile cheesecloth into a 500 mL glass beaker.
10. Add 10 mL of FBS to the beaker to reduce trypsin activity.
11. Add 50 mL of trypsin/EDTA to all remaining tissue clumps and incubate in a 37 °C, 5% CO_2, incubator for 5 min while stirring.
12. Gently decant the remaining cells through sterile cheesecloth into the same 500 mL glass beaker.

13. Using a 50 mL serological pipette, transfer digested cells to a sterile 250 mL centrifuge bottle and bring the volume up to 150 mL with cell culture medium.
14. Centrifuge cells at 874 × *g* for 10 min, 4 °C.
15. Aspirate medium and resuspend cell pellet in 10 mL cell culture medium.
16. Bring the volume up to 100 mL with cell culture medium and centrifuge at 874 × *g* for 10 min, 4 °C.
17. Aspirate medium and resuspend cell pellet in 5 mL cell culture medium.
18. Transfer cells to a 50 mL centrifuge tube and bring the volume up to 30 mL with cell culture medium.
19. To prepare 30 T-150 tissue culture flasks, fill each flask with 30 mL cell culture medium and add 1 mL CEF cell suspension to each flask.
20. Incubate flasks at 37 °C, 5% CO_2, for ~4 days until cells are ~90–95% confluent; after 4 days, each T-150 flask contains about 1×10^7 cells.
21. At this point, cells should be infected or stored temporarily for 1–2 weeks in an incubator set to 31 °C, 5% CO_2. Cells stored at 31 °C can be used for preparing CEF monolayers in multi-well plates for MVA titer and growth analysis.

3.2.2 Preparation of Crude MVA Stocks

1. Remove MVA stock from −80 °C storage and thaw in a 37 °C water bath.
2. Vortex MVA briefly at full power and store on ice prior to ultrasonic disruption.
3. Sonicate virus in a cup horn filled with wet ice for 30 s at 40% power, and repeat three times with 30-s rest between each sonication.
4. Prepare virus inoculum by diluting virus in infection medium so that each T-150 tissue culture flask is infected at a MOI of 1–2 PFU/cell in an 8 mL volume. Below is the general formula used for infecting one T-150 tissue culture flask with MVA at a desired MOI:

$$\frac{(\#\text{CEF cells in one T}-150\,\text{flask})\times(\text{MOI})}{(\text{Virus titer in pfu / mL})} = \text{Total mL of virus / flask}$$

5. Aspirate cell culture medium from 10 to 30 T-150 tissue culture flasks of CEFs and add 8 mL of virus inoculum per flask.
6. Incubate flasks at 37 °C, 5% CO_2, for 2 h making sure to gently rotate flasks every 15–20 min to prevent cell sheets from drying out.

7. After 2 h, add 20 mL of infection medium to each T-150 tissue culture flask and incubate at 37 °C, 5% CO_2, for 48–72 h.
8. When sufficient CPE is observed, harvest infected cells through scraping. Alternatively, cells can be released into the medium by firmly tapping the sides of each flask.
9. Using a 50 mL serological pipette, transfer cells to a sterile 250 mL conical and centrifuge at 874 × *g* for 10 min, 4 °C.
10. Aspirate supernatant and resuspend cell pellet in 1 mL infection medium per T-150 tissue culture flask.
11. Lyse the cell suspension by performing three freeze-thaw cycles using a dry ice/ethanol bath and a 37 °C water bath.
12. Sonicate the cell lysate on wet ice for 1 min at 40% power with 1-min rest between each sonication.
13. To assess sterility, inoculate 2 mL Luria-Bertani (LB) broth with 20 μL purified virus and grow for 48 h in a 37 °C shaking incubator.
14. Aliquot virus into tubes and store at −80 °C.

3.2.3 Preparation of Purified MVA Stocks

1. For purification of MVA, prepare 30–60 T-150 tissue culture flasks containing primary CEF cells 4–5 days prior to planned infection. Infect and prepare cell pellets as detailed in Subheading 3.1.4.
2. Resuspend cell pellets in 1.5 mL 10 mM Tris-Cl, per initial T-150 tissue culture flask, by pipetting cells up and down with a 10 mL serological pipette. Continue to purify as for VACV if needed (Subheading 3.1.5).

3.2.4 Titer Determination of MVA Stocks by Immunostaining

1. One day prior to virus titration, prepare the appropriate number of 6-well tissue culture plates containing primary CEF cells.
2. Remove an aliquot of crude or purified MVA virus from −80 °C storage and thaw in a 37 °C water bath.
3. Vortex virus prep briefly at full power and store on ice prior to ultrasonic disruption.
4. Sonicate virus prep in a cup horn filled with wet ice for 30 s, three times at 40% power with 30-s rests between each sonication.
5. Prepare several tenfold serial dilutions of virus in infection medium.
6. Remove the cell culture media from confluent 6-well plates and add 1 mL of fresh infection medium to each well.
7. Plate out 1 mL of each serial dilution on duplicate wells of the culture plate and incubate at 37 °C, 5% CO_2, for 4 h.

8. Aspirate virus and overlay each well with 2 mL complete methylcellulose.
9. Incubate culture plates at 37 °C, 5% CO_2, for 3 days.
10. Remove methylcellulose and fix cells with 2 mL of a 1:1 acetone:methanol mix for 5 min.
11. Aspirate fixative and wash each well with several exchanges of sterile PBS.
12. Add 1 mL diluted rabbit anti-WR VACV-IgG to wells and rock for 1 h at room temperature.
13. Aspirate and wash each well with several exchanges of sterile PBS.
14. Add 1 mL diluted HRP-conjugated secondary antibody to wells and rock for 1 h at room temperature.
15. Aspirate HRP-antibody and wash each well with several exchanges of sterile PBS.
16. Add 0.5 mL of working substrate solution to each well and rock for 10–15 min at room temperature.
17. Aspirate working substrate solution when foci develop and wash each well with several exchanges of deionized H_2O.
18. Add 1 mL deionized H_2O to each well and count stained foci using a light microscope.
19. Calculate the infectious virus titer by multiplying the foci count by the dilution factor.

4 Notes

1. Individual stock solutions required to make BSS + 0.1% BSA stock are prepared in a biological safety cabinet and filtered prior to use. All chemicals are dissolved in 500 mL of ultrapure water. The iso-osmotic stock solution (stock solution H) is made up by adding 30 mL of stock solution E and 40 mL of stock solution F to 400 mL of ultrapure water. Adjust pH to 7.2 and adjust final volume to 550 mL with ultrapure water.
2. Crystal violet powder easily stains, so prevent spills. Crystal violet solution contains the carcinogen formaldehyde and may require special disposal procedures.

Acknowledgments

This work was supported by the intramural research program of the National Institute of Allergy and Infectious Diseases, National Institutes of Health.

References

1. Moss B (2001) Poxviridae: the viruses and their replication. Fields Virol 2(4): 2849–2884
2. Carroll MW, Moss B (1997) Host range and cytopathogenicity of the highly attenuated MVA strain of vaccinia virus: propagation and generation of recombinant viruses in a nonhuman mammalian cell line. Virology 238:198–211
3. Cotter CA, Earl PL, Wyatt LS, Moss B (2017) Preparation of cell cultures and vaccinia virus stocks. Curr Protoc Protein Sci 83:11–15

Chapter 19

Intravital Imaging of Vaccinia Virus-Infected Mice

John P. Shannon, Olena Kamenyeva, Glennys V. Reynoso, and Heather D. Hickman

Abstract

Intravital multiphoton microscopy (MPM) allows the direct visualization of viral infections in real time as they occur in living animals. Here we describe the routes and considerations for murine infection with vaccinia virus (VACV) for imaging, and the preparation of the skin and inner lip (labial mucosa) of infected animals for MPM. Using different recombinant VACVs expressing fluorescent proteins in combination with transgenic fluorescent reporter mice, MPM imaging can be used to examine the movements, interactions, and functions of virus-infected cells or selected immune cell populations after infection.

Key words Virus, Vaccinia virus, Viral immunology, Multiphoton microscopy, Intravital microscopy, Skin immunology, Mucosal immunology

1 Introduction

While once considered a difficult and expensive technology, intravital multiphoton microscopy (MPM) has rapidly advanced in ease and availability, allowing the development of myriad approaches to image the immune response in the tissues and organs of living animals [1]. Intravital microscopy typically (but not always!) makes use of a powerful multiphoton laser that excites fluorophores with longer wavelengths of light, allowing deeper tissue penetration with less photodamage. Initially applied to increase our understanding of the basic principles of the immune response after administration of noninfectious immunogens, MPM is increasingly being employed to probe the immune response to pathogens [2–4]. We have routinely applied this technology to examine the anatomy of the antiviral immune response in VACV-infected skin [5–7], the route of infection employed for human vaccination with the smallpox vaccine. More recently, we have expanded this technology to understand antiviral immunity within the labial mucosa, a critical site for poxvirus pathogenesis. Here, we describe the experimental methods for MPM imaging of VACV-infected

Jason Mercer (ed.), *Vaccinia Virus: Methods and Protocols*, Methods in Molecular Biology, vol. 2023,
https://doi.org/10.1007/978-1-4939-9593-6_19,

animals with an emphasis on the unique aspects of imaging a BSL-2 virus in live mice. Using recombinant viruses expressing a number of different fluorophores created using standard molecular biological approaches [8], we have successfully imaged the movement of both virus-infected cells and responding innate and adaptive lymphocytes in the skin and mucosa.

2 Materials

All solutions and reagents should be kept sterile and prepared in a biological safety cabinet. All reagents should be prewarmed to 37 °C and stored at 4 °C. VACV stocks are usually at a titer of ~2×10^8 pfu/mL [9].

2.1 VACV Infection

1. Thawed and sonicated VACV (*see* **Note 1**).
2. Bifurcated needles.
3. Syringes (1 cc or insulin).
4. Isoflurane machine with rodent facemasks (or another approved method of sedation, *see* **Note 2**).

2.2 Intravital Microscopy of VACV-Infected Ear

1. Infected mice at the desired time postinfection (*see* **Note 3**).
2. Sterile phosphate-buffered saline (PBS) at 37 °C (*see* **Note 4**).
3. Cyanoacrylate (e.g., superglue).
4. Plastic 6-well plate and lid.
5. Office tape.
6. Athletic cloth tape.
7. Weatherstrip and caulking cord.
8. Cotton swab Q-tips.
9. Isoflurane or avertin for anesthesia (*see* **Note 2**).
10. 27-gauge needle.
11. Butterfly needle with 1 mL syringe and extender if required.
12. Low-flow isoflurane machine with rodent facemask (Somnosuite, Kent Scientific); heart rate and oxygen saturation monitor are helpful. Other isoflurane vaporizers can also be used (such as SurgiVet) with mouse nose cone (Braintree Scientific). We find the Somnosuite the easiest and most consistent.
13. Biosafety Level 2-housed upright Leica intravital multiphoton microscope equipped with an environmental chamber and one or more multiphoton lasers, such as MaiTai DeepSee, Chameleon Coherent, or Spectra Physics Insight (*see* **Notes 5**

and **6**); ultrasensitive hybrid detectors (HyD); and a 25× water-immersion objective.

14. External non-descanned detectors equipped with the following filters: a 495 nm dichroic mirror followed by emission filters of 460/50 nm bandpass and 525/50 nm bandpass; a longpass filter of 560 nm; and then a 610/60 nm bandpass, a 650 nm longpass, and a 685/50 bandpass (*see* **Note** 7).
15. Image-processing software: Imaris (Bitplane), ImageJ and Huygens (Scientific Volume Imaging) (*see* **Note 8**).

2.3 Intravital Microscopy of VACV-Infected Labial Mucosa

1. Infected mice at the desired time postinfection.
2. Water-based gel (e.g., KY jelly).
3. Cyanoacrylate (e.g., superglue).
4. Betadine (or other iodine-based surgical scrub) and 70% alcohol prep pads.
5. Surgical cloth tape.
6. Surgical scissors and dressing forceps.
7. Isoflurane for anesthesia.
8. Rodent warmer/heating pad (Braintree Scientific) (*see* **Note 9**).
9. Stainless steel mini-stage (*see* **Note 10**).
10. Removable microscope imaging stage with standard glass slide-sized insert.
11. Gold-seal coverslips No. 1 (Thermo Scientific) (*see* **Note 11**).
12. Low-flow isoflurane machine with rodent facemask (Somnosuite, Kent Scientific); heart rate and oxygen saturation monitor are helpful. Other isoflurane vaporizers can also be used (such as SurgiVet) with mouse nose cone (Braintree Scientific). We find the Somnosuite the easiest and most consistent.
13. Biosafety Level 2-housed Leica DMi8 inverted five-channel confocal microscope equipped with environmental chamber and dual-multiphoton lasers, such as MaiTai and InSight DeepSee (Spectra Physics); ultrasensitive hybrid detectors (HyD); and a 25× water-immersion objective (*see* **Notes 5** and **6**).
14. Four HyDs equipped with the following filters: a 495 nm dichroic mirror followed by emission filters of 460/50 nm bandpass and 525/50 nm bandpass for SHG and GFP; a longpass filter of 560 nm; and 610/60 nm bandpass, a 650 nm longpass, and a 685/50 bandpass for imaging RFP and far-red fluorophores (*see* **Note** 7).
15. Image-processing software: Imaris (Bitplane), Image J and Huygens (Scientific Volume Imaging) (*see* **Note 8**).

3 Methods

While working with VACV, follow all Biosafety Level 2 (BSL-2) procedures, including appropriate personal protective equipment (PPE) and proper animal handling as designated by your institute. A vaccine is available for VACV. All solutions are warmed to 37 °C unless otherwise stated.

3.1 Epicutaneous VACV Infection in the Ear Pinna and Labial Mucosa

1. Remove VACV stock solution, thaw on ice, and sonicate three times to break up aggregated virus.
2. Anesthetize mice for skin infection using isoflurane. Once mice are appropriately sedated, gently stretch the mouse ear on a solid support (such as the cap of 50 mL tube).
3. Remove sonicated VACV from vial with a 1 cc or insulin syringe and place a drop of virus (~10 μL) on the ventral or dorsal side of ear (depending on the experiment; *see* **Note 12**). Spread VACV droplet on ear using a bifurcated needle (saturating prongs of needle) and gently poke 5–15 times in the ear, taking care not to poke through the thin ear skin (*see* **Note 13**).
4. For mucosal infection use approved injectable anesthetic or isoflurane (*see* **Note 14**). Once mice are sedated, place a drop of virus inside the lower lip using an insulin syringe. Gently poke the mucosa five times with a bifurcated needle, making sure not to poke through too deeply and infect the skin (*see* **Note 15**).
5. Wipe ear/inner lip to remove the remaining VACV. Allow mice to recover from anesthesia. Mice may be kept in the vivarium until desired day postinfection (dpi).
6. Discard unused VACV appropriately—we typically do not freeze/thaw as this can affect viral titers.

3.2 Routes of VACV Infection for MPM Imaging Other Organs

1. For lymph node (LN) MPM imaging, mice may be infected subcutaneously or intradermally. To image the popliteal LN postinfection, we typically inject 10^5–10^8 pfu subcutaneously in the hind footpad in total volume of 20 μL. To image the inguinal LN, we achieve the best results with up to three injections subcutaneously in the flank skin immediately surrounding the node.
2. For imaging the spleen after infection, virus may be injected intravenously and intraperitoneally. High viral doses will result in mortality within a week, so the dose should be chosen carefully based on experimental requirements (*see* **Note 16**).
3. For imaging the lung, VACV is administered intranasally.
4. For imaging infected ovaries, VACV should be injected intraperitoneally.

3.3 Mouse Preparation for Intravital Microscopy of VACV-Infected Skin

MPM imaging of the mouse ear skin has been reviewed a number of times [10–13]. In this methods chapter, we focus on some of the unique aspects of imaging VACV-infected skin.

1. After epicutaneous viral inoculation in the mouse ear, viral replication proceeds as previously described [5, 6]. Select a day of imaging that allows visualization of the desired infected cell populations. For imaging motility of VACV-infected inflammatory monocytes, we recommend days 3–4 postinfection.
2. Remove mouse from vivarium at the selected time postinfection and anesthetize.
3. Following initial anesthesia, the mouse can be maintained with isoflurane using a facemask at a concentration of ~1% or with avertin using a butterfly needle into the peritoneum (*see* **Note 17**). The rate of anesthesia must be modulated in accordance with mouse weight and sex. Take special care to avoid overheating the mouse during anesthesia in the environmental chamber; temperature must be carefully monitored.
4. Place mouse on a solid support that allows transport between the surgical preparation area and the microscope. We typically use a disposable, 6-well polystyrene plate. Gently secure mouse to plate using tape (take care not to tape too tightly (breathing can be restricted)). The tape secures the mouse as well and also limits movement due to breathing.
5. Secure the inner ear to the 6-well plate using a small drop of cyanoacrylate glue. The ear should be very flat with no creases (*see* Fig. 1 and **Note 18**). Be careful not to disturb any scabs present on the VACV-infected skin. Though some protocols for MPM in the skin utilize a depilatory cream to remove hair, we do not do this on inflamed VACV-infected skin.
6. Create a well around the glued ear that will be imaged using caulking cord (*see* Fig. 1 and **Note 19**).
7. Fill the caulking cord well with 37 °C PBS. Virus may be released from the ear into the PBS—take care not to spill. Gloves should be exchanged after the mouse is prepared.
8. Use a cotton Q-tip to gently remove any bubbles from the well (*see* **Note 20**).
9. Transfer the mouse to the heated chamber of the MPM microscope and carefully lower the dipping objective into PBS. The microscope objective will likely come in contact with virus and should be properly decontaminated at the end of the experiment.
10. We typically euthanize the mouse at the end of the imaging session while still under anesthesia. However, as imaging the skin is minimally invasive, it is possible to allow the mouse to recover from anesthesia and image again at a later time.

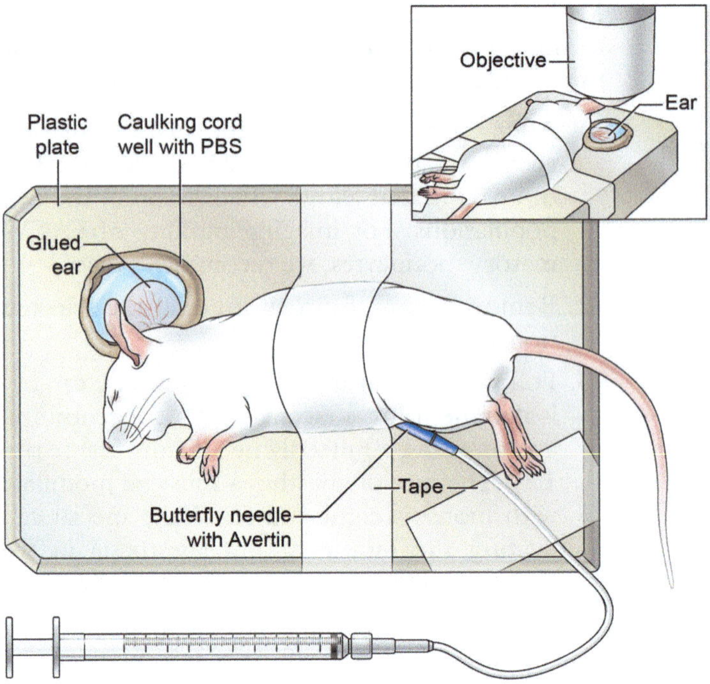

Fig. 1 Mouse preparation for MPM imaging the VACV-infected ear skin. For imaging the ear skin, anesthetize the mouse with avertin and immobilize to a solid platform using cloth tape. The ear is also attached directly to the plate for added immobilization. The ear is then surrounded by a caulk well which is filled with pre-warmed PBS. The prepared mouse is transferred to the environmental chamber of the microscope for MPM imaging

3.4 Mouse Preparation for Intravital Microscopy of VACV-Infected Labial Mucosa

1. Induce animal anesthesia in induction chamber using 2% isoflurane. Following initial anesthesia, maintain isoflurane influx using a nose cone at a continual concentration of ~1.5–1.75%. Adjust if necessary based on oxygen saturation.
2. For surgery and preparation, place the mouse and the microscope stage insert on a heated pad, and secure the mouse face up using surgical tape (*see* **Note 9**).
3. Disinfect the surgical site with three alternating scrubs of Betadine (or other iodine-based surgical scrub), and 70% ethanol.
4. Make 2-mm-long incision on both corners of the mouth, avoiding trauma to the anterior facial vein and other major blood vessels (*see* Fig. 2a.1 and **Note 21**).
5. Expose labial mucosa by flipping mouse lip inside out, and pulling it gently through the opening in stainless steel mini-stage, followed by gluing the external lip surface to the holder using superglue. Extra care should be taken so as not to allow superglue to enter the inner surface of the lip (*see* Fig. 2a.2, a.3).

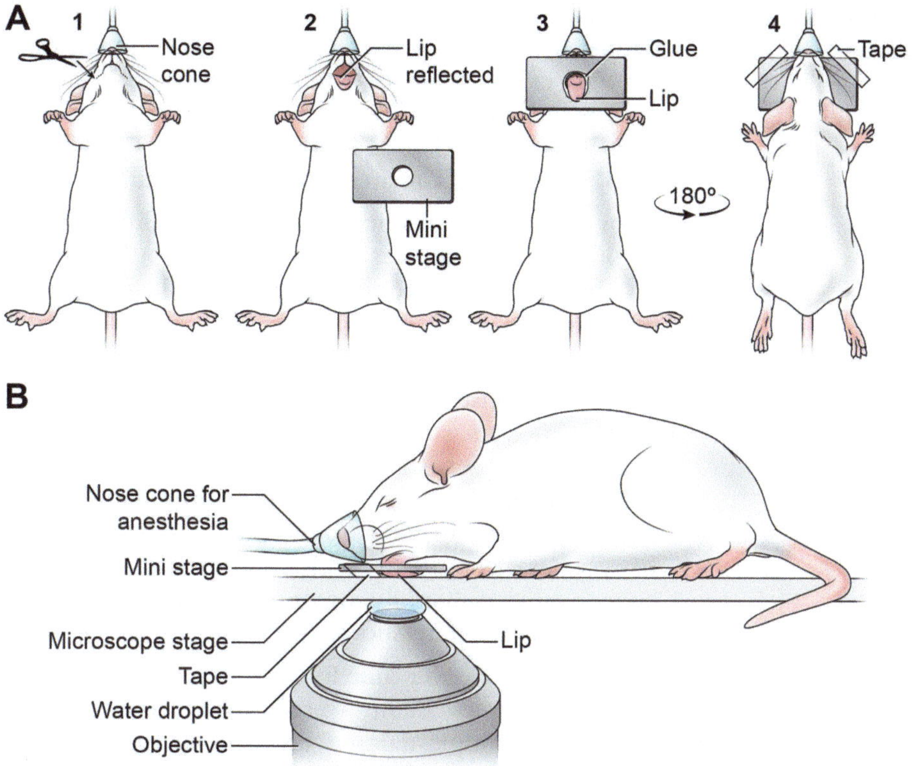

Fig. 2 Mouse preparation for MPM imaging the labial mucosa. (**a**) Sequence of animal preparation for MPM imaging of the VACV-infected inner lip. (1) The anesthetized mouse is placed in the supine position and two 2 mm incisions are introduced at both oral angles. (2) The lower lip is gently pulled down to expose the inner mucosal tissue (the labial mucosa). (3) The lip is pulled through the opening of the stainless steel mini-stage taking care not to damage the tissue in the process. A drop of superglue is applied to the external lip surface in order to attach the lip to the mini-stage. (4) The mouse is now placed in prone position on the microscope platform, and the mini-stage is stabilized using surgical tape. (**b**) Side profile of a mouse prepared for MPM imaging of the labial mucosa. On an inverted microscope, the 25× water-immersion objective faces the exposed labial mucosa of the anesthetized mouse

6. After complete attachment of the lip to the mini-stage, immerse exposed labial mucosa in water-based gel.
7. For MPM imaging on an inverted microscope, flip the mouse face down over the coverslip of the microscope imaging stage, and immobilize the corners of the mini-stage using surgical tape (*see* Fig. 2a.4, b).
8. Carefully place a drop of water on the inverted objective and then transfer the secured mouse into the heated microscope chamber. Slowly raise the objective to come in contact with the coverslip (*see* **Note 22** and Fig. 2b).
9. Imaging the mouse labial mucosa requires minor surgery; therefore mice are euthanized without recovery from anesthesia.

3.5 MPM Image Acquisition

1. Set the multiphoton laser wavelength to 900 nm for simultaneous eGFP and dsRed excitation (if using a single MaiTai laser to excite these two fluorophores). For far-red fluorophores tune InSight DS laser to 1150 nm. Set to other wavelengths as appropriate [14]. We have imaged VACV expressing tagBFP, eGFP, tdTomato, mCherry, and turboRFP [5–7, 15].
2. Allow the mouse to warm to the temperature of the chamber before acquiring images for analysis of motility. Small changes in body temperature can drastically impact cellular movement.
3. Determine and set the imaging parameters as needed. Image acquisition must balance the area scanned (*x*-*y*) and the image depth (*z*) versus the time it will take to acquire each 3-dimensional set (stack) of images. If using an inverted scope with motorized stage, tiled scans of multiple regions of interests over time can be acquired (up to 2 mm^2). The interval between stacks must be adjusted depending on the parameter being analyzed. Rapidly moving cells (such as T cells and neutrophils) will require shorter time intervals between stacks.
4. Acquire z-stacks over time for the desired time range. To determine cellular movement, we recommended tracking cells for at least 1 h. Multiple datasets (containing many hours of data) may be concatenated in order to keep individual file sizes down.
5. Monitor the depth of anesthesia of the mouse during the imaging period (*see* **Note 17**).
6. Analyze datasets using the appropriate software. Typically, we make maximum intensity projections of each stack over time and use these to quantitate cellular speed and directionality using software such as Imaris, which contains several automated spot-counting detection and tracking algorithms. Data image analysis depends largely on experimental design. For previous reviews, *see* refs. 16–18.

4 Notes

1. Sonicate VACV prior to infection (either in vitro or in vivo) to dissociate aggregated virus.
2. Depending on the experiment, anesthesia can be administered either as an injectable (avertin or ketamine/xylazine) or inhaled (isoflurane). Each anesthetic has advantages and disadvantages for imaging. While isoflurane anesthesia typically yields the most consistent results, the requirement for delivery through a nose cone can impede imaging of some areas.
3. Mouse strains should be chosen carefully for imaging the skin with selection of nonpigmented mice whenever available.

Pigmented mice (such as C57Bl/6) have melanocytes and melanin-containing dermal cells that can absorb light resulting in heat and photodamage [13].

4. PBS should be warmed to 37 °C prior to application to the mouse. The addition of cold buffers can alter the movement of cells within the tissue.
5. Anesthetized mice do not maintain body temperature, so an appropriate heat source must be provided during imaging. A heated blanket with a feedback thermometer will work in most cases; however an environmental chamber that keeps the entire microscope stage and anesthetized mouse at a constant temperature provides the most accuracy.
6. Many options exist for turnkey MP microscopes. The first selection criterion is the use of an upright (objective comes from the top) *versus* inverted (objective comes from underneath) microscope. Upright scopes do not have to image through a cover glass like inverted scopes; however the addition of a cover glass can act as a stabilizer for some tissues with high water content. Multiple MP laser options now exist as well; here price can be a major factor. Many scopes are equipped with more than one MP laser, allowing fluorophore excitation at multiple wavelengths simultaneously.
7. Filters should be optimized for the fluorophores being used; our most common set is listed.
8. Multiple choices exist for image-processing software.
9. Imaging the labial mucosa requires a minor surgery, which increases the preparation time. Since mice are unable to regulate body temperature under sedation it is important to maintain physiological temperature during this time.
10. This is a custom-built 2 cm × 5 cm stainless steel plate with a circular window in the middle. After animal preparation, the window will face water-immersion lens on the inverted microscope. The mini-stage was designed to stabilize a small region of interest within a larger organ (here the labial mucosa) by gluing the organ to the stage and allowing exposure of the region of interest through the opening.
11. It is essential to use the proper coverslip thickness based on a given objective's numerical aperture (NA). Most objectives are designed to use #1 coverslips (0.13–0.17 mm thickness). Using an incorrect coverslip thickness may result in decreased signal intensity and resolution.
12. We typically infect on the dorsal ear pinna containing hair follicles. If imaging pigmented (such as C57Bl/6) mice, it can be helpful to switch to the ventral ear to avoid some of the autofluorescence (as this side has less pigment).

13. Infections are performed in areas that maximize MPM imaging ability. Avoid areas too close to the mouse's head, as this will impede imaging on an upright scope due to the objective. Likewise, avoid the ear edges, as these areas have dense hair and can prove more difficult to image.
14. Infection of the labial mucosa is difficult using an isoflurane nose cone. Injectable avertin can be used if approved by institutional committee. Intraperitoneal injection can result in inflammation over time, which should be considered for immunological studies. Avertin injection volumes are based on weight. For most consistent results, make fresh stocks of avertin. Avertin can be stored at 4 °C in the dark for ~14 days.
15. For MPM imaging of mucosal VACV infection, viral inoculation between the inside floor of the lip and outer skin maximizes the ability to image; however it is essential to prevent penetration of the outer lip or infection of the skin (unless analysis of skin infection is desired).
16. When selecting virus dose, one must balance the ability to visualize virus-infected cells and physiology of infection. Typically, a rather high viral dose is required to find fluorescent protein-expressing VACV-infected cells, particularly when animals are imaged soon after injection (such as in the LN). In the skin and labial mucosa, lower virus doses allow the visualization of infected cells after VACV replication.
17. Attach the facemask delivering isoflurane to the mouse securely using athletic cloth tape. We advise the use of a heart rate/oxygen saturation monitor. This allows noninvasive determination of the level of anesthesia during imaging. We have found that an animal with a heart rate of 300–450 bpm and an oxygen saturation of ~90% can be imaged for >4 h [19].
18. We find it helpful to gently press the entire ear with your finger. The glue will fluoresce if accidentally transferred to the skin so avoid contamination from glue adhered to gloves.
19. We find window caulk to be the easiest material to make a well holding PBS around the skin, but other material can also be used.
20. The hair will trap air bubbles after addition of PBS. Bubbles will interfere with imaging.
21. Small diagonal cuts at the oral angles of the mouth result in increased exposure of the labial mucosa for imaging with minimal bleeding.
22. While using water-immersion lens, the water drop between objective and the MPM imaging stage must be reapplied every hour. Sequential 1-h time-lapse videos can be combined later into continuous 6-h time-lapse during post-acquisition data processing using Huygens (SVI) software.

Acknowledgments

This work was supported by the Intramural Research Program of the National Institute of Allergy and Infectious Diseases, National Institutes of Health. Ethan Tyler (NIH Medical Arts) created illustrations.

References

1. Secklehner J, Lo Celso C, Carlin LM (2017) Intravital microscopy in historic and contemporary immunology. Immunol Cell Biol 95(6):506–513
2. Halin Cornelia MJR, Cenk S, von Andrian Ulrich H (2005) In vivo imagining of lymphocyte trafficking. Annu Rev Cell Dev Biol 21:581–603
3. Qi H, Kastenmuller W, Germain RN (2014) Spatiotemporal basis of innate and adaptive immunity in secondary lymphoid tissue. Annu Rev Cell Dev Biol 30:141–167
4. Hickman HD (2017) New insights into antiviral immunity gained through intravital imaging. Curr Opin Virol 22:59–63
5. Hickman HD, Reynoso GV, Ngudiankama BF, Cush SS, Gibbs J, Bennink JR et al (2015) CXCR3 chemokine receptor enables local CD8(+) T cell migration for the destruction of virus-infected cells. Immunity 42(3): 524–537
6. Hickman HD, Reynoso GV, Ngudiankama BF, Rubin EJ, Magadan JG, Cush SS et al (2013) Anatomically restricted synergistic antiviral activities of innate and adaptive immune cells in the skin. Cell Host Microbe 13(2):155–168
7. Cush SS, Reynoso GV, Kamenyeva O, Bennink JR, Yewdell JW, Hickman HD (2016) Locally produced IL-10 limits cutaneous vaccinia virus spread. PLoS Pathog 12(3):e1005493
8. Wyatt LS, Earl PL, Moss B (2017) Generation of recombinant vaccinia viruses. Curr Protoc Protein Sci 89:5.13.1–5.13.8
9. Reynoso GV, Shannon JP, Americo JL, Gibbs J, Hickman HD (2018) Growth and purification of vaccinia virus stock. Methods Mol Biol 2023
10. Goh CC, Li JL, Becker D, Weninger W, Angeli V, Ng LG (2016) Inducing ischemia-reperfusion injury in the mouse ear skin for intravital multiphoton imaging of immune responses. J Vis Exp (118). https://doi.org/10.3791/54956
11. Gaylo A, Overstreet MG, Fowell DJ (2016) Imaging CD4 T cell interstitial migration in the inflamed dermis. J Vis Exp (109):e53585. https://doi.org/10.3791/53585
12. Egawa G, Kabashima K (2016) In vivo imaging of cutaneous DC's in mice. Methods Mol Biol 1423:269–274
13. Li JL, Goh CC, Keeble JL, Qin JS, Roediger B, Jain R et al (2012) Intravital multiphoton imaging of immune responses in the mouse ear skin. Nat Protoc 7(2):221–234
14. Drobizhev M, Makarov NS, Tillo SE, Hughes TE, Rebane A (2011) Two-photon absorption properties of fluorescent proteins. Nat Methods 8(5):393–399
15. Hickman HD, Li L, Reynoso GV, Rubin EJ, Skon CN, Mays JW et al (2011) Chemokines control naive CD8+ T cell selection of optimal lymph node antigen presenting cells. J Exp Med 208(12):2511–2524
16. Beltman JB, Maree AF, de Boer RJ (2009) Analysing immune cell migration. Nat Rev Immunol 9(11):789–798
17. Benson RA, Brewer JM, Garside P (2017) Visualizing and tracking t cell motility in vivo. Methods Mol Biol 1591:27–41
18. Sharaf R, Mempel TR, Murooka TT (2016) Visualizing the behavior of HIV-infected T cells in vivo using multiphoton intravital microscopy. Methods Mol Biol 1354:189–201
19. Ewald AJ, Werb Z, Egeblad M (2011) Monitoring of vital signs for long-term survival of mice under anesthesia. Cold Spring Harb Protoc 2011(2):pdb prot5563

Index

Jason Mercer (ed.), *Vaccinia Virus: Methods and Protocols*, Methods in Molecular Biology, vol. 2023,
https://doi.org/10.1007/978-1-4939-9593-6,

S

T

V

W

MIX
Papier aus verantwortungsvollen Quellen
Paper from responsible sources
FSC® C105338

FSC
www.fsc.org

If you have any concerns about our products,
you can contact us on
ProductSafety@springernature.com

In case Publisher is established outside the EU,
the EU authorized representative is:
Springer Nature Customer Service Center GmbH
Europaplatz 3, 69115 Heidelberg, Germany

Printed by Libri Plureos GmbH
in Hamburg, Germany